אֱלֹהִים הַתּוֹפָעָה
תּוֹרָה וּמַדָּע וּמָתֶימָטִיקָה

Elohim
Phenomenon
Torah, Science and Math

Written by
I. D. McClain

Order this book online at www.trafford.com
or email orders@trafford.com

Most Trafford titles are also available at major online book retailers.

© Copyright 2013 I. D. McClain.

All rights reserved. No part of this publication may be reproduced, stored in a retrieval system, or transmitted, in any form or by any means, electronic, mechanical, photocopying, recording, or otherwise, without the written prior permission of the author.

Printed in the United States of America.

ISBN: 978-1-4669-8056-3 (sc)
ISBN: 978-1-4669-8055-6 (e)

Trafford rev. 02/13/2013

 www.trafford.com

North America & international
toll-free: 1 888 232 4444 (USA & Canada)
phone: 250 383 6864 ♦ fax: 812 355 4082

Dedicated to:

Y'Shua HaMeshiach Ben Yoseph OoVen Daveed
(Jesus the Christ son of Joseph and son of David)

Who as "Son of Joseph" died for our sins as the Lamb of G-d,
And Who is returning as the "Son of David," the Conquering King.
Amen.

Thanks also to All those who endured with me
To get this published.

Special Thanks to
My Parents for their Support
And to
Rabbi Robert & Brenda Benbow and Michael Harbert
Who Helped during this Time.

בָּרוּךְ אַתָּה יהוה אֱלֹהֵינוּ
מֶלֶךְ הָעוֹלָם,
אֲשֶׁר בָּרָא אֶת הָעוֹלָם:

Ba.ruch A.tah A.do.nai E.lo.hei.nu Me.lech
Ha.o.lam
A.sher Ba.rah Et Ha.o.lam!
A.mayn!

Blessed are you L-rd, our G-d, King of the Universe,
Who created the Universe!
Amen!

Preface

Throughout history the human race has sought out the foundation of the universe. Today, we have gathered enough technical data to assemble the answer. Even so, the answers have always been present with us. They were described in the Torah long before today's technological advances, but we were unable understand them. Secondly, in our attempts to understand, we uncover seemingly contradictory information to the Torah. Some of the problem is that the information was interpreted to exclude Elohim and the Torah. However, all the facts concerning the universe fit into a neat interlocking pattern within the Torah.

Science is often used to explain away supernatural occurrences. Supernatural energy is just that, super natural. It is the nature underneath maintaining natural phenomena. The only time we see it is when something occurs outside the created laws of nature.

This book answers the following questions:

> How did the physical universe not exist?
> What is the smallest granule of matter?
> How are they designed?
> How do they function?
> Why do quarks exist?
> How was the earth formed?
> How did plants survive before there was a sun?
> How much water covered the earth during Noah's Flood?
> How did the ice ages come into existence?
> How did time become so confused?
> Which came first the chicken or the egg?
> Why are there quasars and neutron stars?
> For what purpose was all of this done?

And many other questions are answered, even some of those not asked. The intended audience is to those who do not know science and math, but want to know. The point is: What good is knowledge if only a few understand it? Lastly, the accuracy of this book equals the knowledge of Elohim through Yeshua (Jesus) subtracted by my input. Finally, extreme effort has been taken to bring my input to zero.

Nomenclatures

The labels given presented within this book, obviously, are not the original Hebrew labels given by Elohim. Even so, effort has been made to give names that are descriptive of their nature at least in English. This omission of the Hebrew labels is not by personal volition, but by absence of knowledge caused by decay of such knowledge within the human condition. Therefore, the nomenclatures given can, at best, be a translation of the actual Hebrew word.

There is a reason why that Hebrew is so important. As stated in the Torah, the Tower of Babel caused the languages to be confused. The purpose was to stop them from continuing in their building process. However, not all people were involved in this deed. For an example, Noah was alive during this time. His language, along with others down the lineage to Eber, was not confused during this occurrence; because they were not involved. Since our foreign languages are a result of confusion, there had to be a language in which they originated from. Hebrew is this original language. It is the language that "crossed over," as the word, Hebrew, implies. Even so, within Modern Hebrew many word roots have been lost. Only the Torah and the writings of the prophets have preserved the words that we have for guiding us to HIS Plan. More about these topics are in the book.

Table of Contents

PART 1: Establishing Physical Space

Prelude .... א

Chapter 1
    Foundation 23
        Elohim 23
        Creationism verses Naturalism 24
        Unified Knowledge 25
        The Beginning 26
        Ramification Fractals 27
        Rose Illustration 27
        Tranquil Beginning 28
        Movement 28
        Waters 29
        Phases of the Universe 30
        Chapter 1 Quiz 33

Chapter 2
    Nulverse 35
        Omnipresence 35
        Basic Geometric Understanding 36
        Finite Limits 36
        Expanding Sphere 38
        Infinite Reality 39
        Slope of a Line 40
        Area under a Curve 41
        Elohim's Unity 44
        Nulverse Creation 46
        A Slice of Nulverse 46
        Chapter 2 Quiz 49

Chapter 3
    Inertverse 51
        Shape of a Geometric Point 52
        Three Different Geometrical Systems 52
        Volume of Planes 54
        Dimensional Integrity 54
        Creation of Inertips 55
        Body, Soul and Spirit 56

Table of Contents

  Cleaning the Information 57
  Adjusting Infinitesimal Space... 59
  Chapter 3 Quiz 61

PART 2: Creation of Xyzenthium Crystals

Chapter 4
Gravverse..65
  Discontinuous Infinitesimal Structure.. 66
  Resulting Gravverse 68
  Quebbrix (Cube-Bricks)... 69
  Beginning of Time 70
  Chapter 4 Quiz 73

Chapter 5
Kineverse..75
  Vibrations... 75
  Transfer Pattern of Kyntips 76
  Unified Fluttering 78
  Quubium 79
  Time in the Kineverse.. 80
  Chapter 5 Quiz 83

Chapter 6
Thermaverse 85
  Clear to Infinity.. 85
  Invoked Light... 86
  Xevim Angles 87
  Energy Types.. 87
  Xergotips Planes and Plates 89
  Xentrix.. 90
  Juttorial Response 91
  Shrinkage via Xergotip... 92
  Phaseverse Final Form... 94
  Chapter 6 Quiz 95

Chapter 7
Xyzenverse 97
  Loop Directions 97
  Zeeds and Exozeeds... 99

Table of Contents

Xergopath Sectors 99
Xentrix Fragmentation... 100
The Thermtip Pulse... 101
Diflohexius 102
Xyzenthium Crystallization 103
Final Xyzenverse Crystals 105
Chapter 7 Quiz 109

PART 3: Creation of Neutrons

Chapter 8
Magneverse 111
Dividing Light from Darkness 111
Energy Types 112
Division by Dimensional Reference 112
Loop Interruption 114
Miortex 115
Quarquid 117
Microweak verses Microstrong 118
Energy Shapes 120
Hierarchy of Energy 121
Final Magnetic Xyzenthium Crystal 122
Chapter 8 Quiz 127

Chapter 9
Neutronverse 129
From Cube to Sphere 129
Finding the Perfect Square 131
Crystal Annihilation Phenomenon 132
Landscaping the Neutron Cube 134
Xyzenscape Sequence 136
Neutron Reformulation 137
Chapter 9 Quiz 141

Chapter 10
Quarks 143
Quark Classifications Overview 143
Splitting Submatter 144
Hadrons 144
Leptons and Neutrinos 145
Nucleon Equation 146
Antiquarks and Mesons 147

Table of Contents

    Strangeness 147
    Isobaric Spin 148
    Bubble Chamber Experiment 151
    Cold Neutrons 152
    Chapter 10 Quiz 157

PART 4: Creation of Atoms

Chapter 11
Neutron Breakup 161
    Nightsod Evaporation 161
    Formulation of an Electron 162
    Accounting Mass 164
    The Neutrino Experience 165
    Volumass 166
    Neutron Prime 167
    Xyzenthium Crystal Count 170
    Cubical Earth Illustration 172
    Pluto-Electron Illustration 173
    Chapter 11 Quiz 175

Chapter 12
Plasmaverse **177**
    Subatomic Forces 178
    Hydrogen and Helium Formations 179
    Lithium the Isolated Isotopes 181
    Beryllium and Boron Isotopes 183
    Carbon and Nitrogen Isotopes 184
    Oxygen and Fluorine Isotopes 185
    Neon, Sodium, Magnesium, and Aluminum Nuclei 188
    Silicon, Phosphorous, and Sulfur Isotopes 189
    Chlorine, Argon, and Potassium Nuclei 191
    Calcium to Vanadium 193
    Few More Single Isotope Elements 195
    Mass Shrinkage 198
    Nuclei Analysis 199
    Chapter 12 Quiz 203

Chapter 13
Bondverse **205**
    Magnetic Lines 205
    Shell Levels 206

Table of Contents

Ks (1s) Sublevel 208
Ls (2s) Sublevel 210
Other s Sublevels 211
Sublevel p 212
Sublevel d 214
Sublevel f and Beyond 215
Orbital Limits 215
Atomic Radii 217
Bonding of Atoms 218
Electron's Anatomy 220
Anatomy of a Photon 226
Redshift by Photons 231
Color 231
Forming Magnets 234
Thermodynamics 236
Friction 238
Fire 239
Reforming Plasma 240
Seven States of Matter 241
Chapter 13 Quiz 245

PART 5: Creation of the Macrocosm

Chapter 14
   Creating Galaxies 249
      Firmament Subset Phenomena 250
      Firmament- the Implosion Phenomenon 250
      Isotope Coagulation Shells 251
      Particle Elimination by Collection 253
      Neutron Snow 255
      Isotope Rings and Islands 259
      Multigalactic Shell Breakup 259
      Gravitational Divergence 264
      Antishell Subset Phenomenon 266
      Quasars and Neutron Stars 266
      Novas 268
      Chapter 14 Quiz 271

Chapter 15
   Creating Earth 273
      Formation of Our Solar System 274
      Natural Hydrogenation on Other Planets 276

Table of Contents

    Earth's Plasmatic Storm..278
    Forming of the Crust 284
    Formulating the Felsic Layer..286
    Finishing Felsic Formations..290
    Initial Atmosphere...293
    Forming of the Ocean..295
    Chapter 15 Quiz..299

Chapter 16
    Initial Formations 301
    Initial Land Features...305
    First Signs of Life..306
    Igniting the Sun..307
    Entrance of the Moon...309
    Complex Gravity..311
    The Process of Creating Life Continues 319
    Adam and Eve...320
    Chapter 16 Quiz...327

PART 6: Post Creation Activity

Chapter 17
    Old World Phenomena 331
    Cursing of the Ground 335
    Resulting Solar Features..336
    Solar Flipping Deceleration...339
    Pre-Flood Atmospheric Conditions..341
    The Flood 342
    Hudson Comet Impacts upon Earth 345
    Post Flood Activity...347
    Chapter 17 Quiz...349

Chapter 18
    Peleg's Earthquake Overview.. 351
    Analysis of Magnetic Plasma...351
    Earth's Magnetic Polar Phenomena Set..352
    Earth's Shell Phenomena Set 355
    Rotational Phenomena Set...357
    Jupiter Arrives 358
    Time Measurements During the Interactions...............................362
    Magnetic Poles of Today 368
    Chapter 18 Quiz...371

Table of Contents

Chapter 19
   Kadummagen's Demise..373
      Mafic Plain Moves 374
      Australia Breaks Away..375
      Antarctica Captured..376
      South America Forms 377
      Alaska Crashes..378
      Siberia and Greenland Separates 380
      Compression of the Magnetic Poles 382
      Americas Remove 384
      Africa 386
      Decompression Influence 386
      Other Influences and Factors 387
      Chapter 19 Quiz..389

Chapter 20
   Ice Ages...391
      Foundational Information about the Ice Ages....................391
      Ice Ages Scenario..396
      Historical Accounts by Life 401
      Changes in Time 403
      Chapter 20 Quiz..405

Chapter 21
   Resulting Atmosphere 407
      Polar Pressure Phenomenon 407
      Jet Stream Phenomenon..408
      Tornadoes 412
      Atmospheric Layers..414
      Chapter 21 Quiz..415

PART 7: Logic of Creation

Chapter 22
   The Big Ending 419
      Why Our World? 421
      The Dreamer and the Dream..422
      The Rebellion in Heaven 425
      Yeshua 426
      Repentance..427
      Abominations Generally Described.....................................427
      Standing Fast 429
      Mikvah (Baptism)...429

Table of Contents

Welcome 430
Epilog- Tav Judaism 430
Chapter 22 Quiz 433

Appendix
    Antiderivative 437
    Pluto and Beyond 443
    Life Continuum 445

Resources 447

Glossary 449

Answers 469

Index 471

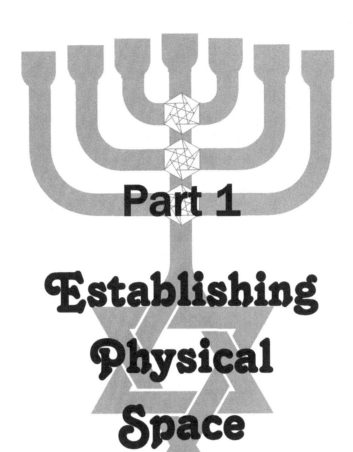

בְּרֵאשִׁית

א בְּרֵאשִׁית בָּרָא אֱלֹהִים אֵת הַשָּׁמַיִם וְאֵת הָאָרֶץ:
ב וְהָאָרֶץ הָיְתָה תֹהוּ וָבֹהוּ וְחֹשֶׁךְ עַל־פְּנֵי תְהוֹם
וְרוּחַ אֱלֹהִים מְרַחֶפֶת עַל־פְּנֵי הַמָּיִם:
ג וַיֹּאמֶר אֱלֹהִים יְהִי־אוֹר וַיְהִי־אוֹר:
ד וַיַּרְא אֱלֹהִים אֶת־הָאוֹר כִּי־טוֹב
וַיַּבְדֵּל אֱלֹהִים בֵּין הָאוֹר וּבֵין הַחֹשֶׁךְ:
ה וַיִּקְרָא אֱלֹהִים לָאוֹר יוֹם וְלַחֹשֶׁךְ קָרָא לָיְלָה
וַיְהִי־עֶרֶב וַיְהִי־בֹקֶר יוֹם אֶחָד:

ו וַיֹּאמֶר אֱלֹהִים יְהִי רָקִיעַ בְּתוֹךְ הַמָּיִם
וִיהִי מַבְדִּיל בֵּין מַיִם לָמָיִם:
ז וַיַּעַשׂ אֱלֹהִים אֶת־הָרָקִיעַ וַיַּבְדֵּל
בֵּין הַמַּיִם אֲשֶׁר מִתַּחַת לָרָקִיעַ
וּבֵין הַמַּיִם אֲשֶׁר מֵעַל לָרָקִיעַ וַיְהִי־כֵן:
ח וַיִּקְרָא אֱלֹהִים לָרָקִיעַ שָׁמָיִם
וַיְהִי־עֶרֶב וַיְהִי־בֹקֶר יוֹם שֵׁנִי:

ט וַיֹּאמֶר אֱלֹהִים יִקָּווּ הַמַּיִם מִתַּחַת הַשָּׁמַיִם
אֶל־מָקוֹם אֶחָד וְתֵרָאֶה הַיַּבָּשָׁה וַיְהִי־כֵן:
י וַיִּקְרָא אֱלֹהִים לַיַּבָּשָׁה אֶרֶץ
וּלְמִקְוֵה הַמַּיִם קָרָא יַמִּים
וַיַּרְא אֱלֹהִים כִּי־טוֹב:

B're.shyit

A.leph   B're.shyit ba.ra El.o.him et ha.sha.ma.yim v'et ha.a.retz.

Bet   V'ha.a.retz hai.ta to.hu va.vo.hu v'cho.shech al-p.nay t'hom
V'ru.ach El.o.him m'ra.he.fet al-p.nay ha.ma.yim.

Gim.mel   Va.yo.mer El.o.him y'hee-or, vai.hee-or.

Da.let   Va.yar El.o.him et-ha.or ki-tov.
Va.yav.dayl El.o.him bayn ha.or u.vayn ha.cho.shech.

Hey   Va.yi.kra El.o.him la.or yom, v'la.cho.shech ka.ra lai.la.
Vai.hee-e.rev. vai.hee-vo.ker yom e.chad.

Vav   Va.yo.mer El.o.him y'hi ra.ki.a b.tok Ha.ma.yim.
Vee.hee mav.deel bayn ma.yim la.ma.yim

Za.yin   Va.ya.ahs El.o.him et ha.ra.ki.a va.yav.dayl
Beyn ma.yim mee.ta.chat la.ra.kee.a
U.veyn ma.yim a.sher may.al la.ra.ki.a, vai.hee ken.

Chet   Va.yi.kra El.o.him la.ra.ki.a sha.ma.yim
Vai.hee-e.rev. vai.hee-vo.ker yom shay.nee.

Tet   Va.yo.mer El.o.him yee.ka.vu ha.ma.yim mi.ta.chat ha.sha.ma.yim
El-ma.kom e.chad v'teh.ra.eh ha.ya.ba.sha.

Yod   Va.yi.kra El.o.him la.ya.ba.sha e.retz
U.la.mik.veh ha.ma.yim ka.ra. ya.mim
Va.yar El.o.him ki-tov.

ב

B'reshyit (Genesis) 1:1-10
(Inorganic Creation)

*In the beginning **Elohim** created the heaven and the earth. And the earth was without form, and void; and darkness was upon the face of the deep. And the Spirit of **Elohim** moved upon the face of the waters. And **Elohim** said, Let there be light: and there was light. And **Elohim** saw the light, that it was good: and **Elohim** divided the light from the darkness. And **Elohim** called the light Day, and the darkness HE called Night. And the evening and the morning were the first day.*

*And **Elohim** said, Let there be a firmament in the midst of the waters, and let it divide the waters from the waters. And **Elohim** made the firmament, and divided the waters which were under the firmament from the waters which were above the firmament: and it was so. And **Elohim** called the firmament Heaven. And the evening and the morning were the second day.*

*And **Elohim** said, Let the waters under the heaven be gathered together unto one place, and let the dry land appear: and it was so. And **Elohim** called the dry land Earth; and the gathering together of the waters called HE Seas: and **Elohim** saw that it was good.* (KJV, modification: Hebrew for English- Elohim means G-d)

The correct rendering of the Hebrew for "In the Beginning" should read "In a Beginning." The formulation of this universe out of nothing is a beginning. This implies that it is not "The Beginning" as there are other beginnings and the eternal past has no beginning. However, for our physical universe, this is its beginning.

This should not be construed to mean that these universes are created by trial and error. Contrarily, it means that this present version of the physical universe and Heavenly universe were created for a specific purpose. As we shall observe, this purpose expresses itself within the very patterns by which the universes were created. This is true even for the development of inorganic material out of nonexistence.

With this in mind, let's begin by examining the creation of inorganic material…

ד

Chapter 1

Foundation

Imagining a time before physical space existed is nearly impossible, unless we become aware that some of our basic premises are not totally true. We cannot be expected to accept this concept because someone says it is true. However, we can come to grips with this information when we look at it from a mathematical perspective. Laying a foundation of understanding requires first to understand the nature of Elohim mathematically. Then, we can examine the construction of this physical space. In this, we will examine the first two phases of this universe.

Establishing the foundation for this material is easy in one sense. Creation has it recorded from the beginning of time unto now and will continue to the end of time for this universe. Confusion comes from our fallen human nature. We often assert our partial understanding above other partial understandings. Scriptures tell us that we should not lean on our own understanding, but only on the Knowledge of G-d. Even in doing that, we sneak in our own understanding. We can only trust His Word.

Another issue to address is the notion that Judaism and Christianity are as most religions ethnic centered. This is not true. When Noah left the ark, everyone knew, as there were only eight, who Elohim was by Name. After the Tower of Babel, even before, people drifted doctoring the truth to their personal design. Abraham was the only descendant of Eber within the descendants of Shem that clung to the truth at his time. Noah, buy the way, was alive when Abraham was born. Because of the sins of the forefathers, we exist with different beliefs in different regions. These beliefs are always a variance of the one truth. For an example; we could say 1 + 1 = 2 or 1 + 1 = 3 or equal to any other number. Obviously, two is the right answer; but, two is a number out of many. In logic, this fact makes the truth as one opinion among many.

Elohim

G-d is **YHWH** (spelled with Hebrew Letters: Yod Hey Vav Hey). Other Roman spellings are YHVH, JHWH, JHVH and IHVH. **YHWH** is the name of the original, all encompassing uncreated being who created everything, owns everything and came to earth to die in our place (by our acceptance). There are reasons for the vowel omission

Part. 1: Establishing Physical Space

in the Tetragrammaton of HIS Name. The original name of **YHWH** had no vowels originally assigned in the written Hebrew language, as vowels were understood to be present. Moreover, the name is HIS actual personal name. Just as we teach our children not to call their natural physical father by his name, with the same respect, we do not call Elohim by HIS Name. Lastly, the usage of the word of Elohim instead of "G-d" occurs because many religions say G-d without knowing who HE is, making this word generic. We omit the usage of the Hebrew Tetragrammaton transliteration because it entices us to attempt a pronunciation. The usual rendering is Adonai (L-RD) or HaShem (The NAME). HE gives us only the earthly name of Yeshua (Jesus) to use.

Elohim refers to the characteristics of HIM as the Creator. The Hebrew word, "Elohim," is not only in an honorific form but is also plural in the truest sense. Despite this plurality, Elohim is a singular awareness. If Elohim does not have a plural state, then HE could not simultaneously interact with each person individually as HE does.

Creationism verses Naturalism

Perhaps the greatest mistake made by the free-thinking scientists is that of not realizing that the laws of nature are created, let alone, having any purpose other than accidentally forming life. Inadvertently by this error, they feel safe in assuming that Elohim does not exist. On the other hand, the occulting practitioners attribute the natural phenomenon as gods. Many religions worship nature as being Elohim for the same reason. Some even try to worship Elohim as a nameless and often mindless force of life that can be brought under the dominion of human beings. They believe that they can learn to access this mindless force and use it for their own causes. However, Elohim does exist, has a real name, and is the creator of nature. Secondly, the laws of nature do have a purpose by His design.

Another point is that the Creator does not abandon His creation. The misguided concepts that human beings have concerning the laws of nature contradict this truth. The laws of nature were created to give forms for life. The automatic mechanical processes are not a sign of abandonment but of love, it provides a set of rules in which the universe can exist without violent upheaval to its occupants.

Consider a puppeteer. As long as the puppeteer is moving the puppet, the puppet is not moving by its own will. The idea is that of making a puppet that is able act by its own will. Then we arrive at the concept of a robot. A robot has programming in it to cause it to function in a certain way. As long as the program functions, the robot moves in the desired manner. The ultimate robotic dream is to have the robot function in the desired manner by its own will. By the same reasoning, Elohim created the laws of nature to facilitate our being with a body subject to our will. There is so much more detail to expound within this topic; however, we would be getting ahead of ourselves.

Chapter 1: Foundation

The intended implication is that the laws of nature, themselves, did not evolve by some mindless natural progression. Each new phase of the universe required Elohim to intervene in order that the development of the universe was able to continue to attain His desired aim. Otherwise, the universe would remain inert to the forces required to attain any new development. In other words, the universe is incapable of naturally developing new phenomena beyond the capabilities provided within the established parameters previously set without aid from Elohim. This will become evident as we continue through creation.

One final basic principle to note is that Elohim is sovereign over His creation. He can add energy or take it away (unlike individual finite beings). This is seen in healings that are accomplished in the name of Yeshua (Jesus). Either matter is being removed out of existence as in cancer, or it can be inserted into existence like a restored limb. The same applies to His acts of creating the universe. However, unlike in times past, the sovereignty of Elohim is not used to blanket lack of knowledge. It should be noted those in times past had to blanket their lack of understanding with bind faith; thus, preserving their stance with Elohim rather than being swayed by the presumable wisdom of scientific unbelief. For this, they should be commended.

Unified Knowledge

Two basic mindsets exist within modern western understanding, the Hellenistic and Jewish mindsets. The Hellenistic mindset fragments knowledge into compartments. Sometimes within this mindset the linkage between compartments becomes obscure. Unlike the Hellenistic perception of reality, the Jewish mindset recognizes all truth is of Elohim. Because of distortion within the Hellenistic view, mathematics and theology are two separate unrelated entities. Conversely, these are viewed as manifestations of G-d within Judaism. The Torah tells us Elohim created the Heavens and Earth, but does not illustrate in scientific detail and all the ramifications of His spoken command. Both the Hellenistic and Jewish mindsets are two sides of the same coin; it is our fallen nature that prevents us from truly unifying the truth.

The Torah sets guides (a foundational structure) by which creation is to be understood. Geometry tells us that a triangle has exactly three sides. Science tells us there is magnetic energy and photons. They are undeniable truths. These are a direct result of Elohim's living awareness. Elohim sustains the existence of all created phenomena within nature.

Along with this, we have the created laws of nature. Gravity, heat, inertia, kinetic laws, and the set of subatomic energy are all created. The manner in which they function is a created truth by Elohim. Therefore, how can we separate the laws of nature from its creator? Note again: the objective is not to prove but to present and recognize the Glory of Elohim.

Part. 1: Establishing Physical Space

In Genesis, Elohim spoke the Heavens and Earth into existence, but the Torah does not tell us the physical processes that occurred to accomplish the command. Actually, this fact does not need to be explained to be accepted. The Torah then defines the truth in which the unwritten laws of nature find their existence. We will find that the Torah provides the missing information in the unwritten laws of nature.

One last statement concerning the unification of truth, there is a hierarchy of our understanding of the laws of the universe. The highest is the Torah itself, then the created laws of nature, and lastly the world as seen today. Any understanding that does not align itself with the Torah or any part of the Tenach is wrong.

The Beginning

In the beginning, Elohim created the heaven and earth. And the earth was without form, and void; and darkness was upon the face of the deep. And the Spirit of Elohim moved upon the face of the waters. B'reshyit (Genesis) 1:1-2

From this text, it is easy to assume that the word "earth" refers to the planet as a physical sphere as it is known today. However, as we read on into the next sentence, we see this is not the case. The words read, "without form;" we could interpret this phrase to the minimal meaning stating merely that there were no life forms. This is correct because there were no life forms. However, the words, "and void," that follow afterward shows that there is a deeper absence than that of just being the absence of life. While these words seemingly depict a Hebrew reiteration of the previously stated condition, at a fundamental level there is a reason for the Hebraic expression. As we shall observe later, there are two immeasurable manifestations sustaining the finite manifestation within our universe.

The Torah was not ignorantly written. It is the law governing all knowledge, given by Elohim. HE has given us information in these words that unlocks the mysteries of the universe, both seen and unseen within its physical nature.

The Hebrew word used for void is Bo-Hu. The actual letters are Beth, Hey and Vav. It means in general an indistinguishable ruin. Another interesting facet of the Hebrew language is that each letter represents a picture. These pictures actually describe the word that they spell. Often they give a message concerning the word, but not always. Elohim revealed this particular concept to Dr. Frank T. Seekins (the address is found in the list of resources). This understanding provides a very useful revelation tool. Even though our book does not use any quotes from his books, it does work within the guidelines established. The Hebrew letters correspond in the picture language as House-Window-Nail. The central figure is a window. The house modifies the image of the window. A window, in essence, is a hole in the house or a vacant location. The nail gives the attribute of unconditional attachment to the condition. This gives the final meaning of being an indistinguishable location with no chance of change within itself.

Chapter 1: Foundation

If we were to move from point A to point B within this void, we would not know if any movement truly occurred or if that we performed our actions within one location.

Ramification Fractals

Despite the apparent complexity of the information within the Torah, the information is not hard to understand. The nature of the information is as a fractal. A fractal is a pattern that repeats itself within itself. Theologians call it, "shadows and types." The pattern is defined. All, of its levels of meaning, follow the lines of the defined pattern or "Law." The above information illuminates the need to realize the fractal nature of the Torah (Law). A person could believe just by faith that Elohim created the universe by the given pattern without examining all the fractals and be correct. However, they would be incorrect in denying the fractal nature of the design.

We also note that Elohim created two universes- a heavenly one and an earthly (physical) one. The focus then shifts entirely to the nature of the physical universe. As we shall observe, the creation of Heaven exists independently of ours. Therefore, the formulation process of Heaven is of no consequence or concern to the physical realm

Rose Illustration

In the interest of establishing an understanding of the Torah in the context of being "void and without form," we present the following illustration: Imagine being an observer looking upon a blank sheet of paper. After picking out a certain minute nondescript spot, we continue to focus on this location as an artist paints a picture upon the paper's surface. Let us say that our spot of attention ends up painted red and was a petal of a rose, particularly if the process was to stain the white paper red. We could say, "In the beginning, there was a rose," as it is our focus, "and the rose was without form and void," as it has no shape or color before being painted. Then we could also say, "and white was on the face of the deep," as the entire painting exists upon a white piece of paper (recall, "darkness was on the face of the deep" (B'reshyit Gen 1:2)).

While it may seem silly to make that statement because the painting of a rose is only a representation of a created real rose, it does provide a perspective of the observation. Relate this to the statement that the earth was without form and void. To say that it was without form means that there was no shape; the area designated to be the earth was not yet formed. Being void means that it also had no make-up as in being **elementless** (having no atomic composition); this relates to the absence of red paint that composes the shape of the rose.

Having a form without substance is impossible within our physical universe, just as a solid metal coin cannot have one side missing. The relationship conveyed by the Hebrew language is the acknowledgement of both sides. Form appears from the outside

Part. 1: Establishing Physical Space

toward the object. Substance observed from within the object supplying material composition outward to produce the form of the object. In a more general mathematical sense, it is viewing finite values from both the infinite outside towards the defining attributes and from the infinitesimal inside outward to the defining attributes.

Tranquil Beginning

Our universe was not contrived out of chaos as understood in today's terminology (meaning disorder via upheaval), because there was law and order established from the very beginning. However, chaos also means indefinable, indistinguishable and defused. The Hebrew letter for chaos is Mem. Mem is, in the original picture language, water. While it is true in today's world, water waves in a storm are wildly destructive; this is not the "ground state" of water. It has no form of itself, hence formless. The meaning is then that of being undefinable shape. The purpose here is not to play semantics, but to negate the present claim that the universe was originally in upheaval (referring to its supposed precedent violent existence being exclusive of Elohim). Because of the presumption attached to the word, "chaos," to be indicative of upheaval only, we need to consider the overlooked meaning of this word: being indefinite in nature.

If the universe were to originate in total upheaval by energies out of control, this would require velocities, which attain their properties by acceleration. This implies that someone or something had to accelerate them. Moreover, the energies themselves have defined attributes. To have any defined attribute, there has to be order. If there are no defined attributes, then we are returning to our definition of chaos.

The basic premise to establish is that all the soon to be created forces of nature are at a balance. The resultant vector of force at any location in space is zero. Note: the composition of a vector is of both direction and magnitude. The statement then reads: no energy and no active direction exist. This means that all the properties of nature are at a balance in such a manner that they annihilate the manifestation of an opposite created property if it was to exist. For example, if a negative charge and a positive charge were to share the same center, the result is no charge. Another example of self-annihilation of a given property of energy appears in gravity. Imagine gravity pulling upon a point with an equal opposite force. The result is zero movement. The very first state of the physical universe, then, has the attributes of being stable and motionless.

Movement

Within Hebrew there are other words exist that also translate to both move and shake. However, in those words there exists an undertone of destruction or fear. The actual word in this text is rachaf (רכף). The composition of this word is Reysh (Head), Chet (Fence) and the Pey (Mouth). Note: the absence of the dogesh (dot) in the final Pey gives the "f" sound. The central theme is the fence. There are many ways to look at a fence. Within this word, we are looking at the nature of a picket fence in contrast to a wall. The idea is that every other plank is missing giving the alternating effect. Reysh (head) is the primary modifying image the word giving the image of the primary, first or initiated "fence" or the initial alternating effect. Further modification of the picture is the mouth that speaks words. Whether we look at the moving mouth or the vibration of words, the resulting image is that of an initiated alternating movement, hence, the enacted vibration. This word will take on more meaning as we examine the nature of the most fundamental essence of the physical universe.

There is another attribute to the statement in B'reshyit (Gen 1:2), "Elohim moved upon the face of the waters." Elohim exerts this force outside of the physical universe. The image of moving upon the face is that of something moving over a flat plane. This means that the source of motion is not part of the flat plane. However, the flat plane represents a three-dimensional volume that encompasses the entire physical universe. The conceptual image being that Elohim existed outside the physical volume, yet moved upon the volume as if it were from another dimensional direction. As it will later be seen, this dimensional direction is a superimposed condition created by Elohim.

Waters

The next objective is to establish a connection of the ancient term "waters" with the high resolution of modern scientific terminology acquired by humankind through knowledge of the physical universe. The term, "waters," could easily translate into our term of liquids. However, even this term inadequately describes the wide variation incorporated in the word, "waters." This is despite the fact that often the term, "waters," means precisely: waters, liquid, or liquids. While the term's large contextual mask seems to be a reflection of crude scientific knowledge; it is also descriptive of the nature of the substance. The large contextual mask actually preserves the attribute conveyed.

Consider for a moment: the problem created if we were to observe an alien substance that has no earthly equivalent. We could invent a word to describe it. However, the word would be meaningless to others unless it had a root, which people could identify to some property of a known substance. Without doing that, the substance would remain totally unknown. Even by using a familiar root word, people would not know the exact nature of the substance. The usage of a familiar word would allow them to recognize the substance by the described attribute. That is, if they did not become too dogmatic about the meaning of the chosen word. The term, "waters," becomes a contextual fractal that gives us the understanding in the basic nature of the substance involved. In this instance, it describes the nature of the **preatomic** universe as being, "waters." The reason this term would not translate well into being a liquid is that a liquid needs atoms to exist within a certain state of activity for such a definition. Even

Part. 1: Establishing Physical Space

in modern terminology, the term atomic plasma (a step above vapor) is not an accurate term. Although this term, within our present contextual mask, depicts a liquid, it could also represent some other unknown non-hydrogen-hydroxide (water) primordial state. However, we are not without data; the term does state that it has a fluid like quality.

Consider this preliminary evaluation of a time before space and matter. The preatomic universe was not as empty as might be perceived for matter of the subatomic particles to form from it. Remember that the physical universe, at that time, already formulate into existence out of nothing via the creational directive by Elohim. The substance here is not referring to protons, neutrons, or electrons, but to a substance more fundamental than the concept of a quark, which supposedly compose the subatomic modules mentioned. We should note that the preatomic universe acquired many different states before reaching the state of being a subatomic universe. Imagine a substance by which both matter and space come from. The state of matter between a solid and a gas is a liquid. Liquid becomes a term appropriate for the preatomic universe, even though, there are no atoms to satisfy the scientific qualifications of being a liquid. Elohim chose hydrogen-hydroxide (water: emphasizing that water is an atomic compound and not a primordial element) as it is the state between ice (a solid) and steam (a gas or vapor) that people during that time understood. Phenomenally, it is the union between something and nothing. Even though this sounds absurd on the surface, we shall observe it to be true by further examination.

We should note at this point, that scientists have concluded that matter is energy via Einstein's Theory of Relativity. In this, Einstein states that if we accelerate matter to the speed of light, matter then would become energy. Therefore, rather than concerning ourselves with the creation of matter, per se, the focus should be upon the creation of physical energy. The term physical energy differentiates the energy of the physical universe from among the energy of the Spirit of Elohim (which permeates the physical creation), the energy of the universe of Heaven, and the energy that Elohim selectively connects to living organisms. While we know that Elohim created motion in the physical universe, the objective is to relate all the scientific ramifications this had upon the physical universe.

𝔓hases of the 𝔘niverse

The different states, which the universe transforms into before being the universe that we know today, are phaseverses. The etymological formulation of **phaseverse** is from combining the words **phase** and uni**verse**. Each phaseverse is a unique state of the universe, as Elohim transforms nothing into something. The first eight phaseverses occur in the "first Day" of creation. Each action Elohim takes in forming the "earth" generates a phaseverse. In each phaseverse, there is an addition of energy or the universe becomes universally different in fabric composition. It is not until the sixth phaseverse that there is anything finitely definable. In the seventh phaseverse,

Chapter 1: Foundation

subparticles (sub-submatter) finally formulate into separate stable self-sustaining entities. Even so, there are distinct attributes that distinguish previous phaseverses.

An illustration: an artist desires to make a painting. Imagine having to weave a canvas. The canvas is the first visible phase. Next, a frame is built, which is another phase. Afterward, the canvas is stretched over the frame. Then the image drawn (if the artist chooses) on the canvas is another phase. Paint application to the canvas creates the central phase. Then applying a protective coating and framing of the picture become two other phases. Then there is a setting in which the painting is displayed.

Additionally, the usage of the suffix, "verse," is to indicate that it is a type of universe. It engulfs the entire expanse of physical space on all levels, infinitesimal, finite and infinite. Below is an overview of the different phaseverses of creation as they occurred:

Phaseverse	Name
1	**Nulverse**
2	**Inertverse**
3	**Gravverse**
4	**Kineverse**
5	**Thermaverse**
6	**Xyzenverse**
7	**Magneverse**
8	**Neutronverse**
9	**Plasmaverse**
10	**Bondverse**

These phaseverse definitions increase in detail as we go through the formation of the universe, but first we will give a brief etymological description. All but the sixth phaseverse have a known scientific base to their names.

The **nulverse** obviously derive from the words of null and verse. Null means more than zero in mathematics; in a solution for an equation, null means no solution or the "empty set", as zero is a numerical answer. For an example: an integer less than one and larger than negative one is zero. Zero, here is the solution. However, if we look for a number less than one and greater than two, there is no number to satisfy the conditions. Null then is the empty set. In the same way was our universe- nonexistent.

Part. 1: Establishing Physical Space

The etymology of the **inertverse** follows the same pattern, as does the other phaseverses. In this phaseverse, we observe the manifestation of inertia. The **gravverse** finds its etymological composition from gravity. Although gravity is the most well-known effect of this force, the root of gravity formulates in this phaseverse, which is a primary force of attraction. The **kineverse** derives its prefix by the word, "kinetic". Kinetic energy is the energy of motion. Recall the earlier quotation stated, "Elohim moved upon the waters." The **thermaverse** derives its prefix from the word for heat. Again, this word derivative stems from a specialized effect of the force involved, as Elohim creates "light."

The **xyzenverse** pronunciation sounds like "size n verse." It would be meaningless or lengthy to describe the meaning other than the hint given in its name- size nth. One interesting point about this nearly quanta sized particle is that the size can vary upon different conditions. This refers to the most fundamental finite granule of submatter that has a definite volume selected by Elohim.

The **magneverse** is the point in creation that Elohim separated light from darkness. In this phaseverse, magnetic energy plays a vital role. The **neutronverse** refers to a universe of freshly formed neutrons. The **plasmaverse** concerns itself with the formation of atoms into various plasmas.

The last phaseverse, **bondverse**, derives its nomenclature from a term in chemistry- bond. This refers to the bonding of atoms into molecules and consequently the final composing of planets and stars within systems of galaxies, which are beyond the scope of defining submatter. Today's universe is a modification of the original bondverse. Strangely enough, or maybe not, each, of these ten phaseverses, line up with the Ten Commandments given to Moshe (Moses).

Chapter 1 Quiz:

1. How many answers are there to this equation: 1+1?
   A. 2, not counting 10_{two}
   B. Many, but only one of them is right.
   C. The answer is an empty set.
   D. The answer is relative.

2. Define waters:
   A. H_2O
   B. Liquids
   C. Union between solid and vapor
   D. All the Above

3. What is the Unified Knowledge?
   A. All answers are equally right.
   B. All answers are equally wrong.
   C. G-d, Science and Math unified.
   D. A myth, there is no absolute truth.

4. What is a phaseverse?
   A. A shift in a verse.
   B. Transition phase of the universe.
   C. A stage between nothing and something.
   D. Both b and c.

5. How many phaseverses are there?
   A. 10
   B. 7
   C. 12
   D. 3

6. (T/F) The Name of Elohim can be spelled JHWH or IHVH.

7. (T/F) A nonrepeating decimal is an expression of a fractal.

8. (T/F) The opposite of $1 + 1 = 2$ is $1 + 1 = 11$.

9. (T/F) The thermaverse occurs before the gravverse.

10. (T/F) The bondverse occurs when molecules form.

Chapter 2

𝕹ulverse

The nulverse is a comprehensive label for physical nonexistence. Physical nonexistence is a complete statement of the entire universe. The key words here are- physical existence; this differentiates between the absence of existence and the absence of the physical universe. Since Elohim is eternal, this statement means, only, that this physical universe was not present in terms of the eternal past.

To understand the phenomenon by which the physical universe came into existence, we must have a working knowledge of where it came from. Since Elohim is the Creator and all things exist within HIM, we must come to terms with the nature of Elohim. The following information derives itself from the attribute of Elohim known as omnipresence; which as we know, it means Elohim exists everywhere.

Omnipresence

When we define omnipresence, we commonly attribute the contextual meaning of being infinite. We then often think of omnipresence as something in which no finite distances are large enough to incorporate the infinite nature of HIS Presence. However, there is an opposite side to this expression; there is no distance too small to be outside HIS Presence. This may seem absurd until we realize that human beings have not acquired the definition of the smallest finite distance possible.

Example, Let us say that a certain piece of subparticle measures to be a decimal number with about 30 leading zeros meters in diameter ($1/100 = 0.01$; this number is said to have one leading zero). Then, we discover that this too is composed of some material, and this material is composed of yet another type of smaller substance. This could go on and on and on! However, there is an end. This is mathematically known as a geometric point. Since it has no dimensions, we cannot divide it into smaller particles.

This brings us to the concept of geometrical space. Geometric space is composed of geometric points in all three-dimensions creating a volume that encompasses the entire universe. Since geometric points have zero dimensions, they exist in space but occupy no space. Lining them up into a single line, it would take an infinite number of them to amount to any finite distance. Imagine adding zero, as many times as we would like; the answer is always zero when added a finite number of times. The same is true with lining up a finite number of geometric points; the result is that we have not reached into finite

Part 1: Establishing Physical Space

space. It is by the very existence of finite space that we know that zero can be added an infinite number of times and derive something different from zero. This observation occurs throughout mathematics.

This leap also illustrates itself in the decimal notation for 1/3; we can write this in decimal notation as 0.3333... The three dots mean that the threes continue indefinitely. (There are other notations, but this will suffice.) Multiplying one third by three, we know the answer will be one. However, the multiplication of the decimal gives an answer of 0.999... By the analysis of the two mathematical statements, we observe that, in fact, a decimal followed by an infinite number of nines equals the value of one.

Basic Geometric Understanding

Before getting started with the actual portrayal, we should introduce some geometrical terms. These will prove useful in describing the three-dimensional directions of the material universe. The dimensions provide width, length, and height. Each dimensional direction describes a line. A horizontal line represents the dimensional width. This is commonly labeled as the x-axis. The vertical line represents height. This dimensional direction defines the y-axis. When these two lines extend far enough toward each other to cross on a vertically flat surface, they are said to intersect each other. This intersection then defines the origin. From this origin, it is possible to construct another line directly toward the viewer and directly away from the viewer. This line will be perpendicular to both the x-axis and y-axis, and it is labeled the z-axis. These axes, as all lines, have only one dimension; the other two dimensional attributes measure zero units by any unit of measure. The origin is then observed to have zero dimensions.

A plane is a two-dimensional object formed by any two axes. We can more accurately define this as a surface inscribing any three nonlinear points (points that do not form a line). At this point, we are only concerned with those that are perpendicular to each other within the reference to the dimensional directions. For example, the xy-planes are parallel to the plane formed by the x-axis and y-axis. The other two sets of planes that are perpendicular, we know as the xz-planes and the yz-planes. As proposed, these are parallel to the axis described by the first two characters of its nomenclature. Being a two-dimensional object, these planes have one dimension that measures zero units.

Finite Limits

From this, we can formulate a geometric progression from one geometric point to a three-dimensional projection encompassing all space. During this interval, we are not concerned with the space-time continuum. Our focus is the fundamental construction of the volume of space from the perspective of a geometric point. Note: the purpose is to show that a continuum of geometric points sustains the entire volume of physical space, not that it started from a single geometric point.

Chapter 2: Nulverse

For conceptual purposes, this distance achieved by the addition of an infinite number of zeros will be valued as an absolute one. Note that this absolute value represents the entire finite existence per dimension, and we cannot assign it any kind of measuring unit of distance. This is because we can divide any unit into smaller finite units or multiply it into larger finite units. The finite universe then appears as having two limits. The lower limit is greater than zero and the upper limit is less than infinity.

Further observations indicate that each finite location in space circumscribes an infinitesimal universe consisting of geometric points. Secondly, any finite distance has the same relationship to infinity as the infinitesimal universe has to finite space. Just as in the addition of zeros by any finite number of times, equals zero. Similarly, the addition by a finite number of finite distances creates merely a larger finite distance. This larger finite distance is no closer to infinity than a million zeros added together is closer than adding a thousand zeros is to reach the value of one.

Let us take an unfathomable finite distance of one-googolplex light-years. A googolplex is a number with googol zeros following a one. A googol is a number with 100 zeros following a one. Each light-year is approximately 5.9 trillion miles. Multiply this distance by a googolplex, and we are still no closer to infinity than any other finite count. We are given the following: 1 mole = $6.02*10^{23}$ molecules, 1 mole Hydrogen-1 = 1 gram, 1 mole water (H_2O) = $2*1.00794 + 15.9994 = 18.01528$ grams, and 1 ounce = 28.3495 grams (avoirdupois), 128 ounces = 231 cubic inches, which defines 1 gallon. We then get $(6.02*10^{23} * 28.3495/18.01528)(231/128) = 1.711*10^{24}$ water molecules per cubic inch. Next, we calculate cubic inches per cubic light year. We are given: 1 mile = 5,280 feet and 1 foot = 12 inches giving $2.54*10^{14}$ cubic inches per cubic mile, then $(2.54*10^{14})(1.711*10^{24})$ gives $4.3504*10^{38}$ water molecules per cubic mile, a light-year = $5.875 * 10^{12}$ miles giving $2.02802*10^{38}$ cubic miles per cubic light-year. The number of water molecules within a single cubic light-year = $(4.35*10^{38})(2.028*10^{38}) = 8.823*10^{76}$ molecules. Since: $100 - 76 = 24$, $24/3 = 8$, and $10^8 = 100$ million, $(1/8.823)^{1/3} * 10^8 = 4.8*10^7$ light-year cube gives one googol molecules. Imagine each molecule forms a zero following a one. That is just for a numerical representation of a googolplex!

There is one last side note on unfathomable finite numbers. Imagine being in heaven so long that it would take two centuries to express the number of years of being there if each digit was different! It will happen because it is a finite number. We probably would not want to do all that continuous talking, but we will know it by experience.

At this point, the objective is to illustrate that the fundamental state of the universe is a physical manifestation from the characteristic of Elohim. As has been described, there are two limits to the finite universe. By this, we understand that the finite universe is surrounded by something that is beyond its reality. On the same note, we observe that finite space owes its existence to that which is beyond it. We can further state the following: it is because the geometric points enumerate out to infinity that finite space exists. Physical space is another issue, as will be expounded. Viewing our example as

Part 1: Establishing Physical Space

physical space will not destroy the concept. However, it would be a mistake to assume that physical space is Elohim.

Expanding Sphere

Now that we have established all the needed background knowledge, we are ready to examine the mathematical portrayal of Elohim. Imagine starting at one point in space and reaching out to infinity in all directions simultaneously. An image of a sphere emerges. Now, try to imagine it growing out to infinity. First, we count inches, then miles, then light-years... Then it becomes a matter of conceptually expanding. The problem is that no matter how far we travel in finite space- there is always space in front of the sphere. In this, we experience the edge of the sphere as a sensing device. Up to this point, it has experienced only the space in front of it. As long as we remain in finite space, we will always experience space outside the sphere. Imagine the qualifications necessary to encompass all space unequivocally. Consider this sequence of questions and simplistic but real answers:

> **Q:** How would the sphere know that it had encompassed all space?
> **A:** When it could not sense space outside itself.
> **Q:** How could the sphere realize that there was no more external space?
> **A:** When it could go no further.
> **Q:** How would the sphere experience that it could go no further?
> **A:** When the spherical surface sensed itself.
> **Q:** How did the sphere come to sense itself?
> **A:** When it returned to its origin.

The entire sphere returning to the origin without changing directions accomplishes this. If "the returning to self" came to a point outside the origin, then it would not have sensed only itself but a counter point. Elohim expresses this totality by saying: "there is no other Elohim besides me." Our scenario is not a portrayal of the creation of Elohim. We can know this, for Elohim already defines the totality of space. Just as the geometric points existed outside the path of the growing sphere, so does Elohim exist. We are only examining to understand the reality.

Another interesting observation to make is that Elohim does not have to move from location to location as HE is already there. It is for this reason that the geometric points do not move. Note: the mathematical expression of HIS Essence cannot be used to prove that HE is a living entity. However, as stated earlier, this is not the objective. The point is to describe the characteristics of HIS Geometrical Existence.

By being able to go in two opposite directions and returning to the origin without contact, we get our ancient symbol for infinity. Imagine being in the center of the symbol and moving out in two opposite directions simultaneously. We would return to

Chapter 2: Nulverse

the origin just as we did with the spherical imagery. The loopholes observed exist only in our finite bend representing the process. The greatest argument against this symbol is that it is humanly contrived. In other words, the Torah does not mention it and the symbol appears in occult uses. Often occult usage stems from natural phenomena. Example: Mars was an actual planet before gaining a pagan value. Two mathematical phenomena occur that show the symbol for infinity does have a natural expressions.

Before continuing, there is another interesting observation derived from the growing sphere experience; imagine seeing in all directions from this geometric point out to infinity. This point would observe itself in a giant canopy image. In essence, the geometric point would see itself in its fullness. Every angle referenced upon the surface of the geometric point magnifies. Another sideline, the totality of infinity or the fullness of infinity can be illustrated as 100 percent in terms of being all that is possible. An example of this illustration is answering the questions in an examination correctly. Nobody can get more correct answers than there are questions.

The first mathematical observation found is in plotting a sine or cosine wave using polar coordinates. For analyzing the symbol, we will position it using the cosine function. In polar coordinates: instead of measuring degrees of an angle along the x-axis, degree measurements express the angle from the origin. The amplitude of the angle defines the distance from the same point of origin. The second example is that a sine wave placed on a sphere of the same radius as its amplitude will produce the same symbol.

As preposterous as the concept of returning to the origin may seem, there is even more mathematical phenomenon to back such a claim. There are several typical examples to examine. One is the equation of $y=1/x$ and another example is the concept of a curve that has a radius that is infinite.

Infinite Reality

Graphing the equation, $y=1/x$, upon the xy-plane, we make the following observations of the phenomenon. From the extreme left (negative infinity), a nearly horizontal line veers below the x-axis and then sharply bends downward as it approaches the y-axis. In finite space, this line becomes nearly parallel to the y-axis and does not touch the axis. Meanwhile, on the right side of the y-axis, we observe the line coming down from an extreme height being nearly parallel to the y-axis and bending out to a nearly horizontal line as the x-values get larger. The reason for this is the larger x-values will produce an answer close to zero, but not zero since they are finite numbers. Near the y-axis, the small fractional values of x will divide into one many times creating large y-values on the graph. The finite x-values hold the line away from the y-axis. The left side bends down showing negative values of x, while the right side bends up showing positive values. In math, we say the line is discontinuous at zero, the reason being there is no finite link between the two sides. However, reality does not quit in finite space.

Part 1: Establishing Physical Space

Reality does not quit on the equation and say: "it is too much to continue on; I'll quit in finite space somewhere." No, if it did, we could find such a location. Since reality does not quit, it continues into the infinite realm and finds itself again on the opposite side of finite space approaching zero. We can first assume that the line crosses at the backside of the universe; as we describe the inertverse later, we will see it does much more than that. Right now, our objective is to show that an infinite limit does not exist beyond reality, and is in continuity to any finite realm of reference.

Geometry offers another easy visual illustration of the relationship between infinity, one and zero. In this exercise, we are inscribing different polygons within an equal sized circle. Moreover, each side of the polygon inscribed in the circle is equal in length with the rest of the polygonal sides of the given inscribed polygon. As we increase the number of sides in the polygon, two distinct phenomena occurs: each side forms a smaller portion of the shape and the angle between two adjacent line segments becomes more obtuse forming larger interior angles. To see the phenomena in action more clearly, inscribe the polygons within the same circle. As the number of sides increase, their sides approach the edge of the circle. Eventually, the sides of the polygon will become visually immeasurable even though they are finite in number. As the number of sides approach infinity, the size of each side approaches zero. When the number of sides is infinite, the size of the sides becomes zero creating a single (one) surface.

Slope of a Line

Another point to acknowledge is the relationship a straight line has to infinity. Take a piece of paper and do a wild scribble without using jagged lines. Imagine that this figure is a three-dimensional line that never touches itself. In calculus, we can state the following two items: at any point upon a curved line, there is only one line that can be drawn tangent to the line. The slope of this tangent line can be computed. Moreover, this tangent slope equals the slope of the line examined at that finite location. This is only possible because the line forms a straight line in infinitesimal space. Similarly, any finite twist (angular variation) becomes nonexistent in infinitesimal space. Therefore, a straight line in finite space can create any strange shape in infinite space and still be finitely straight. However, Elohim does not arbitrarily create a fundamental design without purpose. This shows only that the finite universe does not impose any restrictions upon the infinite manifestations. However, the finite universe finds relationships within its structure relating to the infinite structure (seen in $y=1/x$).

Geometry, again, provides the initial equation for finding the slope of a curve by its tangent at any given point. The equation of a line is **$y=mx+b$**. It reads, the value of **y** is equal to a **m**ultiple value of **x** plus the **b**ase displacement (y-intercept, where the line crosses the y-axis). Within a line segment, the difference between the two end values of y (the rise) divided by the difference between the two end values of x (the span) determines the slope (m). In essence, we are dividing the rise by the span.

Chapter 2: Nulverse

Now we are going to use this concept to find the slope of a curved line at any given point. Instead of looking at two isolated points as in the first example, we are now going to look at two end-points that relate by a given equation. We are viewing y as a function of x; hence, the reference to y is replaced with f(x). This reads as: the function of x. In our example, we chose y as the function of x^2 or x squared. Remember that for the rise we took the difference of the two end-values of y; now, we have an equation to determine these values: **$f(x) = x^2$**. By selecting and applying different values for x within the equation, we are able to plot the line graphically.

Our next objective is to define the equation, which will give us the slope of the curve. First, we note that any location on the curve corresponds to a specific location on both the x-axis and the y-axis, and appears as the intersection of these two axial locations giving us the (x, y) coordinate. Now, we need another end-point to determine the slope. We need to acquire this value within the framework of our equation **$f(x)=x^2$**. Therefore, we add some distance to x that is h units long giving (x+h) and the y value of **$(x+h)^2$**. We can draw a line between these two points and get a slope that is close to the actual slope of the curve at (x, y) or (x, f(x)). To get closer, we must choose smaller values for h, and ideally bring h to zero. Algebraically, using the slope equation is the first example dividing the difference of y by the difference of x, but y is determined by the equation of f(x). Therefore, we substitute y with f(x) which we can replace with the equations from y_2-y_1/x_2-x_1 with $(x+h)^2-x^2/(x+h)-x$ giving $x^2+2xh+h^2-x^2/x+h-x$. After doing the multiplication, we find terms that will cancel giving $2xh+h^2/h$. One note that we should bring out is that the lone division by h before canceling could never be zero, for the equation would "blow up." Dividing by zero gives an indefinable answer hence the concept of limits, which we will not get into. Since $2xh/h=2x$ and $h^2/h=h$, we finally can bring h to zero using 2x+h giving our answer of **2x**. This means that the slope of the line of this curve is twice in value as the distance from the origin upon the x-axis.

The conclusion reached is that if there were no true zero or geometric point, we could not achieve this answer. We would be forever stuck with a close but not an actual answer, or that space is a continuum of some finite granules. In addition, any finite line, except that which we can be described as an intersection of two lines (a jagged line), cannot disturb the nature of infinitesimal space. Similarly, we cannot assume the nature of a straight line in infinite space by its nature within the finite space.

Area under a Curve

Even more demonstrative than the derivative of an equation is the antiderivative in which the area under a curve is computed. For this purpose we are looking at the same equation: $f(x)=x^2$. Since we are looking for a definite area, we will limit the x-values to be from zero to three. In this context, the best way to describe an antiderivative is that of adding up columns (thin vertical rectangles) defined by an equation.

Part 1: Establishing Physical Space

The method of determining the antiderivative has two components, the selection of rectangle size and of the adding of sequential sums. Regarding the size of the rectangles there are two factors, one governs the height and the other is the width. These are important, as the formula for a rectangle is the following: Area is equal to length multiplied by the width. In selecting the length, we will let the value farthest from the origin to be the determining value. We could have chosen the closest value or some middle value. However, all would eventually lead to the same answer. Using only one column forms a rectangular area of 27 units, since the rectangle is 3 units wide and 9 units high. Note that 9 is 3 squared. More generally, we will use w for the width making the length the measurement of the width w squared. The value of 27 units is far greater than the actual value as most of the rectangle exists above the curve.

Next step: we divide the width into n segments; hence, we get w/n as the width. The greater in number the segments are, it follows that the width of each segment becomes smaller. Furthermore, the smaller the width segments become, we get an answer closer to the actual value. Notice that each rectangle base has the same width, and the length of each rectangle is different because of our equation. These differences in length make it necessary to address each increment individually. For this purpose, the letter "i" is used to represent each increment that we are going to add to get the total area. Let us divide the width of three units into six equal parts and reference the length of the fifth rectangle. To do this, we set w=3, n =6 and i=5. The width of the rectangle is always going to be w/n, and in this case 3/6 or one-half unit. We determine the height of the rectangle by the number of 1/n increments traversed away from the origin and apply this value to the equation to get the y-value. In our case, iw/n becomes 5*3/6= 5/2 or two and a half, so applying this value to the equation of x^2, we get 25/4 for the length. Then we multiply the width by the length giving (1/2)*(25/4) = 25/8 or 3.125 units squared. That is the process just for one rectangle. This process continues to become more horrendous as we refine our resolution for the answer.

We are now ready to look at the process of summing values. In approaching the actual answer, we need to add large quantities of increments together. Imagine adding 1 through 100, and then try adding 1 through 1,000,000,000. The first one would be hard enough, the second nearly impossible by time itself. However, as in many cases, there are mathematical shortcuts. In adding 1 through 10, consider: 1 plus 10 and 2 plus 9 and 3 plus 8 and 4 plus 7 and 5 plus 6. We get 11 five times giving a total of 55. Therefore, for adding 1 through 100, all we need to do is add the first and last number half as many times as there are numbers, or multiply by one half of the number of elements. The result is (100+1)100/2 giving 5050. We can put the equation in general terms: (n+1)n/2 or in better math form, n(n+1)/2. For i^2, it becomes more difficult. This information is in the appendix: "Antiderivative" of this document. After adding the various summations, we get, as one form of the results, $(n^3/3 + n^2/2 + n/6)$. Even though the denominator is not uniform throughout the terms, the equation is more uniform in form between terms.

Chapter 2: Nulverse

Now we are ready to solve our problem. Returning to the summation of rectangles having the value of w/n (total width divided into a number of equal segments) for a width and a height of $(iw/n)^2$ (the individual increment multiple of w/n squared). Since w and n become as constants within the equation, as we are not going to change the width or the number of width segments midstream within the calculation, i^2 becomes the only real variable. We are going to delay our calculation, and we will move all of these values in front of the summation or Sigma symbol. This shows that we are going to multiply the summation results by these quantities. The resulting "external constant" is the standard dimensional alteration of the rectangle: w/n (width) times w^2/n^2 (length) giving w^3/n^3. Then we substitute the summation of i^2 with $(n^3/3)+(n^2/2)+(n/6)$ and finish our delayed calculation to include the increment. This results in the equation of $(w^3/3)+(w^3/2n)+(w^3/6n^2)$. After doing all of this, we let n become infinite bringing the value of the last two terms to zero, leaving $w^3/3$ as the answer. This is the basic principle of the antiderivative as seen in finding the area under a curve defined by x^2.

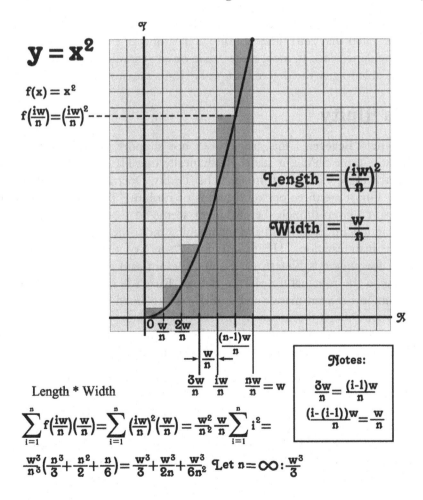

43

Part 1: Establishing Physical Space

The actual evaluation of this integrand is 9. If we had chosen to evaluate the definite integral from 1 to 3, we would arrive with 8 2/3. We subtracted 1/3 from 9, since 1^3 is still 1.

There is another variable involved in our equation that is invisible to our situation but in calculus is always considered. Every time we do an antiderivative, there is always a chance of another variation. Recall the formula for the slope of a line: y=mx+b. Our calculations have only concerned themselves with the m part of the equation, but there is a b. In our case, the value is zero hence not considered. Just as the line does not always pass through the origin, we could have a variation in placement of the curve giving a constant c. Therefore the correct form of the antiderivative, in our case, is $w^3/3 + c$ or presented in the standard mathematical equation: $x^3/3 + c$.

While this illustration might seem like a partial review of basic calculus to some, this is an illustration of a basic concept of the relationship. Notice that when we take n to infinity that the width of the rectangles becomes zero. This makes the areas of each rectangle to be zero. We are multiplying zero by infinity to arrive with a finite answer. Recall that we stated that zero times infinity equals one and that the value of one unit could be of any dimension: an inch, meter, mile or any variation in between. Therefore, the equation is able to give the determining value to its measurement.

Elohim's Unity

Along with other "Believers" this article does not support the Trinity Doctrine (See Devarim, Deut 6:4, the Sh'ma, it states emphatically that HE is one), but supports the data from which the conclusion formed. The Trinity Doctrine states that the Father, Son, and Holy Ghost are three personifications of Elohim. The root of the problem is the word, person. Even though the argument seems so semantic, it is the difference between drifting toward polytheism and maintaining monotheism. While Elohim is one being by three manifestations, HE is not three "sets" of three manifestations. The Hebrew word for person is Nefesh (or Soul). Elohim said that HE is **Ruach** (Spirit) when referring to the root nature of HIS Being. Ruach is not Nefesh; Nefesh is the result of the union between matter and spirit. As explained earlier, the finite universe owes its existence though the union of the set of geometric points with the totality of the universe. HIS "Nefesh," we know as HIS only begotten Son. The Son came to earth via HIS Ruach overshadowing Mary and became known to us as Yeshua (Jesus).

Along the same line, there is a belief that the soul and the spirit are the same thing. Looking at the words in Hebrew does not help much as they have interpretations similar in meaning. However, going through the Tenach (Old Covenant Scriptures) and B'rit' Hadeshah (New Covenant Scriptures), the reference to soul and to spirit have different contextual behaviors. The soul remains with a particular body until death. A spirit can enter into a body or leave at will without death. Finally, there is a scripture

Chapter 2: Nulverse

addressing this issue in Hebrews 4:12. It reads, "...piercing even to the dividing asunder of soul and spirit, and of the joints and marrow..." This was in reference to the word of Elohim being sharper than a two-edged sword. While it shows them as two different distinct objects, it also shows that they have attributes difficult to separate. Another observation is that this is written to Hebrew people, not to a Hellenistic society. This illustrates that the differentiation between soul and spirit being objects beyond a Hellenistic understanding.

Nefesh is composed of the following Hebrew letters: Noon (Fish), Pay (Mouth), Sheen (Teeth). The Fish represents the life of the waters. The Mouth speaks expressions. The teeth generally devour; however, we should note that it aids in forming words. While breath can be attained with the mouth, the central theme of the word is expression. Since the fish picture modifies the mouth picture, it is an expression of life inside as the teeth are in the mouth. Another avenue of meaning for Sheen describes a transformation quality. In this, the Sheen indicates that the soul is in transition. In other words, its attributes are transitory and can fluctuate often in characteristics between good and evil. As we shall observe, the spirit cannot alter its nature from being evil to that of being good.

Ruach is composed of the Hebrew letters: Reysh (Head), Vav (Nail), and Chet (Fence or Sanctuary). The similarity between a fence and sanctuary provides an image of a place where we seek refuge from an enemy. Later, we observe a tabernacle erected as a covered fence. The central theme is the nail picture representing that which holds in place or sustains. The head picture modifies the image giving a meaning of a primary sustaining force. The Chet refers to this force as being held separate from the external influences. This is especially true with the human spirit being separate from Elohim's spirit. Note: A sanctuary can become defiled or compromised. Once destroyed, it cannot regenerate for its resources are depleted. By this, we observe the fall of Adam as a ruin beyond his means to reconstruct. Hence, we find evil spirits unable to change themselves or change their intent via external manipulation. Yeshua (Jesus) said in Mat 9:17 (KJ), "Neither do men put new wine into old bottles: else the bottles break..." This refers to being born again. Elohim gives us a new spirit and not a refilled or patched up human spirit. Hence, a new sanctuary is given- HIS Holy Spirit. This is the fact that "Lucifer" (Heilel in Hebrew, root meaning bright and praise) did not understand when he caused man to fall. He thought that human beings would become unredeemable, as he is. It was not until Yeshua (Jesus) rose up out from the grave, that people began to understand the plan of Elohim was to redeem humanity through HIS death as a human.

The last point to make is the requirement of a Virgin Birth of the Messiah. There is destruction of the human spirit by the sin of Adam. Yeshua (Jesus) came to die for the sin Adam committed which thrust the world into perpetual sin. In order to do this, HE had to be born with a spirit intact as Adam once had. He chose not to transfigure HIMSELF except for the occasion in which he was conversing with Moses and Elijah.

Part 1: Establishing Physical Space

Finally, the spirit and soul both have a mind. It is not until we are born again with the Holy Spirit that we can renew our (soulish) minds (see Rom 12:2). If the spirit were solely responsible for the mind, born again believers would have no need of renewing their minds. The mind of the Holy Spirit is infinite. This gives each of us the ability to achieve a virtually infinite IQ (Elohim willing).

Nulverse Creation

The creation of the nulverse is more an act of allocating rather than an act of creating. In computer terms, allocation is defining the location in which information is to be stored within the computer's memory or some kind of storage medium such as a disk. Nothing formulates in that space, only a selection of location. It is roughly the same as a builder selecting a site to build; nothing is built, but a location is selected.

The same happens in the creation of the universe. Elohim selects a location in which HE will place the physical universe. Now, we are ready to explore the process of this selection.

A Slice of Nulverse

Imagine a geometric plane, a set of geometric points that forms a flat surface with a depth of one geometric point. By touching the surface of the plane, we do not touch the opposite side of the plane, as it is physically impossible. If the plane were opaque, the fingerprint would appear on one side of the plane, but not the other. By this, we observe that the geometric points have sides that provide physical separation while each point has no dimension in itself.

The next observation of this plane is that the geometric points have sides that touch the adjacent points that are not available to the touch on either side of the plane. Even though all of these sides exist, there is only one geometric point. Notice also that these sides have no surface area in them. From this analysis, we understand that a geometric point is a surface that is, in its totality, also the core of the same object. An image of a sphere appears. This gives an infinite number of sides to a point without having a side constituting as a geometric point in itself. The sides are merely an angle of reference to the surface of the same geometric point in question.

The reference angle to the geometric point that is part of the definition of the physical universe is an **alephtip**. The term of alephtip stems from the Hebrew letter, "aleph" and the English word, "tip." The term aleph refers to the beginning, as this letter is often used. The term "tip" signifies that the geometric point is not engulfed but merely referenced by an angle. However, the alephtip does not totally formulate until the inertverse. The alephtip is a general term alluding to the relationship that the reference angle of the physical universe has to a given geometric point. Therefore, we observe

Chapter 2: Nulverse

that the physical universe does not own the geometric points that represent the actual Spirit of Elohim, but merely reference it. We can then say that the entire physical universe or any other universe (i.e. Heaven) is comprised of "virtual" geometric points. For all practical purposes, they are the geometric points within that universe.

These unformatted alephtips are in a diffused state known as the **nultip** (null tip). Actually, three nultips per geometric point ascribes to the physical universe as will be seen shortly in the description of the inertverse. The actual contact of this angle to the geometric point is the **alephfield** (a-lef' feeld). The alephfield is the field of force exerted upon the point as an interface with the dimensional presence of the universe and Elohim.

Imagine this same plane having a pattern on it. This pattern represents the existence of the physical universe. Remove the pattern; now, we have the nonexistence of the physical universe, hence the nulverse. Because there is an absence of a pattern to distinguish a side of the geometric point from the rest of the sides, the side is indefinable from the rest. By this, the physical universe is unformed and no different from the rest. The nulverse truly is as a sheet of paper before the picture, with the picture being the physical universe.

We have one last important observation before discussing the inertverse. Since the physical universe only references the geometric points of space, the physical universe does not contain the geometric points. It only exists in reference to the geometric points. Therefore, there is no way in the physical universe to examine Elohim's Spirit scientifically. However, this does not excuse us from acknowledging HIS Existence as HIS Handiwork appears throughout creation. Moreover, the act of assigning the geometric points to the Holy Spirit does not assign lifeless physical space is to HIM, for HE is the one sustaining physical space.

Chapter 2 Quiz

1. Omnipresence is presence in what?
   A. Infinite Space
   B. Finite Space
   C. Infinitesimal Space
   D. All the Above

2. The Expanding Sphere illustrates:
   A. Balloons expansion.
   B. A sphere thrown into outer space
   C. The nature of an infinite radius
   D. None of the above

3. The purpose of illustrating an antiderivative is:
   A. Show the intelligence of mathematicians.
   B. Finite Space is the union of Infinite space and Infinitesimal space
   C. Show the annihilation of a derivative
   D. Practice

4. What is a Nultip?
   A. A tip that comes to a very sharp point.
   B. An object that has no tip.
   C. Side of a geometric point.
   D. A waitress not getting a tip.

5. What is an Alephfield?
   A. Contact of physical space to a geometric point.
   B. A field in which alephs are harvested.
   C. Energy field generated in a single instant.
   D. Endurance of an energy field

6. (T/F) Omnipresence incorporates the two opposite extremes.

7. (T/F) The Expanding Sphere purpose it to experience inner space.

8. (T/F) Slope of a Curve is an example of an antiderivative.

9. (T/F) There are precisely 7 nultips per geometric point.

10. (T/F) This Physical Universe has always existed..

Chapter 3

Inertverse

Thus far, the alephtips of the nulverse, by definition, are without any type of physical existence other than allocation. However, "data" of existence is about to be introduced into the system. In this, the alephtips will become transformed into another state of being. The comparison between the nulverse and the inertverse can be seen as the difference between a set of binary zeros (absence of energy) being replaced with a set of binary ones (energy) within a data field in a computer. There is a process by which this is accomplished.

The quality that distinguishes a physically existing alephtip from its nonexistent state is the presence of inertia. Inertia is the energy to resist motion. This should not be mistaken for matter. Matter has a specialized volume that houses other energy attributes as well. The inertia fills and holds the universe in the state of rest. While the inertverse is as a blank picture, it is not as a blank piece of paper. The inertverse is formed as a blank picture on one side of the geometric plane. For argument sake, Let us say the color red is displayed upon this side. Unlike, our illustration of the painted rose, the color engulfs the entire medium, hence a blank picture of substance. This image then requires that the same reference angle of each geometric point to have a red attribute.

As we have observed, there are two sides to the plane. This demonstrates the availability of Elohim to create Heaven and earth (two separate universes) simultaneously, which would have opposite polarity. Even though the Torah does not specifically state one way or another, it does not state contrariwise. Since the Torah does not specifically state one way or another, the development of an opposite universe is not an important issue concerning the development of this physical universe. Even after allocating both universes, HE still could develop one side before another. If this were so, we could assume Heaven was developed first as the Torah states "Heaven and earth." This can be stated because Elohim is the true author of the Torah and knew that as a result of sin, we would be "groping" for information. There are other hints of Heaven having been created before earth. Consider the fall of Lucifer. Also examine the purpose of the Tree of Good and Evil in the Garden of Eden, and the method which Satan used the serpent to seduce Eve before the sin of Adam.

Let us continue onward. As stated earlier, the comparison between the inertverse and the nulverse then can be symbolized through examining the two computer bit

Part 1: Establishing Physical Space

conditions. The difference between a computer bit being a binary zero or a binary one is whether or not the bit is holding magnetic energy. In this manner, Elohim connects HIS Energy into an involvement with the physical universe. HE utilizes part of HIS "Manifesting Expression" to hold the physical universe in position. This primary holding phenomenon is inertia.

Inertia is then seen as the first primary energy of the universe. Because of inertia the physical universe becomes tangible. The alephtip is then transformed into a tangible reference angle called an **inertip**, (inert tip) from the words of **iner**tia and **tip**. Inertips are the first expression of the **bethtip** (beth-tip) phenomenon or "house with energy."

Shape of a Geometric Point

The shape of a geometric point becomes a focus of interest as we probe into the nature of infinitesimal space. Looking at a piece of graph paper, we might see the geometric point as a cube. It is neat and occupies all space. One problem with this idea is that there would be variation in distance depending upon the angle. Another is that the access to the geometric point becomes limited to the six dimensional directions. The idea that a geometric point is shaped like a sphere stems from the idea that it can be referenced by an infinite number of angles. However, this is not entirely accurate. If we examine a set of spheres surrounding a single sphere we see that only six can surround the seventh central one. Another problem is that there are empty spaces between the spheres generated by their curvature. The problem in both cases is that we are using finite objects to represent an infinitesimal object which has no volume. The solution is found in the following: despite the fact we cannot touch two opposite sides of a geometric point simultaneously. There is no such thing as a fractional point as 0/2 is still 0 and not some other value. We then see that the geometric point has properties of both a cube and a sphere without being able to generate a finite equivalent.

We see a granular arrangement that is impossible to depict with finite shapes. We can only illustrate these concepts in ways that point to the true nature. Our illustration is that of linear geometric points packed by spherical-like shells of geometric points. The concept is that a straight line grid can be drawn in any radial direction giving a crystalline view of space without forming jagged jumps from one layer to another unlike an angled line on a computer monitor which needs (without using fading techniques) to jump to another layer of pixels to show an angled line on the screen.

Three Different Geometrical Systems

The three different types of systems are derived from the analysis concerning the nature of parallel lines. The initial system (Euclidean Geometry) is built upon the finite observation that parallel lines never meet. This is readily observable as no one can draw

Chapter 3: Inertverse

two parallel lines long enough to meet via the curvature of the earth. However, that isn't the real differentiation concerning the formation of the other two systems (non-Euclidean Geometry). We can visualize a flat plane with no curvature and imagine parallel lines drawn as far as we can finitely imagine. The primary observation of these lines is that they remain the same distance apart.

The first hint that parallel lines might meet is found in examining triangles. On this same flat plane, a person can draw a triangle of any dimensions. However, when the angles of any triangle exist upon a flat plane, the angles always add up to 180 degrees. Imagine an isosceles triangle (triangle of two equal length legs). Using the same base length, increase the inside angle of both legs equally. In doing this, the length of the legs must also increase in order to meet. The angle at which they meet becomes smaller as their interior base angles approach 90 degrees. The idea then becomes; if a triangle has two base angles of 90 degrees; the top angle of the triangle meets at zero degrees. Being consistent with the nature of triangles, it adds up to 180 degrees.

The problem of this triangle's existence is that having its linear base at 90 degrees to the two triangular sides (legs) is the same definition of two parallel lines. It should be observed also that the length of the legs is beyond any finite measurement possible to imagine. In fact, it is infinite or indefinable in length.

Another branch of Geometry studies lines that separate further apart as they approach infinity. This is the study of hyperbolas. Of course, it could be argued that these too would meet at some angle as the lines project beyond finite space.

Our primary concern is focused on the base of a triangle having parallel sides that meet at zero degrees (Elliptic Geometry). Note: the width of the triangle's base does not matter; it will always meet at a zero degree angle. This may seem unimportant until we consider a group of parallel-sided triangles of different base measurements having the same center will become the same line as the parallel line meeting at zero degrees. By this, we observe a finite bounded plane being reduced to a single line.

This concept has yet another unexpected attribute. Consider this group of triangles being rotated around its base's center as a dimensional axis. Another image emerges; imagine the symbol of infinity being placed upon a given xy-plane dimensional axis such that the crossing of the symbol is aligned with the axial origin. After rotating it around the vertical axis, we observe two fountain-like figures. Looking straight down, the center is the focus of the fountain. Now, imagine seeing the same fountain center for each dimensional direction. This will create a final image for lines extending out from one origin or geometric point.

Now, imagine the same for each geometrical point. At this point, we have the complete structure of infinite space relating to the finite universe. We should also note that the center of finite space is finitely undefinable. This reason is that finite space is the union

Part 1: Establishing Physical Space

between infinitesimal space and infinite space. This makes all finite space as a thin sandwich between the two nonfinite natures of space. This is not to negate our unique relationship with Elohim. This spatial relationship will become important when we get to the creation of Adam and Eve. However, in examining the nulverse, we will find that it is structured as our flat plane illustration, due to its crystalline structure.

Volume of Planes

It is easy to imagine one side of a plane being different from the other side. Now, imagine planes stacked on top of each other. The focus of confusion becomes on the ability of transferring physical energy from one plane to another without disturbing the Heavenly plane or with any other plane that may exist. This is no problem because the energy is passed from point to point as a response between corresponding reference angles between any two adjacent geometric points. In this, a given reference angle of one geometric point is enabled to respond to energy transmitted by another adjacent geometric point utilizing the same set of reference angles.

Another aspect of concern is the possibility of inter-dimensional conflict by the three perpendicular planes being expressed as one infinite plane. This is an extension of the basic concept found in the symbol of infinity. Starting with only a two-dimensional "xy-surface," we can observe an object going outward on the x-axis and returning upon the y-axis. In three-dimensional space, the same process exists. Imagine being on an xz-plane traveling down either of its axes out to infinity. Add to this image that of an unrolling wide red carpet upon a blue square tiled floor. First, Let us roll the carpet in the direction indicated by the x-axis of the xz-plane. We observe ourselves coming down the wall on y-axis of the yz-plane. Similarly, if we were to go down the z-axis, we would come down the "xy-wall" facing us. The pattern is described in the following way: a given plane projected upon one of its dimensional axis will return upon a perpendicular plane on a dimensional axis. This axis replaces the projected axis of expansion keeping the axis lateral to the projection incorporated. The result is that the set of three intersecting planes, which define three-dimensional space, are but a single plane.

Now, we have the background to examine the line generated by the equation: $y=1/x$. When x approaches zero (the y-axis), the line races to infinity on the xy-plane and returns on the xz-plane in union with the z-axis. Perhaps the best method to reach the required understanding is to examine the assumed condition that states that the line $y=1/x$ crosses the y-axis on the opposite side of the universe. This concludes: adding to an extreme infinite negative value, we would get an extreme infinite positive value. In terms of good and evil, this would be saying: if an individual strives for the most extreme evil value (sin), that some extremely good value (righteousness) is attained; that is obviously incorrect. This brings us back to the concept that the line returns to the

Chapter 3: Inertverse

origin from another dimensional direction. Another interesting phenomenon: when the x-values approach infinity, it also returns on the z-axis, but it arrives on the yz-plane.

Dimensional Integrity

The infinite path does not twist as a Mobius strip. That is to say, if we were to carpet the internal floor loop out to the right on the x-axis, we would not come back down outside on the y-axis. However, the external floor would. This can be seen by drawing a symbol for infinity and placing dots on one side of the line throughout the symbol. The result will be dots on the inside of one loop and outside of the other loop. The point is that the dots will always be on the same side of the line. This provides integrity of a plane having two opposite sides. We can still only touch one side, even with an infinite fingerprint.

Creation of Inertips

Another issue to address is the effect of each dimensional plane upon a given geometric point at the origin. This is amplified, thinking of each geometric point as an origin. The concept is that each geometric point then becomes referenced by three different angles. This is derived from the idea that the reference angle is always perpendicular to the plane. As the plane returns on different axes, a different perpendicular angle forms upon the same geometric point using the same plane.

Imagine for a moment that a given geometric point is a sphere. These three locations on the sphere would be a 90° angle to each other. This appearance is similar to a sphere's center being the corner of the intersecting cube. There are then three locations upon the sphere's surface that the edges of the cube have in common (intersect). If the edges of the cube were to be extended to the opposite side of the sphere we would have six locations. However, we are forgetting that these are reference angles to the geometric point and that information does not slide across, but is transferred as in a computer from one bit to another. By this, we would observe three nultips per geometric point.

Let us extend this examination a little further. Consider the intersecting corner of the cube and the center of the sphere to be identical to each other. We then observe three lines connecting the three nultips. The area inside these three lines upon the surface of the sphere forms a triangular wedge. This wedge is called the alephfield indentation. The indention is the projection of the surface triangle-like shape down to the central point dictated by the cubical structure. Again, aleph pictures a source, and a field of energy. Putting it together, the alephfield becomes the source of all physical energy. Remember that this is only an allocation process. As meaningless this may seem, the

Part 1: Establishing Physical Space

alephfield will become important as we observe various energies unfold. We must make the following note: the alephfield indentation is not the resulting alephfield.

Now, for the last step, the alephfield indentation has a surface center. Draw a line from this center down to the center of the sphere. Next, move the cube along this line outward to the edge of the sphere without twisting the cube's reference to the sphere. The result is that the corner of the cube is now at the edge of the sphere. Let us return to the geometric point: the line that the cubical corner uses to travel outward is the reference angle by which we defined the alephtip. The edge of the corner joining the reference angle forms the alephfield.

The alephfield is then composed of three unified nultips, or three energy vectors, to depict one single reference angle to a geometric point. Since the entry angles do not approach the geometric point at the same angle, the nultip angle functions as input to the alephtip. They provide three different energy directions in which the geometric point is referenced. By this, the diffused physical universe becomes a focused universe. Inertia, depicted by the reference angle, becomes an existing energy by the action of drawing the three nultips into a single focus within this particular finite space.

The inputs by these four energy vectors, the reference angle and the three modified nultips, are called the **xev**im (pronounced se-vim'; singular xev). This word is an acronym for "**X**-fer **E**nergy **V**ectors" the "im" is a Hebrew masculine plural form.

Their three-dimensional finite representation can be expressed as the framework of cubical corners. The initial image of the structure of the xevim is that of three equal line lengths are joined at right angles upon a spherical surface. The reference angle is a line that goes diagonally through the center of the cubical corner. At this point in creation, each xev contributes equal pressure upon the spherical surface. Pressure equates to focus of awareness by Elohim. These three foci move toward each other on the surface of the geometric point thereby changing the angular location of the pressure. The three xevim collide at the resulting central reference angle formulating inertia energy. This in essence "ignites" the existence of the physical universe.

This action of moving the three reference angles to one reference location transforms the nulverse to the inertverse. To get a feel for this phenomenon, position your thumb, index finger, and middle finger about an inch from each other in a triangular fashion; then bring them together to the center of the triangle that they form. It is with the same ease that Elohim transforms the nulverse into the inertverse. Note that each digit of the hand is exerting pressure upon the other two digits. Similarly, the xevim (energy entry angles) exert pressure upon each other in the alephfield at the newly acquired reference angle of inertia. By this, we observe that there is a focus of attention holding the physical universe in existence. All that is required of Elohim to release the presence of the physical universe is to relax the focus and the universe will diffuse into nothingness.

Chapter 3: Inertverse

As previously observed, the pressure exerted by the xevim within the alephfield is at equilibrium causing no motion. The inertia of the physical universe can be symbolized by its stability as a straight line. However, we will use an equal sign instead of a single line segment; the reason is that a single line segment, the minus sign, is used to represent negatively charged magnetic energy. This symbolic design will make more sense as more energy types are identified. The concept of inertia as being a straight line will also become important as we watch other forces being manifested within the physical universe involving the discussion of the space/time continuum.

Body, Soul and Spirit

The next phase of our observations concerns us with the vastness of our finite universe. Does it stretch out to infinity? As contradictory as this seems, an infinite finite universe does exist. The question should be, "In what manner does it exist?" The answer concerns itself with the continuity of finite space on out to infinity. Consider for a moment: two locations separated by a finite distance within finite space. Two locations can be brought down to two separate geometric points. Each is surrounded by an infinite number of geometric points going out to infinity in all directions. In doing so, they form their own little infinite universe of geometric points totally unrelated to each other yet connected. The same is possible with finite space and infinity. We can have finite distances from two different locations in infinity that will not directly touch even if they extend out into infinite space. This means: no matter how far we travel in these two separate finite locations within the infinite, we would be no closer to the other finite location in terms of traveling through infinite space. While the absolute form of infinity is one in nature, it is plural in terms to the finite nature of our universe. Once again in Genesis, we observe "Let US create man in our own image..." This nature is reflected in the fact that the human body is made up of a multitude of cells constituting a single body, yet each cell has a life of its own. Secondly, the human being is composed of a body, soul and spirit.

We can relate the body, soul and spirit to the three manifestations of space. The infinitesimal space represents the spirit. The finite nature represents the soul, and the infinite nature the body. Just as zero multiplied by infinity equals one, so is the relationship among the body, soul and spirit. This was the original design of Adam before the fall, as we shall observe later. His body was cut off from the infinite nature, leaving him naked and ashamed. That was only the external manifestation. The spirit of man was cut off from the infinitesimal nature of the universe. While it reaches into infinitesimal space, it does not connect to the central point. When we are a "Born Again Believer in Yeshua (Jesus: Salvation)," HE places HIS Spirit within us; HE connects us, once again, perfectly within. However, the external manifestation, the immortal body or the connection to the infinite, is yet to manifest. It will manifest after

Part 1: Establishing Physical Space

the Second Coming of Yeshua. An analogy of this soul creation phenomenon appears in the computers. The physical computer exists as a dormant physical hardware (device). When electricity is poured into the system, the computer attains the expressions of its software (programs). Consider the parts of an equation. It has constants, variables, and the result. The constants represent the different factors involved with the body. The variables are the spirit's involvement. The result is the soul. Notice how the result (soul) varies with the variables (spirit) as manifested through the body. We can find the variables (spirit(s)) involved in an equation by the results of the equation, for it is a known commodity.

Cleaning the Information

Let us pause a moment and bring our thoughts under the jurisdiction of the Torah. There are many kinds of conclusions that could leap out of this information. The problem is that not all of these conclusions, no matter how strong, are always in line with the Holy Scriptures. Any such conclusion needs to be cleaned out of our minds. Otherwise, it will ultimately cause us to fall. These ideas are an unfortunate side effect of our sin nature.

The first obvious conclusions are that there are parallel dimensions as well as separate dimensional universes. These are not wrong conclusions, but the concepts that follow them often are. The concept of other universes should not be an excuse to believe that the purpose of our universe is different than that stated in the Scriptures. This information should never be used to disregard Yeshua (Jesus) as the only begotten of Elohim. Along the same line, it should never be concluded that there is more than one Creator. In considering the possible finite universes created, we should realize that the same Creator created them. The Creator is infinite and absolute. The plurality of HIS Nature does not constitute multiple personalities forming a multitude of G-ds. It means the following: HE made us in HIS Image resulting in a multitude of male and female forms. This attribute provides a singular presence among the multitudes of different people stemming from HIS One Being. Using this information, we should more easily understand how Elohim can communicate with each of us individually at the same time even though HE is ONE in nature.

Consider for a moment the existence of a multitude of "virtual infinite spaces." Even though each has its own realm, it still occupies only as an allocated space of totality. Elohim is that totality; not only of space; but of life in each realm as well.

Another type of misconception derivative is that we often carry preconceived notions that are in error into a new understanding. This will cause us to line up the information within the parameters of this notion. Often, this action occurs unintentionally because the information seems to properly fit. This is especially true with some of the pagan

Chapter 3: Inertverse

beliefs. The reason is that the pagan beliefs do have elements of truth in them. However, these truths are distorted to back a lie that the whole truth is often totally obliterated. When we desire only HIS knowledge, then we are able to let go of the rest.

One such misconception is that of animism. While Elohim is sustaining all matter via HIS Spirit, HE does not give awareness to matter as an automatic attribute of existence. Physical existence references Elohim not engulfing Elohim. HE places awareness into vehicles of matter that have the created facility to function in that measure of awareness given. That is not to say a spirit cannot associate with an inanimate object, for it can. But, it does not give the object itself an intrinsic awareness. These objects become talismans by which a supposed protecting spirit wards off evil or produces evil. It just marks the host object as belonging to a certain demonic force telling other demons to back off from its property. In the case of a curse, it marks and gives the demonic force of its kind a right to attack an individual provided they do not know Yeshua and understand HIS Protection. Sometimes the mindless actions of people will produce the same effect being guided unwittingly. However, Elohim ultimately sustains it all for future judgment. Even though Elohim does not give inanimate object's awareness, the object would cease to exist if it were not held together by Elohim.

Adjusting Infinitesimal Space

Suppose, for a moment, that a certain inertip wanted to move out of its location. It would be nearly impossible, as there is already an inertip present in the adjacent location. The only way that it could accomplish this task is for the entire line of inertips, out to infinity in both directions, to shift in the same direction with it. The image is as stones that are sitting next to each other and touching. It would be physically impossible for any enclosed stone to move in any direction because of the other surrounding stones without causing the directionally adjacent stone to move (excluding stone deformations caused by pressure).

From this illustration, we discover a need for discontinuity at a very basic level. Imagine a physical universe that is made up of alephtips that have two conditions of existence: inertips and empty alephtips. In this image, the empty alephtips are not nultips as the alephtip does not diffuse back into the nulverse state. The alephfield remains intact waiting for the return of xevim pressure. This is accomplished by selecting a geometric point in a finite numerical interval to contain the "final inertip."

The second problem for moving one inertip to another location is inertia. There is nothing within the physical universe to move this inertip out of its original location even if the inertip had freedom to move. By its own nature, the inertip would resist motion. In short, the Inertverse is the only phaseverse that can be described as solid space, which is the inertinuum. **Inertinuum** comes from the words, **inert**ia and cont**inuum**.

Part 1: Establishing Physical Space

It is an extremely faint solid, as it is only its continuity that gives inertinuum strength. Note: Elohim said that HE moved upon the waters; not a solid. This implies that a transformation process occurred from a solid like state to a liquid-like state before Elohim introduced motion.

Inertinuum exists in a hyper-inertial state. All the xevim input are isometric; promoting the pressure of inertia. The fibers of the inertinuum exist in pure uniformity. Unlike the fibers of our universe, as described by Einstein, light (if it were possible to exist) would always move in a straight line. There is absolutely nothing to bend its path.

Chapter 3 Quiz

1. How are inertips formed?
   A. By three nultips coming together.
   B. By the introduction of inertia into space.
   C. By the will of Elohim.
   D. All the Above.

2. The Xevim (Xev) are:
   A. External Energy Vectors
   B. Exiting Energy Vehicles
   C. 10 Ending Variations
   D. Extreme Energy Vortexes

3. The number of Geometric Points touching another point is:
   A. 6
   B. 12
   C. Infinite
   D. 0

4. Bethtips are Alephtips:
   A. Housing energy
   B. Moving parts
   C. Replacing Attributes
   D. None of the Above

5. Inertinuum has a problem:
   A. Too much inertia
   B. Frozen
   C. Occupies everything
   D. All the above

6. (T/F) Physical space came into existence by annihilation of inertia.

7. (T/F) There are 4 reference angles defining physical space.

8. (T/F) We need to keep conclusions at all cost.

9. (T/F) The xy-plane returns from the infinite in the z-axis.

10. (T/F) Inertinuum occupies the entire physical universe.

Part 2
Creation of Xyzenthium Crystals

Chapter 4

Gravverse

Gravity is a force of attraction seen in physics along with some magnetic forces, but the force applied within this phaseverse is not only the root of gravity but of all forces of attraction. While the energy root has a different specific name, the usage of gravity provides us a word root that gives a familiar reference to mark the nature of this phaseverse.

In order to establish intervals between the inertips, Elohim invoked a force of attraction. This force can be thought of as gravity or as the microweak (see page 74 for description) forces since both come from this force. This force is called **jammeria**, pronounced as- jam' area. The contextual intent for the root word of jam is that of cramming together, as its primary function is to pull together. In this phaseverse, it crams external inertial energy into a single unit via attraction.

Within this process, the energy level of the Inertverse is increased and transformed into the next phaseverse called the gravverse. In the uniformly selected places within infinitesimal space, an inertip would eventually receive a greater magnitude of jammerial energy. This permanently alters the universe of inertips into a universe of **gravtip**s. As suspected, the nomenclature of a gravtip stems from the word gravity and tip, giving the pronunciation of grav-tip. It is an inertip with an active jammerial xev. It should be noted that the transformation does not produce an end result of one gravtip per inertip, as the purpose of this step is to create a condition of discontinuity.

Two basic patterns could exist within the gravverse that maintains equilibrium within a cubical pattern. One pattern is to locate the resulting gravtips at the corners of the cube, giving eight gravtips by cubical reference. The second pattern is to locate the gravtips in the middle of each cubical edge, giving twelve gravtips total per cubic reference. Seemingly, the first pattern is the obvious choice until we examine its behavior within the kineverse. As we shall see, the second pattern of twelve gravtips provides the particular pattern required for operation within the kineverse.

In order for these central locations to be centered, the number of alephtips measured is an odd number; otherwise, the centers would be off-centered. Example: when we imagine five blocks in a row; there is a central block. However, if the block count was to be an even number like four, no true central block would exist. We would have to choose one of the two blocks closest to the center. Neither block would ever truly satisfy the required condition needed.

Part 2: Creation of Xyzenthium Crystals

Discontinuous Infinitesimal Structure

Now, we are ready to examine the nature by which the individual gravtips geometrically exist in relationship to each other. If lines were drawn from one gravtip to another so that only the gravtips adjacent to another was connected, a tetradecahedron would form with each edge being the same length. To grasp this image, visualize a cube with each of its corners chopped off. From this nondescript chop, refine it so that the cutting starts at the center of each edge. The resulting shape generates a diamond shape on each face of the cube which connects at the center of each edge of the face. The resulting chopped corners form equilateral triangles. This forms a twelve-cornered object corresponding to the twelve edges of the cube. Once again, this is necessary in order for energy to move in all dimensional directions undisturbed by its local imbalance or be thrown off course by gravity in future phaseverses. This positioning allows the cubical corners to be the intended focus within other phenomena.

As hinted at earlier, our earlier image of the dimensions of the cube holding the tetradecahedron is actually quite small when compared to the actual dimensions. Because of the tiny size involved, we commonly estimate ten to twenty alephtips in length at most. To get an idea of the actual number, think of the ratio between the matter and space in the universe. Then, when we realize number is much larger, we think in terms of hundreds or thousands. Being more adventurous and perhaps more accurate, we think in terms of millions and upward. Even in terms of trillions, there still would be an infinite number of gravtips per any finite measurement.

A side note: From our analysis thus far, we observe that twelve plays an important functional role in Elohim's structuring of our physical universe. The number seven mathematically represents the six dimensional directions plus their intersection. Note: this intersection or origin holds them in place. We also observe the utilization of the number six to deny the existence of the origin. Think of the six faces of the cube covering the cube's center in relation to Lucifer as the covering cherub.

We should note that there are other finite geometric patterns that could occupy the entire volume of space. However, the cubical based pattern is the only geometrical figure that gives all three dimensions equal and minimal input. In encompassing all space, it maintains the integrity of the three physical dimensional directions. In this formation, there is a matrix of gravtips throughout the infinitesimal space of a given location that occupies a tiny fraction of the possible bethtips within that location.

Another side note: jammeria must be distributed to all other inertips; otherwise there would be no annihilation. There are two reasons that it is necessary for jammeria to exist in all bethtips for gathering. One, inertia would not move without a force acting upon it to displace it from its location. Two, gravity can only pull upon gravity. Therefore, the jammerial xev has to be activated in all inertips for this to occur.

Chapter 4: Gravverse

1.1.1.1.1.1.1.2.1.1.1.1.1.1.1

The "1" is read as normal gravtips, and the "2" reads as a gravtip with greater jammerial force. The "." that separates the two numbers is there purely for visual readability. There are no gaps between gravtips, as it was formed from inertinuum.

The gravtips energy will gravitate to the local central gravtip-two. The acquisitions of gravtips by the central gravtips are uniform both in number and in timing. The necessity of collecting an equal amount of gravtips is that it keeps the jammerial attraction in balance. The timing is important; also, to keep the kinetic forces (energy of motion) in balance despite the fact that there is still no finite movement.

The inward flow of gravtips into the central gravtips is not a loose gathering of gravtips as there are no empty alephtips in between them. The image is more like tissue paper lying on top of a water surface. In this image, a stick is pushing down upon the surface paper into the water. This brings the tissue paper into nearly a circular fashion around the stick. The paper immediately surrounding the stick is pulled down into the water with the stick as it continues its journey downward into the water. As the paper is pulled down with the stick, it collapses toward the source of the force.

Recall that inertia can be represented as a flat line. Jammeria appears as an edge of an arrow's head pointing down. We should look at it as an inertia line with its center pulled downward. This, too, will become important when we talk of the space/time continuum. The symbol we'll use for jammeria is a double chevron pointing down, as inertia is represented by an equal sign.

A numerical representation of the drainage of energy into a central location is given below. Each line represents a snapshot of a line segment during this process.

1	1	1	1	1	1	2	1	1	1	1	1	1
0	1	1	1	1	1	4	1	1	1	1	1	0
0	0	0	1	1	2	6	2	1	1	0	0	0
0	0	0	0	0	2	10	2	0	0	0	0	0
0	0	0	0	0	0	14	0	0	0	0	0	0

Note that the central gravtip in our example gains energy from both sides, as we will see later, the gain is far greater. Note also that the zeros by line symbolize the growing empty physical space by the formulating alephtips. Notice the increase of focus.

Even though the basic concept is to divide the entire universe up into cubes, another image emerges. Remember that in cubically dividing physical space, we selected the center of each cubical edge. Imagine from each of these locations that a balloon is being blown up to encompass as much space as possible. The spheres will collide as

Part 2: Creation of Xyzenthium Crystals

they grow and form planar faces that eventually join forming edges. The resulting geometric shape is an octahedron (two four-sided pyramids joined at their bases).

By this, we see the entire universe divided into interlocking octahedrons. Note also, in our balloon visualization, that neither side occupies the corners. The "casing" defining the form of the octahedron also is instrumental for defining adjacent octahedrons. We cannot physically engulf the "casing", as geometric points cannot be divided into fractional values of existence.

After realizing that the external octahedron structure cannot be absorbed internally within any octahedron, we can see it as energy of a finite value, which is divisive. There are locations in which the energy values are divided in half as energy vacates away in two opposite directions indicative of the planar faces of the octahedrons. Energy values, at the locations forming the edge of the octahedrons, divide into fourths as four octahedron edges touch. The cubical corners divide into 1/6 and the centers into 1/12.

Resulting Gravverse

As noted earlier: after the energy from the gravtips migrated to their respective central gravtip, the universe returned to complete stillness. The image is that of a universe formed by layers containing rows and columns of equal spaced dots. Unlike the stillness of the inertverse, the gravverse exists in a balance of tension; this tension is caused by the attraction between the gravtips. The force exerted from these gravtips divides infinitesimal space into an interlocking sequence of octahedrons of three different orientations.

Another observation along this line is the unification of the physical universe. In the inertverse, each inertip existed individually without association. In the resulting gravverse, the existence of a given gravtip location depends upon the location of the other gravtips. As noted, the isometric stillness of each gravtip depends upon the equal distribution of other gravtips. This is the first step by Elohim to establish interacting existences beyond HIS absolute sustaining power. Note: HE is still sustaining the physical universe; it is only that an equation has been established by which these infinitesimal particles will interact.

Yet, another observation to be made of the gravverse is that each finite location is identical to the next. In fact there is nothing to distinguish one finite location from another. Even though the substance of the gravverse is discontinuous at the alephtip level, it is still continuous at the finite level because there is no space between finite locations. Resulting from this fact, we observe it as an ocean of gravtips covering the entire gravverse in an even distribution. The key word is ocean, as the universe is not solid at this point of creation. This space could be manipulated, as there are empty alephtips to comply with the movement. Since it is not a true liquid by definition, it is called **uniquid** and is pronounced as u-neek'kwid. It is derived from the words unified and liquid. This term can be remembered as a unique liquid. As we shall observe, the uniquid undergoes many transformations before the subparticles that create the submatter known today become a reality.

In the gravverse, the uniquid is in the state of being **duoquiet**. This means that the gravtips are not moving in finite space or in the infinitesimal space of the alephtips. In the formation of the gravverse from the inertverse, the forming uniquid was **monoquiet** as there was movement in infinitesimal space. In the phaseverses to come, the uniquid will become monoquiet once again and then progress into **nonquiet**. For the universe to be nonquiet requires both infinitesimal and finite movement.

Quebbrix (Cube-Bricks)

There is a major obstacle that frustrates the concept of achieving the smallest dimension of finite space possible excluding infinitesimal space. The problem is any unit devised within finite space can always be subdivided into smaller segments. It is then seen that the smallest measurement of finite space is indefinable. We can say that it is the smallest nonzero value possible. Inversely, we could say it is the largest finite number of geometric points before becoming infinite. Only Elohim, having both the infinite and infinitesimal attributes of awareness, can assess the determination of this number. The length of a quebbrix actually cannot be considered as a real finite number or a distance, as a number or distance can always be divided into smaller finite segments or distances. However, we need to account the ramifications of bethtips movement upon finite space in smaller units than the smallest particle or cell generated by Elohim. As we shall see: even the smallest cell of submatter contains a framework of activity just as a biological cell. This "cell" is beyond being organic or inorganic. Nevertheless, it is the measurement of a set of resulting gravtips established by Elohim.

For the accountability of the relationship between energy and space, an artificial division of the uniquid locations becomes necessary. The quebbrix is the transliteration of the spelling of cube and links. The letter "Q" equates to the Hebrew letter "Quf." Within the schema of the Hebrew picture language, this letter represents the least; hence, defining the least in size as a distance. Naturally, the word "bricks" implies that two different finite locations are connected together by an infinite number of alephtips, per dimensional direction. Despite the number of alephtips being infinite in count, there is a definite set. Example, the alephtip of one quebbrix does not become a component on another quebbrix. Because of this nature, we can examine the effects of infinitesimal alterations upon the finite universe. Since the quebbrix is a finite measurement, its dimension is far beyond the size of any finite count within infinitesimal space.

Before discussing time and establishing the kineverse, let us explore the composition of a quebbrix. Consider the impact of the spacing between the gravtips or any future energy-filled alephtip (bethtip) arrangement will have upon finite space. The primary

Part 2: Creation of Xyzenthium Crystals

function of a bethtip is that it contains inertia which gives the energy its reality and the ability to manifest within this universe. Alephtips can only transmit the manifestations.

Initially, let the quebbrix consist of a uniform set of bethtips. If the bethtips were to be removed from their original location and became twice as far apart throughout the quebbrix by inserting an alephtips between each bethtip, it would form a discontinuous line segment that is two quebbrix long. From this example, we observe the infinitesimal activity of geometric points altering finite space. Even though the quebbrix of inertverse doubled in length, the line decreased in half in energy per quebbrix from the original quebbrix. Therefore, it can be stated that the expansion and contraction of infinitesimal space alters not only the finite length, but the density of energy altering the magnitude of energy as well. In other words, finite volumes can change by infinitesimal alterations, even though each additional alephtip adds only zero to a given dimension.

Within further observance of the phenomenon, we find definite values of infinity. For an example: Let us examine a line segment that is one unit of measure long. The number of geometric points forming this line would be exactly one half the number of geometric points forming a line two units long of the same measure. Both quantities are infinite in nature but have different values. Similarly, there are many infinite realms composing the unity of infinity. The full value of infinity exists in the inverse manner as that of the zero unit or a geometric point, as two opposite absolutes. As we progress through the creation process, the plurality beyond a finite realm will become important.

Beginning of Time

Before finite time, there was only eternal time. Eternal time can best be described as the eternal present. We observe the eternal present as the moment that we constantly live in. However, eternal time is a more complete experience: all the past and all the future are completely integrated into the present experience. Actually, the future and past are elements purely of finite time. The best description of finite time is mechanical time. We can know that mechanical time is created because it can be altered by acceleration.

Another observation is that inertia is ultimately responsible for finite time. In the famous Einstein equation, which establishes the relationship between mass and energy, we observe acceleration as a factor of transformation. The equation reads: energy equal mass multiplied by the velocity of light squared. The term, "squared", refers to acceleration. Acceleration involves overcoming inertia. The equation also states that there is a limit to acceleration before mass becomes energy.

From this information, we get the Lorenz Transformation. The Lorenz Transformation measures the effect that acceleration has on mass as it approaches the speed of light. From examining the following information, we shall see the effect that the constant battle between inertia and acceleration has upon mass.

Chapter 4: Gravverse

According to the Lorenz Transformation, an object accelerated to 90 percent the speed of light will shrink by one half in the direction of the motion. Along with this, time slows down by one half. When an object moves 99 percent of the speed of light, matter will shrink 1/4 of its dimensional definition in the direction of the motion. Time responds accordingly to 1/4 of that which is experienced upon the earth. The process continues in the same fashion as mass is being accelerated. Along with this, mass grows by a factor of four. Out of this equation, only the effect upon time has had any observable proof. While inertia increases the effect upon time, the three dimensional directions are seemingly unaffected. If this shrinkage did exist, then there would be no wavelength variation in photons as they move at the speed of light and have mass.

From his viewpoint, we can establish a direct link between the resistance of inertia to motion and slowing of time. The strength with which inertia is activated by acceleration determines the retardation of time. Inversely, we can state that the decreasing of acceleration upon inertia increases the motion of finite time.

If we stop for a moment and think about it, inertia and time are logically linked. Motion requires time to manifest; inertia does not. Moreover, inertia resists motion, hence, resists conversion from eternal timelessness into mechanical time. The energy required to accelerate mass, annihilates or converts the energy of time (inertia) into motion. In other words, time decreases as inertial energy becomes tied into the velocity generated by acceleration.

Just as there is an upper limit to motion, there is a lower limit. By slowing mass down to a speed approaching zero, the inverse occur. Accomplishment of this task cannot occur upon the earth as it is moving, but it can in space as a black hole. Time would then speed up causing more vibrations per given time span. These concepts are getting a little ahead of our present historical account, but it shows the pattern.

Putting this information in perspective to the phaseverses that we have covered thus far, we observe that the infinite amount of time is passing through each moment. In the nulverse, time was eternal, as there was no movement. Secondly, despite the movement of the gravtips toward the central gravtips, the resulting gravverse is as the inertverse in time as the phaseverse remains motionless.

At this point, we should note that Elohim is keeping track of time from outside this framework. HE is telling us: within the framework of time existing in the created Heaven, not even a day (approximately 1,000 years in our time) passed before HE began to move on the waters (uniquid).

Another problem with the Lorenz Transformation is the "constant c" or the speed of light. It is assumed to be a constant instead of a variable. However, this does not negate the equation, as there is an upper limit generated by kinetic energy involvement with inertia. This relationship is established by Elohim to generate mechanical time out of eternal time. This is the reason that after destroying this universe, HE creates

Part 2: Creation of Xyzenthium Crystals

another final universe for existence. In the next universe, there will be finite time again to mechanically experience life forever. The actual upper limit of the speed of light, at this point, cannot yet be determined. We do not have enough data, or means to generate such data. However, one item is certain; the speed of light is slowing down. The rate in which it slows down cannot be assumed to be a straight line but a line similar to $1/x$ as x is greater or equal to zero. In this, the speed of light, being an upper limit, has decreased immensely from the beginning of Creation, and now decreases very slightly in an almost linear like manner. There is no danger of the speed of light becoming zero before the end of the universe. Along with this, by the time photons come into their initial existence to express this speed, this limit will have already dropped. The reason for the drop is the release of pressure. This topic will be discussed more as we study the formation of our universe.

Imagine accelerating from a location in which our velocity is 99 percent absolute stillness. Using the inverted form of the Lorenz Transformation, we would have to accelerate 1/4 times as fast to achieve the acceleration of that experienced in our "normal" universe. This is because the distortion of time multiplies our effort. Example: if we were moving 40 mph within the framework of a plane moving 99 percent of absolute stillness. One hour in this framework equals only 15 minutes to an external observer viewing it from earth giving the appearance of moving 160 mph. However, an object moving 40 mph within our framework appears as an object moving 10 mph within this framework. By this, we see reaching absolute stillness is nearly impossible. As we approach absolute stillness, motion is multiplied by our percentage of achieving the goal. To decelerate to absolute stillness would take infinite energy. It is not that anyone would want to do this as our life-spans are finite. The phenomenon would cause us to grow old and die instantly in view of an external observer as all time passes through a single moment for us.

Chapter 4 Quiz

1. What is Jammeria?
   A. Pulling Energy
   B. Devouring Energy
   C. Musical Energy
   D. None of the Above

2. Uniquid is:
   A. Union between something and nothing
   B. First Fluid in the Universe
   C. A Unique Liquid
   D. All of the Above

3. Monoquiet is the occurrence of:
   A. Motion in Infinite Space Only
   B. Motion in Infinitesimal Space Only
   C. Motion in Finite Space Only
   D. All quiet in time

4. Time and Inertia are:
   A. Unrelated
   B. Related via Motion
   C. Related by a Special Energy
   D. Relatives by Family

5. Eternal Time is:
   A. Outside this Universe
   B. When all time exist in a single moment
   C. When there is absolute rest
   D. All the Above

6. (T/F) Jammeria is the force of expansion.

7. (T/F) The purpose of the gravverse is to generate light.

8. (T/F) Quebbrix is the first building block of the physical universe.

9. (T/F) Eternal time in the physical universe exists without movement.

10. (T/F) In essence, time slows down into measurement.

Chapter 5

Kineverse

And the Spirit of Elohim moved upon the face of the waters.
B'reshyit (Genesis) 1:2b

Gravtips within this phaseverse acquire kinetic energy (motion). Energy in pure isometric balance will not suddenly or gradually get out of synchronization by itself. Some external force needs to be applied to the assembly of static balance of energy in order for motion to occur. The kineverse, therefore, must rely on external stimuli to become activated. The added energy generated by Elohim transforms the gravtips into **kyntip**s; it is pronounced as kine'tips to represent kinetic tips. Recall the observation, once again, that there are three dimensional xevim surrounding a central xev of inertia. Kinetic xevim are those aligned with the cubical edge.

The word translated as "moved" is m'ra-Ha-fet in the Torah. The Hebrew root word is rachaf (רחף, ch is hard as in Ba**ch**). The Torah states that Elohim's movement was as a flutter rather than a nondescript motion. We then can further extrapolate from the terminology the concept of a vibration. Now, we need to examine the kind of vibration implied by the scripture. Vibrations are analyzed by the nature of their waves.

Vibrations

There are two basic kinds of waves: longitudinal waves and transverse waves. We generally think of a wave as an up and down motion. This is a transverse wave as the wave is moving in a 90 degree angle to the angle of the motion causing the up and down motion experienced. A longitudinal wave moves in the direction of motion. By using a toy slinky (a large weak coil spring), an analysis of these movements is possible.

Attach one end to the wall; after stretching it out, begin to rapidly alternate the tension of the spring by pulling back and forth on the spring. We can observe waves forming; there are regions in the coil farther apart than other regions. These regions move back and forth throughout the coil. Observe that the up and down motion is not necessary for these waves to carry out their motion.

Examples of these various forms of waves are found throughout nature. The prime example of a transverse wave is light. An excellent example of a longitudinal wave is a

Part 2: Creation of Xyzenthium Crystals

sound wave. An illustration of a combination between longitudinal and transverse waves can be seen on a sea as water waves.

The motion involved in the uniquid vibration is purely longitudinal. It is noteworthy that Elohim spoke the creation into existence as speaking generates sound that is longitudinal in wave nature.

Transfer Pattern of Kyntips

Duocollision means that two (duo) kyntips strike together using only two alephfields during the energy transfer. When we think of two objects colliding together, we visualize that only two objects come into contact, and they either crumble or bounce back. This is a type of duocollision seen in infinitesimal space. However, there are two other considerations that we normally would not consider. Infinitesimal space is a pixel-like universe. Geometric points are like pixels upon a computer screen, in that, they cannot be subdivided. While this alone will not cause an alternate process, the phenomenon does have the added characteristic of involving odd numerical integers in counting the distance between bethtips. They bounce back a different exchange.

Triocollision is defined by three (trio) alephfields involvement in the transference. Within this scenario, two kyntips collide upon an alephtip in which neither kyntip can claim sole ownership of the newly formed bethtip out of emptiness.

Let us examine the uniquid by analyzing a segment of movement within one of the dimensional lines defining a cubical row. To symbolize this line segment of kyntips and alephtips, let us use an alphanumeric character subset for representation of the phenomenon. Let a group of periods (dots) represent the interlude of all alephtips between kyntips. A "1" represents a kyntip. Let an "o" represent the alephtip that forms the corner of the cube structure, giving an odd count. A "2" indicates that the exchange of energy is occurring in a single alephtip location. As seen, these corner alephtips will become the impact locations of the kyntips throughout the uniquid. Sample locations are labeled from left to right: cubical corners A through E. The line segment activity through a given sequence of time then can be symbolized as seen in the table to the right.

Time	Locations
	A B C D E
1	o...1...o...1...o...1...o...1...o
2	o....1..o..1....o....1..o..1....o
3	o.....1.o.1.....o.....1.o.1.....o
4	o......1o1......o......1o1......o
5	o.........2.........o.........2.........o
6	o......1o1......o......1o1......o
7	o.....1.o.1.....o.....1.o.1.....o
8	o....1..o..1....o....1..o..1....o
9	o...1...o...1...o...1...o...1...o
10	o..1....o....1..o..1....o....1..o
11	o.1.....o.....1.o.1.....o.....1.o
12	o1......o......1o1......o......1o
13	2........o........2........o........2
14	o1......o......1o1......o......1o
15	o.1.....o.....1.o.1.....o.....1.o
16	o..1....o....1..o..1....o....1..o
17	o...1...o...1...o...1...o...1...o
	A B C D E

Chapter 5: Kineverse

The kyntips converge to corners B and D until time 5. At these locations the corner alephtips acquire the energy of two kyntips. However, the kinetic motion set by Elohim causes the energy to continue its journey. After this time, they converge to the opposite corners A, C, and E. As the process continues, the kyntips return to their original position with the same vector settings; thus, completing the cycle. From this point, the kyntips continue to oscillate in this path.

The pattern demonstrated is only one-dimensional; however, we are examining three-dimensional space. This phenomenon is occurring in all three-dimensional directions. The activity then could propagate (transmit) in two basic patterns. One: the kyntips of each dimensional direction could reach a designated corner independently. Two: they could reach the corner simultaneously. Pattern two is less complex; however, it is more uniform.

Septicollision (sep-tea-collision) occurs in order for the uniquid to integrate energy between the three dimensional directions, they arrive simultaneously. This means that triocollision occurs in each of the three dimensional directions. Instead of the alephtips becoming bethtips of a double magnitude, they would achieve a magnitude of six. This is due to the kyntips arriving from six different directions dictated by three-dimensional space. This process defines the septicollision- a collision involving seven alephtips.

Now, let us reexamine our model. The kyntips form a momentary bethtip of the sixth magnitude. The visualization then becomes an image of seven spheres. Six are kinetic bethtips and the seventh a stationary alephtip. The bethtips exchange energy as they unite in adjacent locations with the central alephtip. This energy is transferred without any kind of loss by deformation, heat, or any kind of deflection as they have neither volume nor mass.

We can visualize a pattern in which there are only six spheres that meet without the seventh sphere being involved. It is seemingly reasonable that the adjacent meeting phenomenon (duocollision) is correct and not the merging bethtip phenomenon (triocollision). While within one line we can bring two spheres right next to another, this becomes physically impossible when more dimensions are added. Within a three-dimensional axial intersection, we can bring six uniform spheres together creating a central gap having a central length smaller than the spheres involved. However, in doing this, we violate the nature of the infinitesimal universe. The smaller gap length represents a fractional geometric point, which, by definition, cannot exist.

We could also follow the established reasoning within the gravverse to another point of concern. The illusionary impression is that if the two gravtips joined in another alephtip forming a high-magnitude bethtip, the energy would not be permanently bonded within the bethtip. This is contrary to the following observation: the kinetic-like movements of the gravtips into their respective central gravtip were also equal in magnitude, yet they did not cause the inertinuum to reconstruct.

Part 2: Creation of Xyzenthium Crystals

The reason that the reconstruction did not occur in the gravverse is that there is no kinetic energy beyond that generated by jammeria with the adjacent impacting bethtips. The impacting bethtips had less jammeria than the central impact bethtip. However, later, after the next primary force has been included into the phenomenon of creation, the uniquid does respond in the duocollision pattern generally within general regions.

Unified Fluttering

The motion of Elohim fluttering upon the waters gives still other aspects to observe. Thus far, we have been observing the activity of kyntips in a single segment. Let us expand our line segment of the uniquid to include a multitude of kyntips. Note that every other kyntip is moving with the same velocity in perfect symmetry to the opposite set of kyntips. In essence, we have two sets of kyntips moving in unison within their vibratory motion. Now, put that image into a three-dimensional perspective of the uniquid. The resulting image is of an infinite object fluttering in a pattern generated by six overall movements instead of a multitude of individual infinitesimal movements.

This phenomenon can be easily visualized by examining a single line of dots using two 8.5-by-11 inch transparency sheets. First, we establish our initial location upon the surface of one plastic sheet that is 0.75 (3/4) inches right and one inch down from the upper left corner. At this location, place a dot. Move again, one inch to the left and place another dot. Continue the process until there are eight dots. Place the remaining transparency directly over the one that has the dots. On this piece, place a dot where a dot appears. Now, slide the plastic horizontally until the tick marks align again. At this point, slide the plastic back to its original position. Repeat this process again until the pattern of fluttering is experienced. Note: when one dot lines up with another, all the rest do exactly the same, providing there is mark available. This is true for a plane of dots, and true for a vast volume of dots. This phenomenon occurs on sets of lines out to infinity in each of the three dimensional directions throughout the uniquid. As we observe the movement of the bethtips, we sense a breathing-like phenomenon. In this, the expansion from one location is the contraction in its complementary location and vice versa.

Now, imagine that the two sheets are always moving in the opposite directions. The image between dots is the same. The result of this fluttering is that kinetic energy is moving throughout each "selected" (having bethtips) dimensional line in two opposite directions out to infinity in waves. From this, we can truly see that the uniquid has a fluid like quality. Secondly: if it were possible for a human being to exist within such a phaseverse, the experience would be like being in an infinite liquid with infinitesimal waves of kinetic energy moving in opposite directions within the three dimensional definitions. Note: finitely, this movement seems unnoticeable within the continuum. Since the movement occurs in infinitesimal space in isometric sequences per each

Chapter 5: Kineverse

dimensional direction, the impact upon finite space is zero. Darkness is still the image within this phaseverse as light has not yet entered into the equation.

Quubium

The etymological label of the cubical structure containing the tetradecahedrons is the **quubium** (coo-be-um) from the word cube and the letter "q," which in Hebrew means "the least," and the suffix suggesting a multitude of cubes. **Quub** (coob) is a singular cube of the structure. Let us examine the quubium structure in which the bethtips exist. Since the uniquid is a cubical framework formed by intersecting lines, there is an alephtip at each intersection. Secondly, in order for the bethtip to be centered in between any two corners of the quub, the width of the cubical edge has to be an odd number. If the count was even, the distance to each corner would be different.

The **uniquidron** (u-neek-kwi-dron) is the tetradecahedron inscribed the quub. Unlike the general description of a tetradecahedron, the placement of bethtips that form the corners also form equal edge lengths. These lines are shorter than the edge of the quub. Within this phaseverse, these lines represent the strongest "gravitational pull" existing between bethtips. In essence, this is the smallest three-dimensional structure of the physical universe having an initial width about one quebbrix. Even though within this phaseverse the shape is constantly changing, this form will always return to the original form of the uniquidron within the cycle of motion within this phaseverse.

If we were to alter our perspective from within the volume of the uniquidron to the remaining quub volume, another set of images appear. By examining the resulting images, we can visualize the result of the septicollision process within the kineverse.

Examining the space that the uniquidron does not occupy within the quubium pattern, we find eight tetrahedrons, one on each corner of the cube. Each of these tetrahedrons (triangular pyramid) connects to the adjacent tetrahedron in the adjacent cube. The result, within the cubical pattern, is that there are seven other tetrahedrons joined to any given tetrahedron forming an octahedron. Unlike the force crystals of the gravverse, these octahedrons align in a singular uniform pattern in which only their corners touch.

Starting at the equilibrium established in the gravverse, all the octahedrons are of equal size. At each joining corner is a bethtip. As defined by the septicollision, the six dimensional bethtips collapse toward the central alephtip. This cannot occur in each octahedron simultaneously, because the bethtip cannot move in two opposite directions simultaneously. The result is that we observe one octahedron shrinks in size while the adjacent one grows. The pattern observed is that of alternating sets of octahedrons increases in size while the adjacent set is shrinking in size. This process continues until one octahedron set shrinks into being the corner of the growing octahedron set. At this point, the central uniquidron becomes a tetrahedron and the growing octahedron set starts to shrink and the set that seemingly disappears grows in size again. This

Part 2: Creation of Xyzenthium Crystals

relationship between the octahedrons exists throughout the kineverse. In this, the motion initially maintains an isometric stance of the fibers throughout the universe. This is somewhat unique, considering that the images are in constant change.

Any given line that the bethtips move upon is called a **xergopath** (zer-go-path). Etymology of the word is based upon four words exchange, zero, erg, and path. Erg is a basic energy unit, and is combined with zero to form the prefix sound of zer-go. Replacing the z with an x for the word exchange gives the final form, showing energy exchange exists within the path. The xergopaths are retraced by the bethtips as they move back and forth between themselves. This motion is the isometric nature of the energy that holds them in place. These paths formulate lines within this phaseverse, which are identical to the lines forming the initial quubium structure.

Time in the Kineverse

Recall that time prior to the kineverse was infinite per moment, as inertia was motion free. Now, there is motion. This motion generates time. In the kineverse, time decelerates to a constant, as kinetic pressure increases in the inertial xev (inertia's entry angle, which is the reference angle of the physical universe) of the alephfield. The resistance to acceleration by inertia increases pressure against eternal time, causing time to slow down. Recall eternal time exists only in absolute stillness. Instead of infinite time passing through a single moment, we have a large quantity of finite time passing through each moment, i.e. a million years occurring per given second. Elohim, established in a parallel universe, sees the occurrence within a frame of one day. This measurement of a day still is one thousand years in today's terms.

If we were to examine the time it took for two kyntips to complete their cycle of triocollisions, it would be instantaneous in finite terms. However, there is a finite constant generated even though the uniquid is monoquiet (no finite movement). The problem of realizing time disturbances without finite motion can be solved by further analysis of infinitesimal movements by the kyntips. Suppose that the kyntips of each dimensional axis collided separately. Within this, we examine a single line of alephtips without interference from other dimensional directions. Since kinetic energy transfers to another object by contact, energy transfer between kyntips in infinitesimal space passes through finite space and reaches out to infinity. The image is then of a multitude of steel spheres bouncing back and forth in a straight line. Actually, they collide upon another sphere in between, but for simplicity, assume they collide with each other. Imagine a rider existing upon each sphere. Let us say the spheres moving to the right have their riders on top; the ones moving left have their riders on the bottom. When the two spheres collide, the spheres exchange riders. Each rider appears continuing its journey to the other side of the universe. Each rider represents the kinetic energy of the sphere (kyntip). In this, we observe the energy of motion moving out into finite space and ultimately into eternity. The time it takes for "the rider" to move from one location

Chapter 5: Kineverse

in finite space to another determines the limit by which energy can travel in finite space. Even more importantly at this point, is the rate that time occurs in finite space as there is no matter per se. Lastly, we need to realize that even that these "riders" are moving out to infinity in all directions, while the kyntips seems locked in movement within infinitesimal space. If we added all the infinitesimal distances inertia traveled per event within a second, it would reach beyond infinitesimal space. This energy generates time.

As stated earlier, this acceleration is like that of a car accelerating from a position of rest up to a set speed in a certain direction. In order for the car to reach a velocity that has a speed of 50 miles per hour, it first has to accelerate past the point of achieving 30 miles per hour. Likewise, the acceleration within the uniquid undergoes the same velocity phenomena. In effect, time decelerates from being infinite per moment to a finite constant per moment. This constant eventually determines the speed of light in subsequent phaseverses. The result is that the kinetic pressure within the xev of inertia becomes saturated to its fullest potential, hence eternal time is brought to a stop and finite time begins. In other words, there were periods of time that occurred when all time passed through that were slightly larger than an instant until finite time exists.

This phenomenon appears as the uniquid transforms from being duoquiet to being monoquiet. This implies that the kyntip reaches its impact alephtip finite time becomes established. The time distortion begins in infinitesimal space ending eternal time within the finite physical realm. This distinction can be illustrated mathematically. Recall that $0+0=0$ letting zero represent a geometric point upon a number line. It takes an infinite number of geometric points to create finite space and an infinite number of finite distances to form infinite space. By examining the previous statement, we observe two levels of measuring distance: one by counting points on a line, the other counting equal finite distances. Recall that the geometric points sense themselves by reaching out to infinity, thereby encompassing all distance in the process. It takes forever to count out to finite space and forever to measure out to infinite space. The term "forever and ever" takes on a real meaning. Kinetic energy is being transferred at an infinitesimal level. However, the energy reaches through finite space to infinity, finitely unobserved as there are no discernible finite features. Only alephtips and bethtips transform during the motion; not the geometric points, as they are absolutely still. Again, Elohim does not move, nor need to move, for HE is already present at any location.

Chapter 5 Quiz

1. Discontinuity within the Kineverse exists at the:
   A. Infinitesimal Level
   B. Finite Level
   C. Infinite Level
   D. Electron Sublevel

2. Initiation of Movement is caused by:
   A. Gravity
   B. Heat
   C. Nondescript Acceleration
   D. Elohim

3. Quubium exist within Infinitesimal Space in a pattern of:
   A. Cubical Lattice
   B. Hexagonal Lattice
   C. Octagonal Lattice
   D. Triangular Lattice

4. Uniquidron is shaped like a:
   A. Octahedron
   B. Tetrahedron
   C. Dodecahedron
   D. None of the Above

5. Gravtip-2 are Bethtips located at the:
   A. Cubical Corners
   B. Triangular Corners
   C. Center of Cubical Edges
   D. Center of Triangular edges

6. (T/F) Within the Kineverse, gravity sends the bethtips in motion.

7. (T/F) The arrangement of gravtips exists at the corners of the quub.

8. (T/F) Unified fluttering defines all kinetic movement into six motions.

9. (T/F) Quubium is a lattice filling the entire physical universe.

10. (T/F) Time becomes eternal in the Kineverse.

Chapter 6

Thermaverse

And Elohim said, Let there be light: and there was light. And Elohim saw the light, that it was good: and Elohim divided the light from the darkness. And Elohim called the light Day, and the darkness Night. And the evening and morning were the first day.

B'reshyit (Genesis) 1:3-5

Clear to Infinity

Usage of the term of light existed in the Torah because of the effect it had upon the physical universe. Namely, it caused darkness to disappear. However, it was not the bright light that we think of when we look into the sun; even though it becomes the very energy that produces the brightness of the sun. At this point, its effect is that of light entering a dark room. We normally do not see the space in the room as a bright light, but as transparent allowing vision to observe objects in the room. Imagine that the transparent space was all there was to observe. Instead of seeing darkness upon the face of the deep, we observe a transparent substance with nothing on the other side. For our senses this would be unacceptable, we would formulate a color to interpret the sight. As we attempt to conceptualize the union between darkness and light, a grayish color appears.

However, our eyes see in living color. Taking a black and white photograph of cyan (blue), it would assimilate this medium shade of gray. Even though there is a shade of orange that produces the exact same grayish shade, orange is not a primary color. Consider the color of the sky during daytime. Scientists attribute the color existence to the chemical composition of the air, and the curvature of the earth for red sunsets and sunrises. This information is not disputed. However, they fail to recognize that Elohim did not create a blue sky over the earth just because HE arbitrarily thought it to provide a nice color. It represents the union between light and darkness, the first sight that would be seen in the beginning of creation if it were humanly possible. This appears as the light from the sun joining with the darkness of the night canopy forming blue. When we take a black and white photograph of blue, we see this same image of a medium gray.

Part 2: Creation of Xyzenthium Crystals

Invoked Light

Up to this point in the First day of Creation, there were three kinds of energy involved in the physical universe. Introduction of inertia into the universe by Elohim produced physical existence formulating the inertinuum (the earthly universe). Jammeria transformed the inertinuum into uniquid (waters). As presented in the Torah, the uniquid cannot move in itself. Elohim intervenes by "moving upon the waters." By this, we understand that the phaseverse has no intrinsic quality to initiate itself into motion. Note: the Torah does not tell how Elohim created "the waters," only that they both became present as a result of HIS creative effort.

Kinetic energy provided by Elohim introduced a longitudinal vibration and stored in the kinetic xev within the alephfield. When Elohim invoked "light" into existence, the universe transformed from the kineverse into the thermaverse. As implied earlier, the uniquid became transparent. This was accomplished by exerting pressure within the last xev as a force within the alephfield. This pressure increases the kinetic interaction significantly. As mind boggling as this may seem, Elohim created even more kinetic disturbances after creating kinetic energy. It must be remembered that this xev **force subfield** has a separate purpose than that of maintaining kinetic energy. The primary function of the energy connected to this xev is to generate acceleration, whether it is inertial (0), push acceleration, pull acceleration or just kinetic. This will provide additional finite motion needed in the future.

This phaseverse labeling is the thermaverse because of the introduction of "heat" (heat in a broader sense) into the universe. The term heat, once again, does not accurately describe the energy involved just as the term light does not equate to photonic activity. However, there is a reason Elohim had Moshe (Moses) use the term light. It is the ultimate observable expression of this energy as far as humans are concerned. This term of heat describes something present beyond that of visible light. There is even another force just as fundamental; this is known as the **microstrong** force. The microstrong force can exist without generating physical heat. The definition of the microstrong force will be described later in great detail, but for now it is described magnetically as a strong neutral force. Even this force is not the root of the energy expressed by Elohim's call of "light" into existence.

The label given for this force is **juttoria** (jut-tore'i-ah). This word is a cross between the words jut and factorial. **Jut** means to project or protrude. This gives a symbolic term of its graphical nature. Fact**orial** is a mathematical term meaning to multiply a sequence of consecutive numbers starting at 1. Example: $4! = 1*2*3*4$ (Note: "!" is a symbol representing factorial and "*" represents multiplication). We should note that juttoria does not have a factorial behavior pattern in energy increase or decrease. It behaves in the standard form of energy, in that it increases or decreases in the geometric fashion typically seen in radiant energy.

Chapter 6: Thermaverse

The symbol representing juttoria is a double lined chevron with its point up. This is the opposite of jammeria. The reasoning behind this is that these two affects the space/time continuum in the opposite way. This will become important as we examine the various components of the universe. It will provide examples of the effect these energies have upon electrons and photons, the relationship between black holes and quasars, and time. Again, inertia, by the same schema, is symbolized by the equal sign.

Xevim Angles

Finally, we can assign a xev to each of the forms of energy exerted upon the physical universe. We have the three dimensional angles "touching" the geometric point at the same reference angle and the reference angle itself as a xev of inertia. The assignment of juttoria to the bethtip transforms a kyntip into a **thermtip**.

Each of the four xev exerts pressure upon the alephfield, naturally, at different angles. The inertial xev is as isometric tension upon the alephfield defining the angle in which our universe accesses the "real" geometric point, in which the Spirit of Elohim exists. The kinetic xev is as the inertial xev having magnitude but offers no spatial acceleration, at least without contact, maintaining the space/time continuum fiber in the standard straight-line appearance established by inertia. The jammerial xev pulls downward upon the alephfield; its magnitude produces a spatial acceleration giving a downward funnel-like form within the continuum. The juttorial xev pulls upward upon on the alephfield producing an upside-down funnel fashion or peak within the time/space continuum.

Elohim did not increase the energy level just so that HE could observe the physical universe via quote unquote light. HE already knows: the location of each bethtip, the direction of each, and the speed that each is traversing the uniquid. Up to this point, the uniquid finitely appears totally integrated. In other words, the entire finite universe appears as one singular object as there is nothing to distinguish one finite location from another. The purpose for this increase, at this point in time, can cryptically be stated as formatting the xyzenverse.

Energy Types

Another interesting observation is the pattern in which energy develops. There are two kinds of energy being developed in the first five phaseverses. The first is the energy of attraction, and the second is the energy of repelling. Within this, we observe that the development has a secondary pattern of generating the energy in the isometric form and then in the radial or spatial form. Recall that in the nulverse that the xevim existed in a diffused reference to a given geometric point accessing the geometric point at three different dimensional directions. Forming the inertverse, the three nultips were drawn together to form one reference angle per geometric point. This activity is the first

Part 2: Creation of Xyzenthium Crystals

manifestation of the energy of attraction. The energy of attraction exists only within the alephfield holding the three xevim together forming the fourth central xev or reference angle to the geometric point in isometric tension. Within the gravverse, we find that the energy of attraction reaches outside the bethtip. Jammeria radiates through finite space attracting other particles to itself as gravity. The second energy type forms in the kineverse. This is the repelling force generating isometric acceleration. In this, the bethtip is accelerated from absolute stillness to absolute isometric motion as it is constantly moving from its present location to another. However, it does not radiate from the xev to influence other bethtips to accelerate without contact. Juttoria is the manifestation of the repelling force externally causing particles to accelerate away without contact. Putting it all together, we see a pattern of progression in the creation of energy: unfocused, isometrically focused, externally focused, isometrically accelerated, and externally accelerated energy. Notice the alternating pattern between isometrically formed energy pointing inward, and the externally pointed energy affecting the universe around it. Notice also that in the Torah the "earth" starts in darkness (jammeria) and then light (juttoria) enters. Magnetic energy formulates later within a future phaseverse in which Elohim "divides light from darkness." In this, no new xevim form but different relationships and alterations occur within the realm of the xevim.

Dimensional analysis of the resulting thermaverse form: inertia is a zero-dimensional energy type; it does not cause any acceleration or zero acceleration. Kinetic energy is a one-dimensional type. It can accelerate only in one direction, and cannot accelerate another object without contact or zero distance. Both jammeria and juttoria are two-dimensional energy types. Neither inertia nor kinetic energy generates an energy field of itself. Juttoria and jammeria both create energy fields or, more precisely, acceleration fields. Juttoria provides a repelling field and jammeria a pulling field. These fields they create bend space in opposite directions and diminish in strength by distance from the origin generally expressed as $1/x^2$. The two field energies within this phaseverse reach a point of balance, in other words, canceling their bending effect of the space/time continuum per thermtip within the final thermaverse form. Moreover, there is also an inverted application of juttoria, as we shall observe.

The differences among the four xevim types can be described as the following: using the relationship between the kinetic xev and the inertial xev as a reference, we can describe the energy relationships that exist with the remaining two xevim. Imagine looking at a corner intersection of a cube. There are three distinct edges joining together forming right angles ($90°$) in respect to each other. From this corner, imagine a diagonal line passing through the cube's center connecting the opposite corner. Inertia's visualization is the xev that maintains its original position in relationship to this line. Jammeria is the edge that moves inward directly toward the diagonal line generating acute angles (angles under $90°$). Juttoria leans directly away from the line of reference forming obtuse angles (angles greater than $90°$). Through this, the space/time continuum is distorted into opposite directions by the two forces. Note: the

Chapter 6: Thermaverse

supplementary angle (angle added to form 180° or a straight line) of an obtuse angle is an acute angle. The acute angle formed by jammeria lends itself to form the distortion exactly like that of juttoria only in the opposite direction.

A visualization of this phenomenon can be found in the examination of a balanced door hinged to a floor (a horizontal plane). This door is functional in that it can be opened or closed. In the balanced state, the door is nearly vertical. Because of the friction on the hinges, there is a minute variation of angle in which the door can be positioned and still maintain balance. This position represents the isometric pressure of the inertial and kinetic xevim. The door can drop in two directions: to close or wide-open positions. All it needs to fall in either direction is a slight adjustment of its angle to the floor in one direction or another. Let us translate this information of formulating jammeria and juttoria. Jammeria becomes as the closed door and juttoria as the door swung wide-open. Notice that it is the same force that causes the door to swing to rest in the closed position as in the wide-open position. Jammeria accelerates until it reaches the source of acceleration; while juttoria accelerates an object away from the source by the variation of the angle of the xev. In our illustration, the single relational force is gravity; in the xevim, it is the relationship to inertial and kinetic xevim. The angle variation provided by Elohim gives the desired result of juttoria or jammeria.

Geometric examination of energy leads us to the conditions existing at a given geometric point. Physical space exerts pressure upon the geometric point at an angle of reference assigned to the physical universe. Jammeria exerts a pressure downward upon the geometric point. Juttoria exerts a pressure upward away from the point. Inertia exerts an isometric pressure. Kinetic pressure is a directional variable expressed upon inertia and is an isometric pressure at the kinetic xev angle within the alephfield. Finally, inertia is the root of mass.

Xergotips Planes and Plates

The etymological ascription of **xergotip**s (zer-go-tips) is derived from the same prefix as that for xergopath. The "x" represents the transferring attribute of the component. "Erg," a scientifically defined unit of work, indicates that energy is involved. This imputes to the object the attribute of transferring energy. Note: the usage of erg in this word does not include its scientifically defined measurement of work.

Xergotips are not new locations within the uniquid. Prior to this point, we only viewed them as alephtips in which six bethtips ultimately converge. As noted earlier, the paths in which bethtips move (xergopaths) intersect forming a cubical structure known as the quubium. During this phaseverse, the xergotips are used to define the division of the unified finite space into finite granules.

The specially located xergotip has a unique set of attributes assigned. These attributes are not strange to the created environment of the uniquid except for the juttorial energy

Part 2: Creation of Xyzenthium Crystals

formulating the thermaverse. We can naturally assume that the resulting formulation phenomenon was purely a supernatural act, which it was, and be done with it. Despite the supernatural imposition upon the uniquid, its physical manifestation occurs within the confines of the previously created rules of natural phenomena creating yet another rule. The initial manifestation of juttoria occurs during septicollision when the xergotip contains six times the energy of a bethtip. Elohim's action appears primarily as "turning on the switch" (pressure) to the xevim subfields within the phaseverse.

The **xergoplane**s are three sets of uniformly spaced planes that divide the thermaverse into finite cubes. Each xergoplane is two xergotips in thickness. Actually, each xergoplane forms a plane-pair as both planes are parallel at an infinitesimal distance to each other. Within the finite realm, this may seem a moot point until we recall that a set of points creating a certain length will create twice the length if we were to place them in an alternate pattern (every other one) rather than lining them in a consecutive manner. This is the first time that there exists anything distinguishable within the finite realm since creation of the physical universe. The xergoplane then comprises of two separate planes called membranes. Since xergoplanes are comprised of xergotips, they are not a two-dimensional continuum within infinitesimal space. The purpose of these xergoplanes is to apply an additional juttorial input to initiate a separation between cubes so as to generate granules of finite space.

Xergoplates are the membrane faces of these finite cubes formulated by the intersecting xergoplanes. The image is that of many physically individual cubes placed next to each other forming rows and the rows are placed tightly together forming infinite layers and the layers are also stacked forming infinite columns of cubes. In examining this "continuum" of cubes, we see that each physical cube has its own surface. In other words, each cube has a separate set of faces. Xergoplates are these faces. Note: that two adjacent cubes have one face each facing each other. While our physical cubes are composed of atoms, xergoplates are composed of xergotips. If we were able to look at a single xergoplane, we would see xergoplates in neatly aligned rows. Each xergoplate is actually a square area of a **xergotip membrane**.

Xentrix

Xentrix (sen-triks) derives from three words: **xe**rgotip, c**ent**er, and ma**trix**. It is the cubical volume defined by the xergoplates inclusively. The second application of the juttorial force within the thermaverse is the imbalance of input. Just as the imbalance input in the gravverse causes infinitesimal space to become discontinuous, so does the juttorial imbalance generate a finite discontinuity. Interestingly enough, the application of the juttorial imbalance is of an inverse nature. Jammerial imbalance was centrally introduced; juttorial imbalance is introduced at the edge. While jammerial imbalance has a central point (odd number function), this imbalance has no xergotip aligned with the intended center, hence, an even number function (needed by later usage).

Chapter 6: Thermaverse

Imagine the membranes of the xentrixes as being white light shining into a blue field. The white coloring represents an increased juttorial energy level exerted by Elohim. We will examine the effect within a plane parallel to one of the cubical faces.

Unlike the quubium of the uniquid, each cube has a separate membrane. Imagine a blue plane divided into squares by white lines. Now, let the squares shrink. The white line will then move inward with each square. If the white line were an edge of a membrane between two cubes, it would have to move one way or another but not both. It is able to accomplish this, since each xergoplane is two xergotips thick. Therefore, each planeside is able to move away from the opposite side. By this, each square membrane has its own white outline. The white line will increase in size as the square shrinks, as the white light radiates into the blue representing the migration of juttoria into the internal vacuum of xergotips via spatial energy radiation.

Another item that we observe is that the membranes separate from each other, primarily because of their increased juttorial repelling nature, and the jammerial pull from the interior. As the membranes move further from each other, jammerial pulling forces reach a greater imbalance, favoring the center. The reason for the increased imbalance is that the jammerial forces from the adjacent xentrix becomes weaker due to the increased distance from a given xentrix membrane. Despite the imbalance of jammerial pull from within and exterior to the membrane, the jammerial pull between each xentrix remains in balance maintaining equal distances between each xentrix. In essence, the universe transforms in an isometric manner.

Juttorial Response

Despite the spread of the juttorial energy throughout the three-dimensional matrix, the xentrix shrinks. Shrinkage occurs due to the stronger influence of jammeria within the finite realm. If the xentrix were merely a matrix within infinitesimal space, the imbalance of forces would have no effect since there is infinitesimal space between xergotips, each of zero mass. It takes the infinite collective force that gives each xentrix its finite properties as a centroid.

Looking at the phenomenon from a mathematical perspective, we can see the initial imbalance of juttoria energy value is greater than the resulting uniformity. The primary reason is that its "vacuum" is absorbing the energy imbalance. If we have a value of +1 in one location and a value of zero in another and we divide the value between the two locations, we get two equal values of a half or 0.5. However, as it is seen in this mathematic example: After annihilating all the possible internal "void," it reaches uniformity. The process stops and becomes static once again at a smaller unit value.

The rate in which the shrinkage occurs follows a pattern similar to the secant of an angle. The secant curve attains its definition from dividing the cosine of an angle (normally called "theta"- a Greek letter) into one. This function has two correlating

Part 2: Creation of Xyzenthium Crystals

features: it moves quickly toward the x-axis as it leaves infinity, and its rate of slowing increases as it approaches the x-axis to a point of being parallel to the x-axis when y=1 (plus or minus). Unlike the secant function, the xentrix shrinkage does not bounce back to its original position. Like many functions in natural phenomenon, they often do not use the entire curve. Another reason for using the secant function as opposed to using $y=x^2+1$ is that the secant naturally stops at one and x^2 is adjusted by one.

Notice that from $0°$ to $45°$, the y-value only increases from 1 to 1.4142. Yet, in the next interval of $45°$ to $90°$ (equal length in x-values), the y-values increase from 1.4142 to infinity. By this, we observe a dramatic change occurring in the last little distance that these xergotips travel. The juttorial influence becomes stronger causing the rate of shrinkage to slow down to a stop.

Shrinkage via Xergotip

The normal movement between bethtips within the recent phaseverses was uniform and always repeated their movement back and forth forming perfect septicollisions at alternate xergotips. In starting our examination of the two sequences of movement, we used a time when one set has reached a septicollision. Recall, that the geometrical image forms as a relationship between two sets of octahedrons. When the bethtips reach the xergotip, the octahedrons image becomes indistinguishable. As noted earlier in the kineverse, the expansion of one set equals the shrinkage of the other set. The principle is: as a bethtip departs from one xergotip, it is approaching another xergotip (not lost).

Within this scenario, we observe the distance between xergotips shrinking. As this occurs, the possible distance for the bethtip's return also shrinks, generating a smaller sphere of influx. In the ending stage of our sequence, we observe the bethtips reaching the opposite set of xergotips. By the time the thermtips reach their destination, the xentrix has finished its shrinkage.

Another important factor involved within this scenario is the ratio between the size of the original xentrix and its final size. Earlier, we used the apparent ratio to be measurements as one half, one third or even a quarter in size per dimensional direction. Even these measures create volumes of 1/8, 1/27, and 1/64 respectively. However, the ratio is much more extreme than the ratio of an atom to a grain of sand, and perhaps approaching one divided by a googol (10 x 10 x 10 x ... x 10: 100 times). Despite this extreme shrinkage, the xentrix remains finite in measurement.

There is a juttorial increase. It is more than a collective count of xergotips multiplied by some constant of "heat"; the juttorial constant also increases in magnitude after its initial devaluation. The nature of juttorial force can be visualized as a sphere that has a multitude of thin magnetically strong (strength) charged strands several times the diameter of the sphere. By this, each strand projects straight out from a given sphere. Secondly, as the sphere accelerates toward another sphere, the strands are pulled back

Chapter 6: Thermaverse

slightly by their repelling and inertia. Imagine several spheres coming closer together. The result is that these strands are forced to project in a lateral direction to the direction of their mutual approach. This alteration in the strand's direction gives the resulting plate a greater projection of juttoria. Imagine being able to give a flattop the resulting alignment of spheres. In this, we shear in the direction of the movement and place each strand end on end to form a single line. The resulting line would be longer than the chopped strands from an equal number of isolated spheres with the same flattop. These two lengths represent the magnitude of energy. Thereby, the constant increases as the distance between the juttorial energized xergotips decreases. Another example that illustrates this is; the difference between diffused light from a light bulb and the congruent light of a laser reaching the same location.

This process is called the xentrix ignition. When "light" was introduced into the universe, the universe became clear, all the way to infinity in every direction giving the general color perception to us as "blue". Now, at the ignition, juttoria becomes like a visible glow. This would be represented as a vast array of blue lights, if we could see it. It would not be white, as it is still in union with darkness.

There is a struggle that exists between the two acceleration forces within the xentrix. Juttoria promotes a total dispersion of the xentrix by the means of pushing away adjacent xergotips. Jammeria pulls upon each xergotip attempting to cause the xentrix to collapse into infinitesimal space. However, obviously, neither happens.

Perhaps the best image to present the finite response between jammeria and juttoria is the following: Imagine the xergotips as having two sets of arms and hands. One set is shorter than the other set. The shorter set will be used to symbolize the juttorial force and the longer for jammeria. Imagine the two spheres passing by each other. Before the spheres can reach their apogee (closest point), the larger sets of arms and hands come into contact. As they do, they begin the attempt to pull them together. As the spheres move toward each other, they come in contact with the smaller set. They join hands and attempt to push the spheres apart. The image structurally is a set of long arms bent at the elbows struggling with a shorter set relatively unbent. The force needed at the elbows to bring the spheres together is considerably more than the force needed by the short arm's elbows to push the spheres apart. At a point, they reach a balance where the kinetic energy accumulated by the longer arms becomes negated by the shorter arms. Since these two sets of arms have equal strength at this location, neither side wins. In their battle, the two spheres maintain a fixed distance as two separate entities. Note: the small arms do not represent the extent of the influence, only the comparative magnitude.

The bethtip paths during this process can be illustrated using a loom. Our loom has threads that have no thickness and are evenly spaced per dimension of the loom. If we push inward upon the threads toward the center, the pattern tightens and an external angular pattern forms. Since these treads are so thin, we can pull them so close together

Part 2: Creation of Xyzenthium Crystals

that they seem to come to a point. The image then becomes purely angular in appearance as that of intersecting lines from all angles. However, since the process remains in finite space, we can magnify the center and see the crisscross pattern before it goes beyond our sight resolution. This image represents the path of the bethtips during the process.

Phaseverse Final Form

There is one last item that modifies the image of the xentrix. It is the relationship between the surrounding void and the density of the xentrix itself. In completion, the xentrix material becomes concentrated into a very tiny volume; the external void behaves somewhat like a vacuum suction. Not that the void has any pull, but the imbalance of the jammerial and juttorial energy from the xentrix cube and the void distorts the cubical appearance of the xentrix. Jammeria, being the strongest finite force, pulls inward from the center in a spherical manner. Similarly, from the external faces, edges and corners of the xentrix juttorial energy pulls away from the center as there is no external force from the void to hold the equilibrium. The result of these two functioning principles is that the faces of the xentrix cube sink inward while the edges and corners are pushed outward. While we can visualize this formation as a thorn-like cube, the effect is actually quite subtle. Recall that any finite variation is infinite when compared to the infinitesimal alignment of geometric points. This thorn-like cube shape will become important in the future formulations within the void.

When the xyzenthium crystals begin to develop into their final formulation, the thermaverse terminates. Juttoria will have accomplished the task of dividing finite space into uniform energy complexes. Thermtips, however, still have attributes that will continue to alter the nature of the universe into the next phaseverses.

Chapter 6 Quiz

1. What is Juttoria?
   A. Jut of Factoria
   B. Repelling Energy
   C. Attracting Energy
   D. Static Energy

2. Blue is?
   A. A primary color
   B. A union between darkness and light
   C. Color of the Sky
   D. All the Above

3. Xergotip exists as a:
   A. Bethtip
   B. Outside the Uniquid
   C. Corner of a Quub
   D. None of the Above

4. How is juttoria applied within the phaseverse?
   A. Very carefully
   B. Within selected centers
   C. At the cubical external faces
   D. Every other geometric location

5. What is the final shape of the xentrix?
   A. Spheroid
   B. Ellipsoid
   C. Thorn-like Octahedrons
   D. Thorn-like Cube

6. (T/F) Juttoria and jammeria exists in the same bethtip.

7. (T/F) Xergotips are specialized bethtips.

8. (T/F) Blue is a sing of division between light and darkness.

9. (T/F) Juttoria and jammeria are applied in inverse fashions.

10. (T/F) The xentrix occurs in uniform intervals.

Chapter 7

Xyzenverse

The thermaverse predetermines the formulation of the xyzenverse as a result of its internal activity. The xyzenverse granules define the initial finite objective for the introduction of "light." These formulated granules within the xyzenverse are **xyzenthium** (size-enth'i-um) crystals. Even though xentrixes were the first finite formulation, it is only transitory to the crystal formulation. These crystals behave in a similar manner as a cell nucleus is to the entire cell. Xyzenthium crystals are the first finite "cell-like" units formulated in the physical universe. It is interesting that at this point: Elohim observes that HIS Creation was good (Gen 1:4). It is not that HE could not see HIS Creation before, but now there are finite mechanisms existing within it.

In the naturalist version of physics, these crystals are quanta sized particles. The Quantum Theory is a relatively recent development in the science community. Using this term "quanta" could create confusion, even insulting to some (but not our intention). This is not the Quantum Theory, nor is it an extension of it. Even though the final formulation of the crystal behaves in the manner observed in quantum mechanics, there are elements of the xyzenthium composition that are dissimilar, even undefined by the definition of a quantum. There are other differences also; it is not the observation data nor the calculations that are being challenged, but its interpretation.

The etymological significance of xyzenthium is that it describes the nature of the foundational finite particle. Dimensional axes (x, y, and z) are represented and create the phonetic sound of the word "size." Along with the phonetic sound of the tem "size," is the mathematical term, "nth." Nth is used to illustrate the most extreme finite term. Contextually speaking, the "size nth" means the most extreme small finite size. Therefore, it is chosen to represent the fundamental-most finite building block or granule of our universe. The chemical suffix of "ium" is used to reflect the elemental (fundamental) nature of the crystal. Recall the xentrix of the thermaverse. Its original size is the size of the xyzenthium crystal before the xentrix shrunk into the complex formulation described within the previous chapter. The xentrix mechanism is the smallest finite mechanism, and there is only one xentrix per xyzenthium crystal.

Loop Directions

As we shall observe, when the bethtip streams leave the xentrix into the external void, they are immediately drawn into the adjacent streams of bethtips. The forces involved will be described shortly. The basic process involved is that these streams merge with

Part 2: Creation of Xyzenthium Crystals

the adjacent streams, forming loops. For an example: if there were one hundred xergopath strands leaving a given xentrix xergoplate, twenty-five xergopaths would bend toward each of the four adjacent streams forming a loop with that stream. This process is true for each of the facial xergoplates forming the cube of the xentrix. We can safely connect twenty-five xergopaths to a different set of twenty-five without losing one xergopath within the process.

As easy as this might seem, the actual process is somewhat complicated. The problem is symmetry within a three-dimensional object. Consider a cube containing twenty-seven smaller cubes of equal size. The corner cubes have three faces exposed to the surface while a central edge cube has only two faces exposed. The process occurring at the central edge needs to be different than that of the corner. Yet the corner has to conform to the process occurring within the central edge. Two plus two equals four, but it needs to equal three "working within the symmetrical box." In other words, there is need of a shift in symmetry. The corner xergopath can only link with one side of the cube, but our initial image requires that it link with both sides.

However, there is a solution to the problem. The first rule to the solution is that the number of xergopaths gathered has to be a multiple of four making both edge counts even, meaning no central xergopath. The image is of four isosceles triangles with the base on each edge per cubical face. If we match the corners of the triangles with the corners of the cube, we will find that we have created a set of xergopaths that sit at the union between two cubical faces. This, naturally, prevents some of the xergopaths from moving in the intended directions. However, if we make the triangles slightly smaller and shift the triangle to one side symmetrically in respect to the facial center for all triangles, we will have successfully divided the xergoplate into four equal counts of xergopaths. If we have accomplished this task properly, we will observe a small unused central square occupying zero xergopaths. This is only the first part of the solution.

Remember the three-dimensional problem with the corners; we have removed only half of the corner problem. We now are facing another corner problem. We cannot make these triangles symmetrical to the triangles adjacent to it on the xergoplate attached to the xergoplate of origin in all dimensional directions. However, we must make opposite xergoplate patterns match. Once again, we need two plus two to equal three instead of four. Just as before, it is mathematically impossible. Because of this problem, we still have basically four corners with nowhere to go and an alignment problem.

These offset triangular sets must shift slightly laterally toward the center. In doing this, we made our problem of an untied corner and our alignment problem insignificant. In examining our movement inward, we find that the movement is half the distance that exists between xergopaths. The vector arrangement has radial symmetry between the four triangular divisions of xergopaths per xentrix face. The offset designed division provided by Elohim appears to have no natural phenomenon attached other than pure logical resolve of the problem. However, the interaction between jammeria and juttoria

Chapter 7: Xyzenverse

is the process that pulls the four triangular segments together into a new alignment. Before expounding the details, we need to define some more geometrical terms.

Zeeds and Exozeeds

Up to this point, we have each xergoplate divided into four equal parts. Each part contains a specific set of xergopaths. There are twelve sets of parallel xergopaths formulated from the original xentrix connecting two triangular shapes upon the faces of two opposite xergoplates. As we examine the connecting of the loops starting at the edge-most xergopaths, we find that each loop meets at a right angle at the base. The next set of xergopath loops away from this edge also meet at right angles, but connect under the surface. Each succeeding set away from the edge connect deeper into the xentrix. The resulting shape of these xergotips forms a triangle from the edge of the xentrix down toward the center. This set of intersections is a **zeed** from the words "**z**ero" and "s**eed**." The set of xergopaths connected to the zeed is called the **exozeed** using the prefix of "**exo**" to provide the meaning of being external to the **zeed**.

Another interesting observation is that the xergopaths become disjoined leaving only two dimensional directions of xergopaths connected per zeed. This reduces the zeed from a seven alephtips involvement per xergotip to five alephtips. The reason for this disconnection is that there is lateral movement as well. The reason for this movement is to center each exozeed with the center of their xentrix. The zeeds form six distinct pyramid-like shapes with their apex near the center of the xentrix. However, neither the edges nor the faces of the zeeds ever touch because of the movement of the exozeeds away from the xentrix center. Similarly, the apexes between pyramids are not joined as there is no central xergopath. The zeed count per xentrix is twenty-four, as the xentrix "cube" has twelve edges. Each exozeed contains two zeeds. The image of two adjacent zeeds is as two exact triangles next to each other on two thinly separated parallel planes.

Returning to our set of one hundred divided into four sets of twenty-five, we can now determine the region in which they tie together. The surface area of the xentrix that they occupy is one-twelfth. The volume reaches to the centermost xergopath of the xentrix from both dimensional directions. The centermost xergopath per exozeed union does not occupy the center of the xentrix as the center is empty of all xergopaths.

Xergopath Sectors

There are three basic sets of looping patterns that result in this phenomenon: they are the xy-looping pattern, xz-looping pattern and the yz-looping pattern. The first two letters represent the dimensional plane in which the looping pattern exists. Each of these patterns is identical in nature with the only difference being the dimensional direction in which they are facing. Once again, this uniformity is stable and would not change until Elohim intervenes again.

Part 2: Creation of Xyzenthium Crystals

The energy vortex has six expressions as the bethtips leave the xentrix. The image is as if a cube had faces made out of a type of elastic material stretched out over a cubical frame with each facial center pulled into the cubical center. The reason for this phenomenon existence is that the external void exerts zero force upon the fibers of the space/time continuum, while the xentrix is exerting an extreme force upon the fibers of finite space. The concave surface creates a slight curvature in within the xergopaths, formulating the xentrix, controlling the flow of bethtips.

The xergopath curvature within the xentrix structure will become important shortly. However, we will start out with a simple linear model of the xentrix to relate the movement between xergopaths during the transformation process. The reason that we are able to do this is that the curvature does not affect the relationship between xergopaths of different dimensions within a given position. For example, the edge forming xergopaths still connect; the only difference is that the cubical structure is not exactly cubical. We are now ready to examine the process in which the xergopath strands form into loops.

Xentrix Fragmentation

Now, that we have established the need for the xentrix to breakup in a certain manner and all of its terminology, we can now present the forces involved in generating such a movement. After the xentrix shrunk to its limit, the shrinking process continues to exert itself upon the xentrix. The xentrix fractures in a crystalline manner. The manner of this fracture is as previously described, in that, it is not purely from corner to corner as we might assume.

To understand the processes involved, we need to present another illustration. Within this image, we will start with a single square. We will divide this square into nine equal squares (3x3). At the corners of each square, we will place a dot inclusive of the outside corners. The total number of dots should be 16 as the intersections formed by four adjacent squares are allocated only one dot each. Now, we connect the opposite corners of the initial square with two diagonal lines. The result is that theses two lines intersect in the center of the central square as it is a 3x3 square. However, these two lines also contain four of our dots per line. As earlier, this is unacceptable. No dot or xergopath can exist in two triangular sectors. Therefore we will divide the line at the central intersection and move each segment to the right slightly in radial symmetry.

Now, we are ready to examine the process. Each dot represents a xergopath viewed is as looking at a line segment from its endpoint instead of its side. Secondly, there is a central square formed by the diagonal lines positioned diagonally in reference to the original square. Thirdly, notice that the total dots are divisible by four, deliberately. Another observation is that the triangles formed extend beyond the confines of the original square. The focus of our examination will be the small empty square in the

Chapter 7: Xyzenverse

center of the xentrix. For our slight shift of the diagonal line, we will use the central unit square of our 3x3 division of the original square. Starting with the lower left diagonal line segment, we will generate a parallel line that starts at the center of the horizontal edge to the right the diagonal line. In essence, we moved the line to the right one half of distance between dots or xergopaths. We continue this process in a radial symmetry for each line segment in respect to the center of the original square.

Now, we will shift the triangles to the center laterally along the edge of the original square. The result is that the central diagonally created square disappears. Moreover, the central four dots are closer together than they were in their original positions. This is consistent with the gravitational centroid existing within the center of the xentrix cube. Recall that juttoria prevents the total collapse of the xentrix and that its strength is primarily perpendicular to jammeria. Secondly, the finite size of the xentrix is the factor in which jammeria exerts pressure upon the centroid. As the central four xergopaths move into their final positions the external dots or xergopaths follow per dimensional direction. The result upon the edge of the xentrix is that there is no corner xergopath present.

The Thermtip Pulse

If kinetic energy were the only force involved, xergopaths within these exozeeds would shoot straight out from each face of the xentrix to crash into the oncoming xergopaths from an adjacent xentrix to its direction. Recall that the bethtips within each xergopath moves in two opposite directions. In this, the xergopath splits into two sets of bethtips moving in the two opposite directions. These closely knit sets form the thermtip pulses.

When we examine the zeeds a little closer, we discover that the zeeds come into existence only after the exozeeds shifted. The primary reason for this development is that the xergopaths separate from their original alignment defined by the shift and do not rejoin to the same alignment afterward. The new xergotip alignment forms at a different location in which the zeed find their existence.

Another interesting observation, the initial size of the xergopath loops is not as big as we might expect. However, as we shall observe later, they will grow to encompass the entire xyzenthium cube. The reason for their closeness or smallness is the pull of jammeria. We might conclude that the juttorial energy would prevent the connection from ever occurring. This would be true except for one small detail- the geometric requirements intrinsic within the finite pattern of the loop giving jammeria the control.

Moreover, the two pulses of bethtips meet upon the external side of the loop; their xevim are inverted by the nature of making the loop. Instead of juttoria meeting juttoria, it is meeting with jammeria. Since both pull space in the opposite directions, they are drawn to each other by the tension of the spatial fiber formed within the xergopath. The result is the phenomenon aids the force of attraction forming between

Part 2: Creation of Xyzenthium Crystals

any two colliding bethtips. When the looping is complete, the bethtips return to the xentrix via the collective central pulling acceleration by the bethtips within the crystal.

Examining the pattern in which the multitudes of xergopaths form loops, we find that the edge xergopaths travel the least distance. They move to their corresponding symmetrical counterpoint. The reason for this movement is that the outermost xergopaths' bethtips of two separate streams are closer together than the internal xergopaths. As the two xergopaths move away from the stream, they release juttorial pressure for others to "peel" away from the main stream. This process continues until all the xergopaths peel away and join with their counterpart. The result is that the movement remains a parallel journey for the bethtips.

The pattern, in which the bethtips flow within their loops, flows differently than one might expect. The xentrix structure dictates a linkage between four different loops existing within the same plane. The xergopaths maintain their original positions within the xentrix internally, and invert externally by the geometric nature of the loop. The outermost xergopath within a given xentrix becomes the innermost loop maintaining a parallel-like course. There are three different two-dimensional plane sets per xentrix. The xy, yz and zx-plane sets each operate within different sets of exozeeds which are defined by the occupying xergopaths.

Diflohexius

The next attribute is seen in examining the dimensional axes with the origin in the center of the crystal. Each axial direction appears as a geyser of uniquid. The resulting image is that of six geysers moves energy from the core of the cube outward to each of the six facial surfaces. Similarly, energy also feeds back into the core like a funnel. The six xyzenthium dimensional geyser/funnels are called a **diflohexius** (dye' flow hex' i-us), or in single fountain formulation: **diflohexet** (dye' flow hex-et'). Diflohexius is a word-compound formed from four words, **di-** a prefix for two, **flow**, **hex** for six and rad**ius**. The prefix, "di," is chosen to combine with "flow" to create an image of energy flowing in two opposite dimensional directions simultaneously. "Hex" represents the three dimensional axis in their positive and negative directions. "Radius" gives us the spherical nature of the formation. The diflohexius phenomenon is the process that the xyzenthium crystal establishes its finite size. As we shall see later, this will diversify.

Our initial observation of the phenomenon formulating the xyzenthium crystal is found in the loop pattern itself. If we examine a single xergopath strand within the structure, we find that the bethtips flow in two directions simultaneously. While we view it as an even flow, there is evidence that they intensify in number as the bethtips approach the xentrix. The reason is that in converging within finite space the jammerial pull accelerates the bethtip from its outer excursion within the xergopath loop. This expresses itself as somewhat denser nearer the xentrix base. Recall, also, that the zeed is

Chapter 7: Xyzenverse

the location that bethtips from each dimensional direction will return to as opposed to the exterior two-dimensional face of the xentrix.

Next, we will illustrate the fanning effect of the xergopaths by rotating a selected loop around the y-axis. The loops came nearly in union with the axis as the xentrix shrinkage was extreme. We chose only to represent a few of these xergopath loops to illustrate a simplified image of a fountain/ funnel effect, i.e. one loop per a degree of rotation.

The image is similar to an animated geyser of water shooting up out of the ground and then fanning out at the top in every direction. This is very similar to the diflohexius along a given dimensional axial direction. Secondly, we observe the water coming back to the ground after fanning out at the top. The "falling back to the ground" acquires another attribute, as there is no ground to fall upon. As observed, the water has to be constantly fed into the base; otherwise, the water would soon run out.

Within using our geyser example, another possible solution presents itself in the formulation of the diflohexius. We might think that the two adjacent diflohexets between two xentrixes splashed against each other causing the xergopaths to split away from its central stream-like course instead. In this, juttoria would cause the xergopaths to repel each other. The problem is that all the xergopaths coming from the adjacent xentrix would be continuing a path already established. The jammerial energy within a given bethtip controls the frontal movement. In this, they would be drawn to each other instead of repelling. The pulse then would shoot directly into the adjacent xentrix. We would, in essence, generate another kineverse where energy moves only in dimensional directions. The only differences are that the pulses are finite and they travel between xentrixes instead of xergotips. By this, we see that it is necessary that the diflohexius develops into a process from its "roots," at the xentrix itself. However, after the diflohexius becomes fully developed, each diflohexet does exert pressure upon the adjacent diflohexet between any given set of adjacent xentrixes. This will become important as we discuss the final formulation of the xyzenthium crystals.

After assembling all the diflohexets, we find a pattern very similar to the image representing infinite space. This is befitting that the xyzenthium crystal formulation should reflect the infinite formation of the macro-cosmos. However, this form does not remain as the xergopaths continue their expansion out to the faces of the xyzenthium crystal. Remember there is the additional juttorial force added to separate the xentrixes from each other. It continues to function and becomes the catalyst for xyzenthium crystallization.

Xyzenthium Crystallization

Recall that energy moves in two directions simultaneously within the xergopaths. After the xergopaths are cut into segments by the definition of the xentrix size, the bethtips are ejected out of the xentrix cube by their kinetic energy. They divide into two strands

Part 2: Creation of Xyzenthium Crystals

moving in opposite directions. If there were only one set of xergopaths facing the same direction, the strands would move like a bullet fired from a rifle. However; this is not the case, there are two sets moving at right angles to any given xergopath strand. Initially, juttoria forces them to remain apart, but as they get further apart jammerial attraction overtakes their trajectory. The two pulses pull together by the strength of jammerial and juttorial opposing continuum forces present in each bethtip expressed at the leading edge. It is this connection that initiates the crystal formulation.

Initially the xergopath loops were very tight as the bethtips were brought close together. Again, we are looking at finite distances near zero, from our perspective, but not zero or inside infinitesimal space. The collective juttorial influence again pushes the bethtips apart to reach the balanced distance between bethtips before the shrinkage of the xentrix. The shrinkage caused by the imbalance of juttorial energy within the matrix of bethtips has now been returned back into a balance via redistribution of energy by contact by all bethtips passing through the xentrix. The strand of bethtips lengthens due to the continued time of juttorial influence. The result is the xergopaths lengthen forming larger loops. Another interesting observation is that the juttorial constant per bethtip is greater than in the original thermaverse because of the added juttorial influence to divide the thermaverse into cubes. This provides a repelling force greater than that within the original external "ignition," generating larger bethtip spacing.

As the loops expand, the xyzenthium crystal grows in size. Initially, the crystal looks cubical as defined by the xentrix. This form changes into a squarish spherical-like form in nature as the crystal grows, and the entire assembly is held centrally within the void left by the shrinking xentrix. The forces that hold the crystal in place are isometric in nature by the surrounding forming xyzenthium crystals. The spherical nature dictated by jammeria is in appearance only, as the base form is cubical. However, in viewing the outermost force shape, it resembles more of a perfect cube. The xentrix core is almost a geometric point in comparison to the external form of the crystal. As previously stated, the spaces between bethtips attempt to reach beyond their original distance during the expansion because of the added presence of juttoria per bethtip. As for the behavior of the xyzenthium crystal, it resembles the attributes of a perfect sphere. This is because most of the xergopaths flow in a circular path and behave as a fluid rather than a solid via high energy levels per bethtip. Moreover, the collective force of juttoria per crystal repels the adjacent crystals giving an extremely slippery surface. As we shall observe, this is even true in future phaseverses.

Another phenomenon that occurs within the xyzenverse is the gravitational force presented by a jammerial pull upon adjacent xentrixes. This aids in moving the strands further apart than can be justified by their normal expansion. In essence, the pulse would only expand to encompass the entire loop, except that the strands are pulled away from the xentrix by the external jammerial force. After departing from the xentrix, the bethtip positioned within the xergopath loop causes even more bethtips to

Chapter 7: Xyzenverse

be extracted outside the xentrix until a gap forms. The pulse gap increases at a faster rate than the expansion between bethtips. This process is a little more complex than that, as we shall observe later, but for now this will suffice. As the two strands move through the xergopath loop in opposite directions, they come back into phase forming momentarily a single pulse. This always occurs when the gap centers upon the xentrix. While the local distortion upon the space/time continuum moves with the strands, the composite distortion remains centered within the xentrix. The reason for this continued center is that the xergopath loops exist in isometric patterns in all the dimensional directions. This pulse state continues to exist throughout the phaseverse.

Expansion of the xergopath loops finds a limit as the juttorial energy existing within each bethtip within the strands comes into contact with the juttorial energy of strands from the adjacent xyzenthium crystals. The expansion process halts at the point of equilibrium of the jammerial and juttorial energies. By this we see that there is space existing between the crystals. There is another alteration that occurs, but first we need to examine the nature in which the juttorial and jammerial energies exist within the bethtips, which are thermtips within this phaseverse.

Throughout the observations we have seen the juttorial repelling force manifest horizontally to the movement rather than straight into the movement. If we look at an acute angle and an obtuse angle with sides (or legs) of equal length, we can get an understanding of the different natures of jammeria and juttoria. Within this framework, the acute angle is assigned to jammeria, and the obtuse angle to juttoria. The reasoning behind these assignments is that jammeria pulls together, and juttoria pushes away. If we were to place this acute angle on a horizontal line with its pointed side up and the obtuse angle with equal leg lengths pointed up in direct alignment, then we would be able to observe the differences. Notice that the acute angle stands taller than the obtuse angle. The acute angle illustrates the greater magnitude in jammeria in the kinetic direction of movement of the bethtip. Observe the angles' sides. The obtuse angle extends beyond the acute angle; this gives juttoria the greater lateral influence as the bethtip moves through space. By this, we observe that thermtips repel more laterally than straight on, and that the attraction by jammeria is stronger straight on than as a lateral function. Within the xergopaths containing thermtips, it gives less lateral force.

Final Xyzenverse Crystals

One final image of the xyzenthium crystal is that the diflohexets are very thin. If we were to make a glass image of the crystal's design and influence, it would look almost as a thin squarish sphere with six thin poles joined perpendicular to each other at their center. The end of each pole joins perpendicular at the center of spherical surface. However, each would be flared at the end of the pole as it joins with the spherical shell. We now observe that there is hollowness to the xyzenthium crystal.

Part 2: Creation of Xyzenthium Crystals

Another interesting observation of the xyzenthium crystal is that it expresses itself as a finite representation of a geometric point. In this, we find both the spherical and cubical geometry equally expressed. While the shape of a geometric point cannot be finitely expressed as a shape, the finite representation does incorporate the same intrinsic patterns of behavior. It can rotate like a sphere, yet it has the six distinct faces of a cube. This brings to the nature of the cubical corners of the crystal. As expected, the energy of the corners is weaker as the xergopath density is much sparser than that near any given facial center of the crystal. It follows, that the edges of the cubical crystal is also weaker in energy than the central facial regions, but not as weak as in the cubical corners via the cubical geometrical formation in comparison to the spherical formation. Along with this, the distance between each crystal is greater than the original distances between xergopaths in the kineverse as juttoria pushes them further apart, but this is about to change as we enter the next phaseverse.

Beyond the above information, we observe another phenomenon intrinsic to the crystal. Recall that the weakest location of energy is at the corners, this naturally means that the midpoint upon the cubical edge of a given facial side is the location of the strongest energy intensity, as expected. However, there is another reason beyond the spherical nature of energy. It is found in the geometry of its structure. Recall that the xergoplate was divided into triangles which pointed toward the center forming the exozeed boundaries. The uppermost tip of the triangle is the external most xergopath loop, as it is the innermost strand at the xentrix. By this we observe that the intensity also aligns itself with the dimensional axis of the diflohexius streams. As the inner xergopath loops push outward, they also rotate toward the cubical corner fanning away from the central edge by the lateral repelling force of juttoria. This, too, will become important later.

There is another point to make about these crystals. Despite their seemingly ethereal appearance and behavior, they are denser than osmium, which has the density 22 times that of water. Moreover, these forces give the subatomic particles their density, which is far beyond any stone. Despite this density, their dimensional stability is very weak. The magnitude of the forces existing within the crystals causes the extreme dense form to behave as a liquid. Imagine cold iron behaving as liquid moving like water at the whim of every force. Now, imagine a force greater than that. Inside the xyzenthium crystal itself, these forces intensify as we approach the xentrix. By this increase, the diflohexet holds its integrity as the surface moves to fit the shape demanded by external forces. Fortunately for us, the mass within the crystal is very near to zero and there are large empty spaces within the crystal allowing dimensional matter to exist.

Finally, the spherical crystals form perfect lines in respect to each of the three dimensional directions throughout the entire universe. In essence, the physical universe forms a continuous mass of individual xyzenthium crystals. Again, if a human being could exist in such a space, they would see "clear" to infinity.

Chapter 7 Quiz

1. What causes the thermtip pulse?
   A. Kinetic Energy established in the Kineverse
   B. Unstableness of the thermtips interacting with each other
   C. Both a and b
   D. None of the Above

2. How many zeeds are there in a Xyzenthium Crystal?
   A. 1
   B. 2
   C. 6
   D. 12

3. An exozeed exists:
   A. Outside the Zeed
   B. Outside the Crystal
   C. Inside the Zeed
   D. All the Above

4. Diflohexius has how many Diflohexets?
   A. 1
   B. 0
   C. 3
   D. 6

5. Energy within a Diflohexet moves:
   A. Two opposite ways within one Xergopath
   B. Like a Fountain and a Funnel
   C. Symmetrically in reference to the Xentrix
   D. All the Above

6. (T/F) Final formulation of a bethtip occurs in this phaseverse.

7. (T/F) The thermtip pulse separates from the xentrix.

8. (T/F) Zeeds are roots of the xergopath loop.

9. (T/F) Diflohexets act like a funnel and a fountain for energy.

10. (T/F) Xyzenthium Crystals have red and blue bethtips.

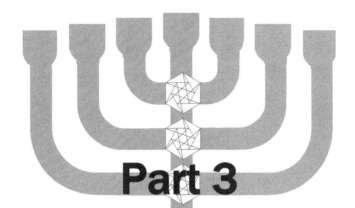

Part 3

Creation of Neutrons

Chapter 8

Magneverse

And G-d saw the light, that it was good: and G-d divided the light from the darkness.
B'reshyit (Genesis) 1:4

Dividing Light from Darkness

The transformations that occur in this phaseverse are described by the historical action by Elohim to separate light from darkness. This brings us to the creation of magnetic energy with all its associated energy patterns. Change occurs at the bethtip level, as well as, in the finite level of creation. The term, "darkness," stems from the initial state of bethtips. Even though there was motion in the kineverse, it did not make the uniquid transparent. It was dark until the entrance or creation of juttorial acceleration in the thermaverse. Within this attribute, the finite granules (xyzenthium crystals) formulate uniformly throughout the universe.

Now we are ready to examine the division process. In the thermaverse, the dimensional fountains moved energy in such a fashion that there were no real distinctions between the six diflohexets within the xyzenthium crystal itself. Each face of the xyzenthium crystal can be assigned the color of blue showing a uniform magnetically neutral state. First item to understand of the process is that the process does not occur naturally. It requires Elohim to interact and alter creation. As stated in the Torah, HE divided light from darkness. Therefore it did not intrinsically occur of its own designed nature. We could emphatically say that it became divided because HE said so. However, we would be missing out on the knowledge of HIS designing principles.

The division occurs during the collision process of two bethtips. Recall that the two bethtips meeting within the loop symmetrically inverted the positioning of the juttorial and jammerial xevim. Instead of both xevim transferring across to the opposite bethtip the trade becomes selective, in that, the xevim of the same type join together in their separating process. Notice that there is no inconvenience in the realignment as the jammerial and juttorial xevim already has two different positions. The process becomes a controlled movement in which no energy was created or destroyed. Actually, it was only a reassignment of xevim vectors associated with the energy during the collision. It

Part 3: Creation of Neutrons

is not just that HE did it, but Elohim works within the created framework. The result is that the jammerial force and juttorial force now are moving on separate bethtips.

Energy Types

The result of this energy exchange, there are two kinds of bethtips instead of one. Each bethtip type has three xevim attached to the inertial vertex. The only common xevim to both types are the kinetic and the inertial xev. The kinetic xev is typically colored blue as it does not bend the time/space continuum. Juttoria is represented by red, and green for jammeria. Letter symbols associated with the forces are the following: "i" for the inertial vertex, "k" for kinetic xev, "u" for the juttorial xev, and "a" for the jammerial xev. The bethtip with two jammerial xevim is called the **hypotip** (high-po-tip) and the bethtip dominated by juttorial energy is called the **hypertip** (high-per-tip). Note: do not confuse these with microweak and microstrong energies.

The final bethtip forms are called **tavtip**s (tav-tips). Tav is the last Hebrew letter of its alphabet, and the tavtip is the last creational change made to a bethtip. Another interesting observation is that the hypertip has no jammerial energy for gravity, and the hypotip has no juttorial energy for heat. The repelling and attracting forces that occur within infinitesimal space has no relationship to their components per finite effect. Within infinitesimal space, it is the bending of the space/time continuum that effect their behavior, as we shall soon observe.

Division by Dimensional Reference

Recall that the bethtip pulse that expanded until the spherical appearance became spherical again. Despite the pressure exerted, the pulse-like format continues to play a role. Consider a time when the set of two bethtip strands expanded to encompass the entire loop when side by side; during this phase there is no interaction. Then they merge into each other continuing their journey through the loop pattern. When they are totally emerged within each other, there is a gap between their existence and the xentrix cube from which they came. The opposite and adjacent loop sets experience the same phenomenon. They will continue until the entire loop pattern becomes engulfed with both bethtip sets continually. However, even as we enter the magneverse, this result has not yet occurred. Even so, it has reached a stage of nearly opening the gap.

The dividing process occurs completely outside the xentrix when both sets of bethtips are completely immersed within the other and making contact. One direction becomes hypertips and the opposite hypotips. This process occurs in all twelve loop sets. Again, there is more designing involved than just arbitrarily setting one direction for hypertips and another for hypotips. The choosing may seem arbitrary and even futile at first, but the pattern has a purpose, as we shall soon observe.

Chapter 8: Magneverse

Within the xy and zy-loops the movement choice is made symmetrical in respect to the y-axis. The hypertips are feeding the xentrix from the top and the hypotips are feeding the xentrix from the bottom. They continue their journey through the xentrix to the opposite side. The horizontal movement toward the xentrix shows like forces moving toward each other. A different process occurs with the xz-loop set. The pattern is symmetrical in respect to the center. Like forces move toward each other at every direction toward the xentrix (center). This brings us to the next attribute of the tavtip arrangement. We know opposite attracts. There is a reason for like energy to move toward each other in a neutral format.

When we visualize a flat plane, we provide it with orientation by our relationship to our planet's surface. Within this orientation, up is away from the surface in the direction toward the sky. Suppose for a moment that two people were visualizing a plane upon two different, even opposite, locations upon the earth. "Up" then becomes two opposite directions. Or, if we were in deep space, up would become only in reference to the position of our body. The idea that the time/space continuum can be bent in two opposite directions within a framework of a single plane and orientation is easy. However, when we are dealing with several planes of different orientations, up seems almost meaningless to an internal observer.

This observation becomes important, as we examine the xev orientation of the tavtips moving through the xergopath loop. Xevim infinitesimal "arms" of two tavtips of the same energy type, hypertip or hypotip, can face opposite directions upon the same plane and bend the continuum in opposite direction even though they are the of the same energy nature. This does not alter their nature as energy, for a hypertip will always affect finite space in the same manner. The difference is in the interaction between tavtips within infinitesimal space. The opposite bending of space will pull the two tavtips together as they try to release the tension of the continuum fibers even though they are of the same energy type.

Again this points to a Creator of the physical universe rather than a mindless natural nature. The dividing of energy, "light from darkness," is a creative act with purpose rather than an arbitrary decision or something that was going to happen anyway with or without thought. The inertial behavior is intrinsic to the universe before the fall, afterward decay became the intrinsic behavior. In both cases, Elohim intervenes on behalf of creation.

After converting thermtips into tavtips, the energy continues its journey back toward the xentrix. If we were to follow the xergopath loop in its entirety, we would return to our starting location after passing through the xentrix four times. We know by studying today's subatomic structure that another process occurs to prevent this continuous inter-looping by tavtips from reoccurring. The last location of the tavtips, before the interruption, equates to the tavtips passing through the xentrix only once, positional speaking. Again, this did not occur by accident, but by plan.

Part 3: Creation of Neutrons

When we examine the tavtip positions after passing through the xentrix once, we see an interesting pattern occurs. The tavtips moving upon the xy and zy-planes present a dramatic change. The hypertips and hypotips become completely separated from each other being on opposite sides of the xentrix. For example, the hypertips could flow to and fro under the xz-plane and the hypertips flow above this plane within a xyzenthium crystal. Examining the xz-plane, we observe a different process occurring. The xz-plane tavtips merely reverse polarity, maintaining the equal distribution of hypertips and hypotips within the xergopath loop system.

Loop Interruption

The interruption of the energy exchange between loops actually closes the loops of the xergopath trapping energy within the external loops. This process occurs when the tavtips are all back into the loops outside the xentrix. Moreover, the closing process occurs for each xergopath simultaneously

The termination process occurs as the tavtips return to the xentrix. Before actually colliding at their respective loop intersection within the loop's zeed, the tavtips pull toward each other. The reason they are able to accomplish this task is that the tavtip's energy bends the time/space continuum more than the thermtips because of the joining by like forces per tavtip. Instead of jammeria and juttoria fighting against each other within a bethtip, now, the forces are joined bending the continuum the same way, adding more strength to the bend. The result is that the severity of the bending nature extends the range of influence by the tavtip to attract another tavtip to cancel the tension. Another observation, this occurs when the returning tavtips at each loop end becomes closer than the adjacent tavtip loop. If the attraction were to occur before the distance was closer, then they would simply join themselves to the adjacent set and become short-circuited.

This phenomenon establishes the magnetic nature of each xergopath loop. Naturally, these xergopath loops are grouped by the sector in which they occur. In general, the loops that we have orientated to be horizontal in nature become polarized into magnetic charges. There are four poles established alternating in charge; as a result, we find that the negative charge and positive charge exist at right angles to each other. The vertical loops become neutrally charged. This can be observed by the hypertip hitting hypertip generating the **microstrong** force on the side below the horizontal loops. **Microweak** neutral energy forms by hypotip hitting hypotips on our orientated xergopath loops above the horizontal loops. Both neutral forces are weaker than the magnetic energy of the horizontal loops. The reason for this weakness is the tension within the magnetic loops are stronger as the hypertips and hypotips bend space in opposite directions and are pulled "down" toward each other. When the two tavtips are equal in nature, there is also an annihilation of spatial bend as the two tavtips approach from opposite directions bending the space/time continuum in opposite directions in respect to each other.

Chapter 8: Magneverse

Another interesting observation is the kinetic pressure asserted upon the base of the loop. As the tavtips move inward to the loop they both have kinetic vectors that form right angles together. Even though the attraction factor generates a "head on" collision, the kinetic movement prior to the attraction interlude continues to push toward the xentrix center. This pressure pulls the xergopath loop into the xentrix. The leaving tavtips from the loop base are drawn back into the "linear" stream via the bend of the space/time continuum generated by the two tavtips. They then collide with incoming tavtips "head on" in the triocollision fashion requiring three alephtips. The central alephtip then holds twice the energy as the regular tavtip.

Secondly, by the identical phenomenon, the xergopath loop maintains its "tear drop" appearance instead of transforming into a circle. This principle can be easily illustrated. Imagine two objects in deep space, meaning no external influences involved, traveling at different angles and being pulled drastically toward each other by their gravitational pull. Because of the strength, they hit straight into each other. All the colliding reactions occur as if the two objects were just pulled together linearly, because the entire phenomenon moves uniformly in the direction represented by the addition of the two original kinetic vectors. Unlike the illustration, this occurs because the original kinetic energy is not destroyed and is separate from all the additional kinetic energy associated with the force that moved them into collision. The tavtips do not stay at the collision site; they have instead, others to replace them as they repeat the same process.

Miortex

As the process continues, the zeed (loop base) triocollision pattern moves toward a central point within the zeed plane. This central point (alephtip) is the base of the **miortex** (mi-or-tex) from the words **mi**rror and **v**ortex. The angle of the collision by the tavtips sharpens as the xergopath loop bases are pulled closer into the miortex "collective." Eventually, the angle collapses into a parallel linear movement toward the miortex called the **zeedpath**. The energy moving down the zeedpath consists of two tavtips approaching the triocollision phenomenon providing them a **tavtip-2** vehicle.

The zeedpath collects tavtip-2s and moves them toward the miortex "collective" by the means of their own kinetic energy and the xentrix centroid. The miortex deflects the incoming tavtip-2, and sends it back the direction at its opposite side. The deflection is made possible by the four zeedpaths leading to the miortex per plane. Recall that there are twenty-four zeeds. Hence, twenty-four triangular zeedpaths planes leading to the miortex. The manner in which the zeedpath collects tavtip-2 is the same pattern used to disperse tavtip-2s. Tavtip-2s perform duocollisions after reaches the miortex instead of the expected triocollision. The duocollision phenomenon occurs in a unique pattern.

Consider for a moment the final pattern of a zeedpath. The image is similar to that found within a checkerboard. In this, we can select two parallel diagonal sets of square

Part 3: Creation of Neutrons

adjacent to each other and of the same color, only to find that they are separated by another set of diagonal squares of the opposite color. The duocollision does not involve the central dividing square. Instead, after reaching a center, it moves energy across the "connecting corners" of the same colored square. In doing this, a duocollision is maintained. However, the manner in which the energy departs from the tavtips is at right angles to the direction of the kinetic movement. To understand the reason for this occurrence, we need to go back to the checkerboard. As before, we are going to select two sets of parallel diagonal squares of the same color. Both sets are of the same color and at right angles to the other set. The opposite color divides both sets and occupies the central square. The result is that the squares that contain our selected color surrounding the central square of the opposite color has energy coming into it from two different angles. Each of the four squares surrounding the central square of the opposite color has energy entering in from their two outermost corners away from the central square by pure geometry. In this, the energy is deflected to the adjacent square rather than using the central opposite colored square, which holds the centroid.

The next alteration that occurs within this phenomenon is that the zeedpath "unzips" within the miortex. To visualize this result, we return to our last usage of the checkerboard. Using the same "impact" squares, we move straight instead of diagonally across the checkerboard. Notice that these parallel lines are off centered forming a vortex image. In this, the zeedpath disappears and the crystal will fully develop.

With the above information, we now can discuss the miortex collective. For this, we will return to our last result at the checkerboard. Recall that the four impact squares that surround a central unused square with its parallel lines moving in line with the dimensions of the board. Now, we are going to connect two adjacent impact squares and project diagonally away from the impact area generating a line of diagonal squares of the same chosen color. We can do this for all the four parallel sets. Notice, that we can duplicate the same outermost pattern along these lines without crossing a single line. In essence, we are stacking loop bases upon each other per zeed direction. The miortex is the collection of collapsed zeeds forming six inverted pyramids that make a cubical shape. The image is as six flashlights shining straight out from each cubical face without a cubical face, but an inverted pyramid interior. This interior is the miortex of the xyzenthium crystal. There are other miortex collectives found in photons.

The result is that four kinds of stable magnetic energy emerge from the process. If we examine the xy and zy-plane loops, we find two kinds of neutral magnetic charges form. The flowing of hypotips in both directions within a given loop generates microweak energy, which is a neutral charge. On the opposite side of the xentrix we have hypertips flowing in both directions within a given loop. The force generated is microstrong, again, neutral in charge.

When we examine the xz-plane loops, we find the energy that we call the positive and negative magnetic charge. The positive charge occurs when the hypertip leaves the

Chapter 8: Magneverse

xentrix and hypotips are moving toward the xentrix. In observing a given loop we see that both occur, that which defines the positive charge at one face (xergoplate) of the xentrix, also produces the negative charge at the adjacent xentrix xergoplate passing through the xz-plane. However; if we examine all four loops, we find that two opposite xergoplates have the same polarity within the xz-plane. By this, we observe that the positive and negative charges exist at right angles to each other within the xyzenthium crystal. Moreover, the magnetically charged xergoplates exists in pairs for each of the polarized active magnetic charges, unlike the single xergoplate for each kind of neutral or inactive charged face.

Even so, the twenty-four exozeed sectors illustrate a different picture of the energy. Each loop, formulates two different exozeed sectors. Looking at the crystal's top, we see four sectors plus four connecting side sectors containing the microweak neutral charge, totaling eight. Then we observe four sets of magnetically charged vertical sectors. On the bottom we observe eight microstrong charged sectors as on the top.

Quarquid

Thus far, we have primarily focused upon single xergopath strands, and examined their transformation. With the exozeed and the xergoplates we have illustrated the effect of these strand transformations into the xyzenthium crystals with regard to their magnetic charge. Now, we need to view the crystal in its finite form. As noted earlier, there are six streams flowing from the xentrix to the crystal facial surface. These also serve as streams moving from the facial surface into xentrix, hence the name of diflohexius. In the xyzenverse, the diflohexius was composed of a single "liquid" of uniquid. Each stream has been divided into four sub-streams defined by the exozeed. Even though, there are three basic patterns of energy on the "top" and "bottom" facial regions of the xyzenthium crystal are two different neutrally charged "liquids." The four streams moving to the sides of the crystal are primarily considered charged despite that half of the energy is neutral in nature. The general description of the charged streams and faces is that magnetic charged tavtip types move in opposite patterns in relationship to their central axis. Notice that in the positive charged sub-streams that hypertips are moving up the axis toward the facial surface and afterward away from the center of the face, the inverse is true when considering the negative charge process. Due to this, the uniquid transforms into the **quarquid** (quark qwid). Quarquid has four forms as illustrated. Even though the word quark is sounded, the idea is that the uniquid is divided into four forms. From these four forms we get the 12 flavors (kinds) of quarks (24 if the antiquarks are counted).

All the charged streams have two neutral sub-streams moving at right angles to the charged. We should note that even the neutral charged sub-streams contribute to the magnetically charge faces since half of the tavtips are moving in the same direction as their charged counterparts.

Part 3: Creation of Neutrons

Even so, another image appears. It is another level of "dividing light from darkness." At the level of the xyzenthium crystal, it appears as if the magnetically charged energy is used to hold "darkness" and "light" apart. "Darkness," here, is represented by the jammerial hypotip input of the microweak force. "Light" is then represented by the pure juttorial tavtip (hypertips) input by the microstrong force. Another observation is that the two neutral sides of the xyzenthium crystal attract each other magnetically holding the two together within the crystal as both represent energy type purity.

Microweak verses Microstrong

As we have observed earlier, two diflohexets (energy fountain and funnel) within the diflohexius promote neutral charges. Each neutral charge has a different composition by energy type. We need to look at them finitely. The jammerial and juttorial energies differ in magnitude by distance in two different processes. The jammerial force has a greater vertical force and the juttorial energy has a greater lateral force. We might tend to believe that jammerial energy is the microstrong force and that juttoria composes the microweak nature. However, we know by the actual phenomenon that the inverse is true. For this cause, we will examine some examples that will illustrate the answer.

Our first example was geometric in nature. Here is another representation that is numerical in nature. Let us look at two different measuring scales of temperature: Celsius and Fahrenheit. The basic formula is $F=(9/5)*C+32$. When the temperature is -40 degrees Celsius, it is also -40 degrees Fahrenheit. When the temperature is 100 degrees Celsius, it is 212 degrees Fahrenheit. When it is -100 degrees Celsius, it is -148 degrees Fahrenheit. By observation, we observe that when the temperature is above -40 the Fahrenheit scale reads a larger number than the Celsius, and below -40 a smaller number even though its absolute value is larger. In this example we are able to see numerically that a process that causes a value to become greater within a given interval will also cause a value to decrease faster.

The comparison of the two scales also shows a location in which their values will be equal. Above this temperature value the Fahrenheit value will be greater than the Celsius measurement value. Below this point, the Celsius scale numerical value will become greater than the Fahrenheit as larger negative numbers are smaller in value.

Another example of another difference between juttoria of the microstrong and jammeria of the microweak is found in examining two light waves. The first light wave we will decrease its size moving it toward the ultraviolet end of the spectrum. We will continue to shrink the light wave past the x-ray, gamma rays, and cosmic rays. We are going to zero. At this point, the light wave becomes a straight line moving upon the x-axis. Our second light wave, we will increase to infinity, first the amplitude (joining the wave to the y-axis) and then the phase length. In this sequence it may be easier to observe the result. This light wave naturally moves in the opposite direction in the

Chapter 8: Magneverse

spectrum of light. It moves beyond infrared, heat waves and radio waves. The increased distance to infinity puts the light wave in union with the x-axis.

When we recall that the fibers of space loop into each other, we are able to observe that the two light waves move on the same line for opposite reasons. This returns us back to the nature of jammeria and juttoria found in our neutral energies. One shrinks down to the x-axis, the other expands and becomes finitely in union with the x-axis. We now can see that one pressure is the opposite of the other, and the necessarily different resulting energy type.

Returning to illustrate the inverted strength, mathematics provides another dramatic picture. Let us take the numbers ten and two by raising their exponent to different powers. Raising ten to the second power or squaring it (multiplying it twice) gives the quantity of 100. Squaring two gives the quantity of four. Look at the following table:

Power	10	2
4	10,000	16
3	1,000	8
2	100	4
1	10	2
0	1	1
-1	0.1	0.5
-2	0.01	0.25
...	...	...
$-\infty$	0	0

Note that the exponent increments or power of ten indicates the multiples of ten in the second column. Similarly, the exponent of two indicates multiples of two in the final column. At zero, both the powers of ten and two are both equal to one. The powers of ten increase at a faster rate when the magnitude of the power increases. Inversely, the powers of ten decrease more rapidly than the powers of two. Actually, all the values of two raised to a negative exponent (power) are greater than any power of ten raised to the same negative exponent.

Zero and negative exponents have to occur in finite space. The logic behind this is that their influence is felt in finite space. The zero exponents represent a point in space that both energies are equal in power. This brings us back to the concept that infinity multiplied by zero equals one. The value of one could be centimeters, inches, miles, light-years or any other measurement. However; when we examined the process in calculus, we found that in using equations, the equation defines the value of this measurement. Similarly, the geometric structure of the xyzenthium crystal defines the

Part 3: Creation of Neutrons

measurement of distance that the two energies are equal. This distance was established by the original size of the xentrix, which is the original size of the xyzenthium crystal. This size as we shall soon observe is much larger than the crystal size found upon the neutron. Even so, it is still extremely small.

The negative exponent values exist in the internal region from the point of equilibrium toward infinitesimal space. When we examine the values of the negative exponents, we find an inverse of magnitudes. Example: when we look at ten raised to the negative power of two and two raised to the same negative power, we get 0.01 and 0.25 respectively. The power of ten is 25 times smaller than the power of two.

Despite the increase differences in ratio between the two forces as their exponents approach negative infinity, they once again become equal. The reason is that the limit is negative infinity for both. When this limit is reached, they both become zero. Essentially, juttorial and jammerial differences within infinitesimal space becomes a relationship of "space bending" and not a matter of finite lateral pulling and pushing. Energy moves in a pixel-like manner maintaining its integrity.

Energy Shapes

The mathematical examples give the impression of disassociation between forces. It is more profound than that. Visualize two blocks of wood with the same rectangular dimensions (two inches by two inches by four inches). One block is lying flat, while the other is standing. The upright one extends above the ground further than the one on its side. While the other is not as tall, it has a larger, sturdier base. The two functions differently even though they are identical pieces of wood. The same is true with the xev of juttoria and jammeria.

Unlike the concept of juttoria represented by the block lying on its side, jammeria is the one standing on end. To understand this better, compare the two pieces to another block of wood. This piece is a cube of 2.5 inches per side. At the base, the piece lying down has a greater surface area than the cube; hence at lower altitudes, the force is stronger. Hence, juttoria is microstrong. However, it is weak in its ability to reach out and influence other objects at large distances. This is because its energy primary objective is to accelerate the particle. Jammeria does the opposite. This force has a smaller base than the cube making it weaker at the base (easily toppled over). Jammeria extends beyond the cube making it a stronger influence at larger distances. By this, we observe that the microweak influence is strong outwardly, and the microstrong is weak in the external finite world.

As observed within the previous information, the microweak and microstrong forces don't correspond well to the strong and weak forces in quantum mechanics. The strong force in quantum mechanics explains for the absence of a lone hadronic quark (up or down quarks from a nucleon standpoint), while the weak force prevent leptons

Chapter 8: Magneverse

(electrons for example) from colliding into the proton. The primary reason for this difference is the viewpoint that infinitesimal space (subspace) has an impact upon finite matter. It is the microstrong force that prevents finite matter from being consumed into subspace by gravity developed in finite space or magnetic energy. By this definition of subspace, there are no gluons holding together hadronic quarks within a baryon (composition of any three quarks) or mesons (two quark objects, explained later) other than internal magnetic energy and externally gravity produced by the hypotip collective expressed primarily in the microweak force.

Hierarchy of Energy

We now have all the different primary energy patterns defined. We can present them in separate classifications. The first classification is Aleph Energy Level being the Ruach Ha Kodesh of Elohim (the Holy Spirit). The source of all energy is the Spirit of Elohim physically defined as a geometric point. However, this geometric point exists outside our universe and is not to be confused with the "virtual" geometric point of the physical universe.

The second classification is the Beth Energy Level. The physical source of all physical energy is found in the alephfield of the geometric point, which is the housing structure of all physical energies. The alephfield is only an angle of reference of pressure exerted within the geometric point defining the physical universe. This energy can exist in an inactive or active state, alephtips and bethtips respectively, as primordial pixels of the universe. In other words the house can be full or empty, but the house is still the same house. The alephfield is the source of the four primary energies.

Jammeria, kinetic, inertia, and juttoria are the four primary observable manifestations of the alephfield. These four comprise the Gimmel Energy Level or the third classification of energy. Juttoria, jammeria, and kinetic xev build up magnitude in three different directions. Inertia forms the unity of the three xev at their "base." The inertial xev is the reference angle in which the physical universe formulates its alephfield. In terms of three-dimensional space, the length of this xev appears only as a geometrical point as its dimensional reference generates no finite movement into physical space.

The fourth classification of energy is the Daleth Energy Level. The collective influences within finite space by the three Gimmel Level energies are the following: from jammeria we get gravity, from inertia we get substance, kinetic energy gives us time, and from juttoria we get heat. The imbalance of heat within the finite physical universe is the result of photonic activity in response to subatomic heat.

The fifth classification of energy is Hey Energy Level. Here energy acquires another dimension of character as they move on a path through infinitesimal space. Kinetic energy is inertial energy transmitted through infinitesimal space. Microstrong energy is juttorial hypertip transmissions through infinitesimal space. Microweak forms the force

Part 3: Creation of Neutrons

acting between jammerial hypotips. The two electromagnetic energies are created by the interaction of transmissions between jammeria and juttorial dominated tavtips through space. Each magnetic transmission is in the inverse fashion of the other.

Final Magnetic Xyzenthium Crystal

As noted earlier, the original space occupied by the xentrix encompassed the entire cubical shape of the xyzenthium crystal before it shrank. The miortex forms as the void outside the spherical xyzenthium crystal begins to pull outward formulating the xergopath loops. The juttorial force is doubled in the hypertips generating a greater lateral repelling force. Remember that jammerial energy is stronger in the direction of the flow within the xergopath loop. The result is that more lateral space is needed for "juttorial tavtips" to pass each other. The imbalance presented by the external void caused the xergopath loops to "relax" into filling the original space occupied by the xyzenthium crystal. In doing this, the spherical crystals press against each other.

Let us return to our colored image of the xyzenthium crystal. By viewing the cubical dimensional relationships in a symbolic color code, we observe four colors. These four provide the crystal alignment within the various surfaces of a neutron. Let the upper face color be night blue representing the weak force. The lower face then becomes as azure (bright sky blue) illustrating the microstrong force. The opposite lateral faces are the same color representing the same magnetically charged force. Let green represent the positive charged magnetic energy and red the negative charge. Notice that the magnetic polarity exists at right angles instead of the expected straight angle or opposite of each other. The reason is that two sets of magnetic poles existing instead of one. Even so, the energy is drawn to the opposite polarity, as experienced with our finite macrocosmic perception.

The snapshot of the final state of the magneverse or the original state of the neutronverse appears as layers of xyzenthium crystals facing in isometric directions. Their alignment determined by Elohim structured the crystals so that like charges were facing each other. Recall that the division occurred in each xergopath loop external to the xentrix, and that Elohim also predetermined the direction of separation. Example: the positive electromagnetic xyzenthium faces are directly facing each other. While the positive and negative directions require uniform alignments, the same is not true with the neutral faces. Even though the faces are neutral, they are composed of opposite tavtip natures. If these natures were next to each other they would interact as magnetic energy. They are neutral only in that their dual flow of the same tavtip type annihilates their charge.

While the juttorial lateral influence of microstrong faces repels adjacent crystals, it draws to the miortex by the shape of the energy and the initiated geometry of the xergopaths within the crystal. Note also the infinitesimal distances between tavtips within the loop.

Chapter 8: Magneverse

This gives the "continuum bend" control of the oncoming tavtip and not to the xergopaths outside the crystal. In this, the xergopaths remain intact.

The final form of the crystal has an orientation determined by the two neutrally charged diflohexets. The action of "dividing light from darkness" establishes the magnetic polarities both in the infinitesimal level and the finite level. The topside has been illustrated, as the microweak side of the crystal, as it is the orientation found upon the surface of the neutron, not yet formulated. The microweak side of the crystal can be viewed as the night view and the opposite microstrong side as the day view of the crystal. This may seem to be upside down, but as we shall soon observe this alignment becomes important in nature of a neutron.

Looking at a cross-section of the xyzenthium layers perpendicular to the xy-plane, we will view a multitude of crystals with their red faces (negative electromagnetic energy) forming the layer's surface. Similarly, the layer perpendicular to the yz-plane will form a green-surfaced layer (positive electromagnetic layer). Dividing the layers in a perpendicular fashion via the xz-plane we will see two kinds of zero charged surfaces. The top surface will appear dark blue (microweak force), while the bottom will appear bright blue (microstrong force) in one layer. Within the adjacent layer on either side the inverse is true, light blue on the top surface and dark blue underneath. This pattern is true for any such division of the xyzenthium layers throughout this phaseverse.

Just as in the gravverse, there is a certain energy dominion of the crystal. Even though it is spherical in nature, it controls a space that is cubical in nature. This is because of two factors, primarily, the physical arrangement of the crystals, and the dimensional axis of the crystal. Each axis has dominion over one face of the cubical structure as each axis is centered within a given cubical face. In this, each face then expresses the magnetic condition of the corresponding diflohexet only.

Our next step is to examine the nature of the magnetic xergopaths as they bend toward the axis of the diflohexet adjacent and underneath within the same diflohexius. Starting with our initial image, it can be acquired by taking a circle and fitting it in a square, such that, their defining lines touch each other without crossing over each other; then, erase three corners of the square back to the point in which they touch the circle, and, lastly, erase the quarter of the circle underneath the remaining corner of the square. This is our original image of the xergopath loop. Now, we need to examine the xergopaths that angle away from the axis of the adjacent diflohexet.

Next, we need to examine this phenomenon at the diflohexet lateral limits. The greatest angle possible in terms of a given cubical face is $45°$ from the diflohexet axis. This can be illustrated by constructing another square and dividing it into fourths by dissecting each edge of the square and connecting the opposite locations; this represents the four axes of the diflohexets existing at the base of the diflohexet that we are examining. Then divide the square again into fourths by connecting opposite corners of the square.

Part 3: Creation of Neutrons

These lines represent the limit that a xergopath can vary and yet be associated with the same facial side of the cube. When we measure the angle between the lines representing the diflohexet axis and the line representing the limit adjacent to the axis, it measures exactly 45°. However, no xergopath exists at this angle, it would form uneven energy.

Unlike the cubical structure of a cube, the edges and corners are the weakest areas of the crystal. The reason is that the radiating energy is the weakest at these regions due to the distance from the xergopaths of their respective diflohexet. Moreover, the "corners" are weaker than the "edges" of the crystal because their distances are even further. The actual crystals are still primarily spherical in nature. This observation will become important later as the neutron develops.

To understand this shape better, let us return to the nature of the diflohexius. Starting at the base or center of the diflohexius at the miortex, we observe six streams of energy shooting nearly straight out from the miortex. As noted earlier, two streams from adjacent crystals seem to splash away from each other at the location that they meet and the "fluid" falls back into the miortex via their respective diflohexet.

However, we need to look at the process more closely. Firstly, the stream formation is not a set of xergopaths that move in a parallel course within a given stream. The xergopaths are near parallel, but not parallel. They actually project slightly away from the axis of their formation as the leave the miortex. Initially, the course of the tavtips is a straight line; they bend away from the central axis before they even reach the crystal "surface" using the zeed as its pivoting reference away from the zeed's bisecting line.

The shapes of the xergopaths that angle away from the adjacent diflohexet axis become modified from their original form. To acquire the image of these xergopaths, we can use either trigonometry or geometric analysis to establish the variation in the finite distances. Without a calculator, the geometric analysis is the easiest. Recall that the original loop can be inscribed within a square that measures one half per side of the xyzenthium cube. If we draw a line from the center of the square to a corner and measure it, we will find that this length is about thee fourths the size of the cubical side, or more precise, this too is only an approximation to the actual value, 0.707 (trigonometrically, the cosine or sine of 45°, this function becomes obvious when figuring other angles). At any rate, this transforms our inscribing square into a rectangle. This in turn transforms our circle into an ellipse, thereby stretching the xergopath. This, in turn, diminishes the density of tavtips per measurement weakening the influence of the structure. In this, the influence of energy is weaker at the corners rather than stronger as we might think.

The actual behavior of these crystals is spherical fluid in nature. Despite their extreme density, they are not solid. The image is that of the magnetic energy of two magnets having the same magnetic charge facing each other repelling; the shape of the energy field alters as the magnets are forced closer together into a more squarish form. Even

Chapter 8: Magneverse

so, the friction between the two fields still is zero. This leads to the concept that the shape of the xergopath can change without destroying the flow as pressure requires.

At this point, we are ready to observe the neutron's formation. The formulation of the neutron ignores much of the theory of quarks; even though, quarks are founded upon the findings by nuclear collisions. The reasoning for this will be explained later as more terms are needed to define the process. The argument is not with the findings, but with the conclusions.

Chapter 8 Quiz

1. What are Tavtips?
   A. Final Form of Bethtips
   B. Hypertips and Hypotips
   C. Result of Division of Energy
   D. .All of the Above

2. What are Hypertips?
   A. Juttoria dominated Tavtips
   B. Jammeria dominated Tavtips
   C. Kinetically dominated Tavtips
   D. All the Above

3. What are Hypotips?
   A. Juttoria dominated Tavtips
   B. Jammeria dominated Tavtips
   C. Kinetically dominated Tavtips
   D. All the Above

4. What is a Miortex?
   A. Contextual Mirror of Meaning
   B. Vortex of Mirrors
   C. Crystal Energy Vortex
   D. Formation of Mirrors holing Energy within

5. The Final form of the Xyzenthium Crystal has:
   A. Three different Charged Faces
   B. Two different Neutral Faces
   C. Spherical Shape with Cubical Influence
   D. All the Above

6. (T/F) Hypertips have two xevim containing juttoria.

7. (T/F) Diflohexets with only jammeria forms the microweak faces.

8. (T/F) Tavtips come in three forms.

9. (T/F) The miortex reflects energy.

10. (T/F) Xyzenthium Crystals have two negative charged faces.

Chapter 9

Neutronverse

From Cube to Sphere

Our movement, from the development of the xyzenthium crystal directly into the development of the neutron without an intermediate step of creating quarks, is neither an oversight nor an act of defiance against quantum mechanics. The primary reason for this omission is that quarks are man-made objects produced by smashing subatomic material or by other contortions such as extreme cold. Moreover, it is necessary that we cover quarks in the next chapter to give answers concerning this stance. In order to illustrate the dismantling process by smashing or extreme cold, we need to first develop the neutron model.

Within the Neutronverse, we will, again, divide space into a continuum of cubes throughout the universe. Each cube contains a multitude of xyzenthium crystals packed in a crystalline cubical fashion. Initially, the resulting **neutron cube**s generated cubical structures that were indistinguishable from each other. The xyzenthium crystals were all facing each other in a symmetry preventing magnetic interactions between adjacent diflohexets. It is from this pre-selected size by Elohim that each phenomenon within this phaseverse manifests which will transform a continuum of cubical neutrons into the spherical neutrons of discontinuous matter.

Two phenomena must occur in order for the neutrons to formulate. Each phenomenon requires intervention by Elohim. The first phenomenon divides the neutron cubes into magnetic layers of various arrangements within the neutron. This is accomplished through flipping xyzenthium crystals into symmetrical patterns in respect to the neutron's center and sets the neutron into layer relationships. The second phenomenon enables the cubical neutron to reconcile its crystal alignment into the spherical form and maintain the integrity of a given layer.

Even though that the xyzenthium crystals are spherical in shape, the diflohexius forces exert energy that form the six square faces of a cube. To understand the properties of the crystals within this phaseverse, we need to view them as cubes. In doing this, we create another visual problem. Rotating cubes in place requires more space than the volume of the cubes. In this, we must remember that their actual shape is spherical in nature. The cubical shape represents only the pressure between crystals in their uniform positions form cubical shapes by each diflohexet. Each cubical face contains the energy of one diflohexet. Along with this, we must remember that it is Elohim that flips the

Part 3: Creation of Neutrons

crystals into the magnetic patterns necessary to from a neutron and not them by themselves by some intrinsic nature.

In order to form a neutron cube, the number of crystals per dimensional measurement must also be equal. Ultimately, the xyzenthium crystals within the neutron cube will form today's spherical-like neutrons, setting aside temporarily the modern triple quark pattern. First, there is a need to establish a small sample cubical set of crystals and transform them into a spherical pattern. For our examination, we will utilize a cube of 27 crystals, which is a cube of three by three by three. Actually, these crystals that we are looking exist at the neutron cube's core. Developing a sphere from a cube of equal sized cubical blocks, at first glance, seems quite simple. We take the cube and alter its defining edges into a sphere. The image is that we take a cube and its blocks and expand its dimension until the limit set by the sphere is reached, or inversely squeeze the dimensions down into a spherical limit.

There is a problem with this solution. Recall that the crystals have different facial natures. We require that only one face per cube to be exposed to the surface. When we reexamine our initial attempt of making a cube into a sphere we observe three different results. The corner cube has three exposed faces in our spherical form. The central edges have two edges exposed to the surface, and only the facial center cubes meet the requirement of a single facial exposure. We can illustrate this by connecting 27 cubes together to form a larger cube and painting its surface, and then separate them to compare their painted surfaces. As we see in our development of the xyzenthium crystals, no two adjacent faces have the same charge. Because of this, it is necessary for each crystal to show one face. If we look at the eight corner crystals, the problem is compounded by three exposed surfaces which are adjacent to each other. The result of any alignment of the cube in this position will expose three different xyzenthium face charges: positive, negative and one of the two possible neutral faces. All of these crystal positions, with the exception of the central facial crystals, challenge the uniform facial exposure needed for each cube contributing one cubical face to the spherical form.

From our sample of the process, we only view one of the nine crystals per side meeting the requirement. In the innermost core of the neutron, this is true. However; as we examine the neutron surface, we find that the number of crystals in much larger. The uniform measurement of distances selected is a finite odd number of xyzenthium crystals per side of a cube. This number is actually 21,601 giving a cube of approximately 10.079 trillion crystals. The derivation of this number will be expounded within the chapter after the next, as we examine the formation of other subatomic material. The necessity of this number is that the number must be a number not divisible by 2 (odd number). The odd number creates a central cube.

Another observation is that the corner crystals are always going to be eight in number, and the number of edge crystals are always equal to 12(n-2) or twelve times the total count of crystals defining the cube minus the two corners forming that edge. The

Chapter 9: Neutronverse

reason for the subtraction is that the corners are already counted. Finally, the vast majority of the crystals only input one face to the surface, as they are centrally engulfed within the facial structure $6(n-2)^2$ per layer ($n=2x+1$ giving 3, 5, 7, 9, 11…).

Thus far: the original neutron state is a large cube containing trillions (this is not a figurative value) of xyzenthium crystals in a multitude of layers with the same crystal alignment. As stated earlier, there are no initial distinctions between neutrons. The xyzenthium crystals are only isometrically stable. This means they exist in a physically delicate, exact, and uniform balance. The surface faces of each neutron are touching each other by energy force, as there is no empty space, meaning a continuum of crystals.

Finding the Perfect Square

Our initial primary configuration problem in the transposition of the cube's surface crystalline layers into a single spherical shell is surface division. We could try and design a way for all the sides of the cube to contain an equal number of crystals; however, this is geometrically impossible. It can be proven algebraically that there will always be at least two crystals left over, no matter the number (within the set of whole numbers) of surface crystals that contribute equally per face. The calculated mathematics is the following.

Step one: Let n equal the number of crystals per side of a neutron cube. Let n-2 equal the number of crystals of the cube buried just under the surface. The -2 comes from subtracting one from the top layer and subtracting one from the bottom layer; naturally, the same applies to all four sides. Another stipulation of the domain of n is that n is greater than 2. The difference between the two numbers cubed gives the total number of cubes of the surface. For an example: if we had a cube of three crystals per dimension there would be 27 crystals in the entire cube; 26 of these crystals would be exposed to the surface with only one being buried. We should note that the number of surface crystal faces exposed equals 54 (9 squares multiplied by 6 faces).

Step Two: Since a cube has six faces the number of cubes forming a face must be multiplied by six if each side is to contribute an equal number of cubes. Using our example consider the following attempts to equal 26 cubes: 6x5=30, 6x4=24, and 6x3=18. Note that the first and last attempts are farther from the actual 26 cubes available, hence unusable. The second is under the exact amount by 2. Observe that the square root of 4 is 2. Our multiple of 6 is also a perfect square of a whole number. Note also that 3-1=2. This observation refers to the square root of four is one less than the number of cubes determining an edge of a facial side of our composite cube. The formula then can be written to say the following: The number of surface cubes (n cubed minus n-2 cubed) subtracted by equal removal of cubes per face (6 times n-1 squared) equals 2:

$$n^3-(n-2)^3-6(n-1)^2=2$$

Part 3: Creation of Neutrons

After formulating the equation, we can do the mathematics to show that the answer is necessarily two. When we cube n-2, we obtain the following polynomial:

$$n^3-6n^2+12n-8$$

When we multiply 6 to the squared quantity of n-1, we obtain another polynomial. Let us accomplish this task in two steps: first squaring n-1, and then multiply each term of that result by 6.

Step 1: n^2-2n+1
Step 2: $6n^2-12n+6$

Now, we need to implement the subtractions from n^3. In order to subtract the two polynomials, we need to multiply each by -1 and add. Multiplying by -1 only alters the sign of the number from minus to plus or vice versa to acquire the subtractive attribute. We now have acquired the following:

$$(n^3-n^3)+(6n^2-6n^2)+(12n-12n)+(8-6)$$

Notice the matched values in the first three sets of parentheses, just as 7-7=0, so do these equations equal zero. The only part of the equation left is the subtraction of six from eight, which gives the value of two.

Crystal Annihilation Phenomenon

Applying the result of the equation to our 27 crystal cube, we subtract 2 crystals from the surface of 26 crystals leaving 24. Now, we need to divide 24 by 6 to give us the crystal count for each face. The answer is 4. To find the number of crystals per side we take the square root of 4 and arrive at 2. This gives us six sets of four crystals arranged in a two by two fashion. Now, we need to be able to remove only two crystals from our cube in such a manner to attain our mathematical description.

Amazingly enough, this task can be accomplished. When we remove the two crystals, we need to remove them in a symmetrical fashion to involve all three-dimensional directions. For our example, we will remove the front-top left one and the back-bottom right one. Starting at the top, we can strip away the remaining front and the right edge crystals away from the crystals used for the top. This leaves four crystals in the two-by-two fashion as needed. From the remaining crystals in the front face, we can duplicate the same process eliminating four more crystals. As we continue the process throughout the cubical surface, we find that we are able to divide the crystals up into six equal sets of four crystals in a two by two fashion without rearranging any crystals. An interesting observation of the result is that each set is offset from each facial center of the cube in radial symmetry.

Chapter 9: Neutronverse

If we were to extend the pattern to the next layer, we will find that we are able to keep with the same pattern as the equation holds true. Instead of 24 crystals, we are working with 96 crystals within a five by five by five cube generating six sets of 16 crystals in a four by four fashion per cubical face. As we go on upward, the numbers become exponentially larger. Two observations can be acquired from this progression. One, the crystal squares are even numbers; and two, the pattern of cubes removed forms a diagonal line passing through the center of the cube. The central cube is also logically removed, as it is part of this line.

A side note concerning the aim of the process: After dividing the crystals into equal squares, we can align them to match sides. Imagine the side faces pulling toward each other. The result is that we have a squarish spherical-like shape that has a circumference of eight crystals. Actually, the process is somewhat more involved than that, as there are crystal alignments to consider and the forces that move them together. But first, we must expound the annihilation process.

Annihilation of crystals within a neutron cube is diagonal in nature as previously alluded. But, there are more factors involved. If all the diagonal lines of annihilation were to occur in a parallel fashion, the created forces could not distinguish between neutrons along the line, as all locations along the diagonal line are identical. Using a checkerboard to examine the process, let each square represent a neutron cube. We can set diagonal lines through sets of squares that are touching in the direction of the diagonal. Example: within a blue and purple checkerboard, we can connect the blue squares diagonally in one direction with a set of parallel white diagonal lines. The purple squares can be connected with another set of diagonal lines that are perpendicular to our previous set. We now have two different sets of diagonal lines crisscrossing the checkerboard diagonally in two directions matching the color of the squares upon which they reside. These intersections provide a reference point to determine the size of a neutron as the intersections mark the opposite corners. This view is only the beginning of the answer. We are working within three dimensions and the corners are not on the same plane as observed in a two-dimensional checkerboard.

Using the 27 crystal cube as our model, we observe four annihilation patterns. Our first pattern starts at the front-bottom-right corner and ends in the back-top left corner. In the process it passes through the central crystal. Our second pattern starts at the front-bottom left and ends at the back-top-right corner. If we were to look straight down from the top, we would see two diagonal moving along the plane in right angles to each other in the fashion seen on the checkerboard. However, there are two more patterns that appear the same from the same two-dimensional view. In our third pattern, we will start from the front-top-right crystal and end up at the back-bottom-left corner crystal. This produces the same top two-dimensional diagonal seen in our first pattern. The fourth pattern will two-dimensionally match our second pattern by starting at the front-top-left corner and ending at the back-bottom-right corner. But it also is the inverse of

Part 3: Creation of Neutrons

the second pattern in that it moves downward from the front-top instead of upward from the front-bottom. Note: All diagonal lines pass through the central cube.

The sequence is not arbitrarily chosen. If we were to make four cube sets using their respective annihilation pattern and stack them with one and two on top with three and four on the bottom, we will observe a hole forming in the middle. The four missing crystals join together to form a larger hole. These holes actually mark the corner of eight neutron cubes, not just the four seen here. When we continue the pattern of crystal annihilation into the neutron cubes in front of "the hole" we see an inversion of the pattern continuing to extend the annihilation lines along their path. Pattern four and three sits on top of the inverted order of two and one annihilation patterns. This occurs throughout the phaseverse.

These holes exist outside the remaining neutron cube material and are named **neutron exohole**s. The neutron exohole set in an offset manner as well. Each hole marks the location in which eight neutron cubes meet. Each neutron cube joins two neutron exoholes, which are set on opposite corners. Recall that a cube has eight corners. Therefore, there are six corners in which there exists no neutron exohole. While the phaseverse is still in isometric hold, locally between neutron cubes they are not. The holes create a void, and the jammerial influence will pull the crystals away from the hole. However, this is not the only influence pulling at the crystals, there is also a realignment that also occurs that causes or necessitates the annihilation process.

Landscaping the Neutron Cube

The neutron is made up of layers; each layer has a particular "landscape" called a **xyzenscape** (size-n-scape). Xyzenscape comes from the words "**xyzen**thium" and "land**scape**." There are five different xyzenscapes that occur within a neutron. Each xyzenscape holds a particular charge resulting of the crystal alignment. Again, these formations did not occur by themselves. Elohim designed them for specific purposes. The image is as a builder interacting with HIS Work. HE gets creation to a point in which it becomes usable for the next step. At these points the creation is helpless in itself to move onward to the next step without HIM intervening. After Elohim made the magneverse bricks occupying the neutron's volume, HE then places them into a pattern that will give HIM a functional neutron. Instead of moving them from one location to the next, HE has to only flip them in their present location. Remember that the crystals can flip without altering or moving their external structure. In essence, they are as spheres shifting in orientation.

Recall that from the magneverse the xyzenthium crystal has primarily three sets of cubical faces. The neutral faces we gave to the top and bottom face. The microweak face is the top and the microstrong force makes the bottom face. The four side faces hold the magnetic charge. The front and back faces are both positive charged, and the

Chapter 9: Neutronverse

two remaining side faces are negative charged. Throughout the description of the xyzenscapes these faces will be assigned different; colors to help us visualize their composition. Light blue represents the microstrong force face; dark blue represents microstrong; green faces are positive charged and the red are negative.

The five xyzenscapes are labeled the following: **daysod** (day-sode), **nightsod** (nite-sode), **electrosod** (elect-tro-sode), **protosod** (pro-tow-sode), and **chexosod** (check-so-sode). Even though there are other alignments possible, these are the initial alignments found in a given neutron. The "sode" endings of these xyzenscapes come from the Hebrew word for foundation, the "o" is long. This keeps with the idea that the alignments are foundations of the neutron.

Nightsod refers to the night aspect of creation. In fact, this xyzenscape forms the surface of the neutron as will be seen later. In this, the dark blue (microweak force) is facing up. Essentially, nightsod is a daysod turned upside-down. The internal facings of the crystal faces are magnetic. The green faces of positive energy are always facing the red faces of negative charge by the adjacent crystal. The microstrong force formulates the faces underneath. While this force is underneath, it is not without function, as we shall observe.

Daysod has day as the root word as Elohim called the light- Day. This is not just a memorial, but it pertains to the essential nature of its function. Since daysod is actually an inverted version of nightsod, its interior crystal faces are identical in nature. The green and red faces are always facing toward each other between crystals. This means that the electromagnetic forces of the crystals are interlocked within the layer. In daysod, the microstrong force is always up and the microweak force is always down. While the function of daysod is more indirect within the neutron, it plays a vital role in the electron, and in the second-generation quarks.

Electrosod stems from the word for electron. The surface of electrosod is red illustrating a xyzenscape displaying a negative electromagnetic charge. This means that the opposite surface is also red. The layer is held together weakly by the positive (green) faces connecting with the two neutral faces (blue). This is like a piece of non-magnetized iron and a magnet connection as opposed to have two opposite magnetic surfaces meet. Actually, the microweak face does have some magnetic interaction with the positive faces as the positive faces are projecting hypertips and the microweak faces are projecting hypotips, both are ingredients for a magnetic interaction.

Protosod stems from the word for proton. The green faces of positive electromagnetic charge are facing up and down making both sides of the xyzenscape "green." In this, it is the negative (red) electromagnetic faces that hold the layer together. Again, we are observing magnetic faces interacting with neutral charged faces. In this instance, we have hypotips projecting out from the negative magnetic face and hypertips form the neutral microstrong faces. This layer is the strongest layer within the neutron structure,

Part 3: Creation of Neutrons

not only by its own internal structure, but by its placement within the scheme of xyzenscapes. It interacts magnetically both internally with hypotips of the next layer holding the layer in place.

Chexosod xyzenscapes have a surface different than all the rest. Just as the root word, "checker", suggest, it has a checkered like surface. It appears as a red and green checked surface on both sides. It has a net magnetic charge of zero just as if it were neutral. This layer seems to be the most unstable. The internal lateral structure of the xyzenscape contains a pattern of the two neutrals and both electromagnetic faces. The electromagnetic faces facing primarily a neutral face (xyzenscape). Even though the number of negative and positive charged faces is equal, the pattern is not completely a checkerboard. The exception to the pattern occurs at the corners affecting four parallel edges. Recall that at each corner only three crystals meet. There is no way to form an alternating pattern using three mutually joining faces. By this, an edge forms with two crystals of the same magnetic charge next to each other. Even so, internal to the structure, only neutral faces touch charged faces of the crystal. This xyzenscape is not really evident until we study the nature of positron's complementary electron outside the framework of electron formulation.

Xyzenscape Sequence

Neutrons exist in a sequence of layers. These layers are the same for each neutron cube. The basic concept is an alteration between a charged layer sequence and a neutral layer sequence. These two sequences form in groups of three layers (three xyzenscapes). The outermost layer sequence is neutral. The external most xyzenscape sequence is neutral composed of nightsod, chexosod and nightsod. The following charged xyzenscape composition from top to bottom is electrosod, nightsod, and protosod. This sequence alternates throughout the neutron placing protosod at the bottom. An observation should be made at this point; daysod never faces outward within a neutron. This cannot be said to be true of second-generation quarks.

As noted earlier, the chexosod xyzenscape is a magnetically neutral charged layer because of its dual charged nature. This gives every other layer a charged xyzenscape containing only red, only green, or red and green checked. During the proton formation, the outermost chexosod xyzenscape crystals flip forming into nightsod, as this xyzenscape surface faces magnetically interact. Within the structure of xyzenscape layers, we observe the dual pattern existing on two levels: the alteration between xyzenscapes, and the alteration between the groups of three xyzenscapes. The number of times the pattern of six layers occurs is exactly 1,800 times as we shall observe.

As stated earlier, chexosod makes a neutral layer as the facial charges of the xyzenthium crystals within the xyzenscape creates a net charge of zero. Even though the net charge is zero, it is magnetically active. As noted, chexosod is the least stable layer, the two

Chapter 9: Neutronverse

charged xyzenscapes are more stable but not as stable as the nightsod/ daysod xyzenscapes. Nightsod has its polarized magnetically charged faces internally aligned. Another interesting item is the pattern alternates between charged and nightsod throughout the neutron. The establishment of this pattern occurs not by chance, but by Elohim's design. As we shall observe latter, HE puts HIS Mark on the design.

Xyzenscapes have a hierarchy concerning the optimum expression. As presented above the least favorable xyzenscape condition to the optimum xyzenscape: chexosod, electrosod and protosod, and lastly the xyzenscapes of daysod and nightsod. The electrosod and protosod xyzenscapes are equal and only stable xyzenscapes. While the optimum xyzenscape is the nightsod/ daysod having the magnetically charged faces joined internally, this makes them unstable. To realize this, we must examine the nature of magnetic energy. Just as two opposite charges will pull together, so does the tavtips between crystals. In this, the internal magnetic energy continuously pulls upon each other devouring crystal volume. In other words, the layer continues to shrink despite the resistance of the mass underneath. Upon the surface of the neutron, it causes the layer of crystals to breakup and eject from the surface. Internally to the nucleon, the nightsod is held in place by the external xyzenscape layers.

Normally during the process of forming a proton, the chexosod layer converts to nightsod as the surface layer breaks up into gamma rays. The relatively gradual breakup of the surface xyzenscape gives the chexosod opportunity for adjacent crystals to magnetically flip toward each other. This starts a chain reaction throughout the layer that terminates before the surface nightsod layer evaporates. The only limit to the chexosod conversion process is the interaction imposed by the covering surface layer before "excavating."

Neutron Reformulation

Once again, we will use our 27 crystal model to illustrate the process. Within this illustration, we will assign the crystals magnetic "coloring" to visualize the phenomenon. Remember that there are three crystals are going to be deleted, and that they form a single diagonal line from corner to corner, such that the diagonal passes through the central crystal.

Using our formula, we know that each face is going to have four crystals positioned in the nightsod xyzenscape format. Secondly, we know that the internal crystal magnetic alignment most be consistent. If we were to examine our pattern closely, we would observe a contradiction occurring at each of these annihilated cubes. Example: the corner crystal needs to have all four side faces negatively charged or positively charged. The central crystal needs to have all six of its faces neutral microweak forming nightsod in all directions. These conditions cannot exist. This makes the elimination process more than a geometrical necessity, but a magnetic necessity as well.

Part 3: Creation of Neutrons

Annihilation of crystals extraneous to the formation of a sphere actually dissolves by absorption into the adjacent crystals. This will not naturally occur; Elohim uses the energy upon the faces of these crystals to link the energy of the diflohexet into the adjacent crystal thereby dissolving its existence. But, it does far more than dissolving the crystals. This increases the juttorial and jammerial energies within the "walls" of the crystal layers. This expands the xergopath loops via tavtip count, and the added jammeria attracts energy from the adjacent diflohexet and linkage occurs unifying the "spherical wall." Secondly, this linkage also increases the shrinkage of the neutral layers.

Another observation is that when the crystals move into geometrical alignment, they also move into magnetic alignment. The magnetic alignment allows the side faces of the nightsod (the outermost xyzenscapes of the neutron) to interact as the interior faces interact as a unified and interconnected spherical layer.

Our next illustration describes the nature in which the neutron cube shrinks and how crystals attach themselves to other crystals. The external most loops of energy are defined by their xergopath placement relative to the surface of the crystal. If we follow these xergopaths toward the xentrix, we find that they are the fibers that reach toward the innermost core, as the loops defining the xentrix's edge form the innermost loops.

Notice the number of xergopath loops that form the exterior-most path is one in number while there are a multitude of xergopaths formulating the inner-most layer of xergopath loops. If we were to divide the xergopaths such that half were external and half were internal within a particular zeed, we find that the ratio is about 3:1 layers. This can be seen by taking an isosceles triangle and measuring the top half and bottom half.

The singular outermost xergopath per exozeed, giving four per a crystal face, serves as feeler for external energy. When they find suitable external energy attraction, they break away from the loop pattern to join the external pattern. These four interchanged energy xergopaths have a stronger hold as their paths become shorter that the loop pattern, generating a greater density of tavtips per a given measurement.

After the first two xergopaths establish contact, the xergopaths in the layer underneath join in with their corresponding xergopaths in the adjacent crystal. If this process were to continue to the innermost xergopath loop layer, the two crystals will become absorbed into each other. If all were involved, they shrink into infinitesimal space and become finitely nothing. However, the microstrong force from within the crystal structural alignment prevents this. Since daysod is always facing down toward the core of the neutron, the juttorial push keeps the neutron from shrinking into nothingness.

As the neutron faces shrink, each face shrinks past the edge the adjacent lateral faces of the neutron cube. This gives the edge faces of both to attach themselves to each other. As this occurs, the faces of the neutron cube unify into one surface. While this creates some structural tension, the tension is not as rigid as we might think. Remember the quarquid is a liquid-like substance, and as such finite structures are not rigidly held. As

Chapter 9: Neutronverse

the neutron cube continues to shrink, the cubical shape becomes more spherical because the microstrong force pushes outward restraining the lateral shrinking process by magnetic energy. The end result is a spherical neutron with cubical like qualities. This will become important as we develop atomic nuclei and as we look into quarks.

The feelers of the external most xyzenscape facing the external void also connect. They connect to the external feelers of the adjacent crystal with its adjacent diflohexet. In this, there are no dangling feelers external to the outermost xyzenscape. Even so, they form surface feeler xergopaths between crystals and behave as self-engulfing geysers shooting out into the external void.

Chapter 9 Quiz

1. Annihilation of Xyzenthium Crystals occurred because:
   A. They were defective
   B. They were extraneous
   C. They were not of the original set
   D. None of the Above

2. Xyzenscapes are:
   A. Layers of magnetically aligned Crystals
   B. Xyzenthium paths of annihilation
   C. Escaping Xyzenthium
   D. All the Above

3. Nightsod is:
   A. Negatively charged Xyzenscape
   B. Positively charged Xyzenscape
   C. Microweak Neutral charged Layer
   D. Microstrong Neutral charged Layer

4. Chexosod is:
   A. The most unstable of the layers
   B. Checkerboard between Positive and Negative Charges
   C. A Neutral Xyzenscape
   D. All the Above

5. Neutrons formed Geometrically by:
   A. Expanding into a Sphere
   B. Shrinking from an Octahedron
   C. Expanding from a Point
   D. Shrinking from a Cube

6. (T/F) Neutron Cubes occupied practically every physical location.

7. (T/F) Neutrons have layers of repeating xyzenscape sequences.

8. (T/F) Protosod has a surface and base containing positive charges.

9. (T/F) The outermost layer has a xyzenscape of nightsod.

10. (T/F) Daysod occupies the opposite side from nightsod.

Chapter 10

Quarks

Quark Classifications Overview

Before continuing on, we must understand the information and reasoning behind a quark. While this article does not utilize the quark concept, it does not disregard its findings. After a brief overview of the quark concept, we will discuss our reasoning behind the omission. Note: understanding this chapter is not a prerequisite before going on to understand the presentation of the creation of the universe. This portion provides information for those interested in the quark phenomenon and get an insight to the behavior of submatter. **Submatter** is anything less than a molecule, i.e. atoms, neutrons, protons and so on down to xyzenthium crystals.

In the quark theory, there are two basic kinds of quarks: hadrons and leptons. There are two kinds of leptons: the larger charged negative particle, i.e. electron, and a neutral particle called a **neutrino**. Both are much smaller than hadrons. Each of these quark types (flavors) has forms in three different generations. For an example: there are second generation hadrons or third generation leptons. The hadrons have more different flavors (kinds) of quarks assigned providing more combinations. **Meson**s formulations require two hadronic quarks, and **baryon**s require three hadrons. Quarks of the first generation primarily form by natural phenomenon, such as electrons and electron-neutrino. Secondly, the initial smashing of the **nucleon**s (neutron or proton) produces the up and down quarks, which are logically back fitted into being the components of a neutron or proton. **Hyperons** are three hadronic quarks that do not form the natural nucleons and are much larger in mass. **Antiquarks** are hadrons or leptons having a notable annihilating characteristic, usually being of opposite charged particles in equal magnitude.

Relatively recent, antineutrino has become the definition of the particle originally labeled as the neutrino, which first reclassification was electron-neutrino. The first reclassification to electron-neutrino was reasonable, as scientists were able to create more generations of quarks. However, the term antineutrino gives the false impression that antimatter is a natural phenomenon, which it is not. **Antimatter** is strictly superficial; they are artificially produced by scientific processes, which have no function in forming the universe. Since the original name of this particle was neutrino, we are continuing with that name. Even though, it is a piece of submatter that will evaporate down to nothing.

Part 3: Creation of Neutrons

Splitting Submatter

While scientists are able to form the different quarks via collision, it does not necessarily mean that these particles aided in the creational process. There are two indications that quarks are post creational phenomenon and not a phenomenon of creating the universe. First, the quarks do not add new dimensions to the formation of a neutron or proton. Example: their formation does not answer any questions concerning the composition of the neutron. It only adds another layer postponing the previous questions of composition. Secondly, the second and third generation quarks have no functional purpose in sustaining the atomic structure of atoms formulating the elements.

Consider the following scenario of a smashed neutron. As scientists apply intense pressure to the neutron, and it begins to split. Its crystalline nature induces a fracture that would split the neutron into halves. The image is that of half shells forming as the split approaches the core. Before the split can reach the core, the two multi-layered shells are repelled from the core. This occurs at the point that the shells have equal mass to the remaining core. The repelling force is juttoria. In this illustration, we see three separate hadrons formed. While it is true, in relationship to the original particle, the three hadrons could join with corresponding formulating hadrons formed by the colliding mass. In this instance, it could form three mesons.

The following illustration further demonstrates this argument. If we threw a drinking glass upon a piece of cement, it shatters. The pieces of the shattered glass can be put back together to reformulate the glass. This does not qualify these pieces to be the original subdivisions that formed the drinking glass; however, sand is. The entire quark phenomenon is a study of broken pieces and assembling them to reform the "original glass" or other "glass forms". Unlike the pieces of broken glass, fragments of subatomic material are self-gluing via the interaction between hypertips and hypotips.

Another phenomenon observed by scientists is that there are subatomic particle fragments that have a half charge. This is in line with the description given in this article. It is even possible that many of the masons are in fact canceled 1/2 charged subatomic fragments instead of 2/3 or 1/3 charged particles.

Hadrons

There are six quarks classified as hadronic flavors. Their labels are the following: **up** (+2/3), **down** (-1/3), **strange** (+2/3), **charm** (-1/3), **top** (+2/3), and **bottom** (-1/3) quarks. Quarks do not vary only in sizes but in sequence of formation called generations. The first generation quarks stem from the original source of nucleons. These quarks are the up and down quarks. The neutron is assigned two down quarks and one up quark giving the equation of $2(-1/3)+(+2/3)=0$. The quarks assigned to a proton are two up and one down or (u,u,d). This gives the result of adding +2/3 twice

Chapter 10: Quarks

then adding a -1/3 bringing 4/3 back down to 3/3 or a +1 charge. The second generation of quarks formulate from the smashing of these quarks together. The down and up quarks of the second generation are the charm and strange quarks respectively. Finally, scientists formulated the third generation. The hardest of these was the top quark, which turned out to be more massive than calculated. The quarks of this generation are the top and bottom quarks. According to calculations, there can only be three generations of quarks. By this, there are only six quarks; this is not counting their antimatter counterparts giving twelve hadronic quarks. For example anti-strange (-2/3) is the antimatter counterpart for a strange quark. This quark classification signifies that they have magnetic "glue" with each other, and there has not yet to be one to exist alone. They form either baryons or mesons. The baryons have two subdivisions: the nucleons and hyperons. The nucleons are the neutrons and protons. **Hyperons** are more massive than nucleons, but they are unstable (having half-lives). A mason is a quark and antiquark pair, generally unstable.

Leptons and Neutrinos

This class of quarks formulates out of hadronic material. They are usually smaller than any hadron. The exception is the tau lepton. The electron, **muon**, and tau all have a -1 electromagnetic charge. The only stable lepton is the electron (does not decay). The first generation lepton is the electron; second generation is the muon, leaving tau as the typically massive third generation quark. Unlike the hadron quarks, these quarks will not collect into groups of two or three. These quarks require that they remain singular quarks as there are no natural occurring quarks of its nature with an opposite magnetic charge. The electron's antiquark is a positron, which we shall expound upon shortly.

An interesting observation is that even through these quarks of the different generations have more mass, they still hold the same charge. Some might attribute it to being a net charge. An example is that -3.5 + 2.5 = -1. Nevertheless, the answer exists with the nature of an electron. The electron has a surface equal, actually slightly larger, to that of a proton and has comparatively very little mass internally. The muon is an electron with more mass internally, same number of xyzenthium crystals upon the surface. The tau lepton has even more mass. Suppose for a moment, there is a fourth generation lepton containing the mass of three nucleons; it would still have the same amount of crystals upon the surface. The reason this is able to occur is that the crystal layers within the neutron cube were much larger before shrinking into a sphere. When we examine the nature of the electron, we observe that its size in volume can vary with the energy involved.

All neutrinos have zero electromagnetic charge. At one stage of quantum mechanics, the following terminology was used. The first generation neutrino is the electron-neutrino. The second is the muon-neutrino; third generation formulation is the tau-neutrino. Each generation progression has more mass than the previous. However,

Part 3: Creation of Neutrons

some initially believed that the neutrino mass might be zero. As seen by their names, all neutrino formulations occur with the manifestation of a given generation of lepton.

Nucleon Equation

The foundation of hadronic quarks is found in the natural occurring phenomenon of the subatomic nucleons. Therefore, the first generation hadrons are defined by the nucleons: a neutron is said to be composed of two down quarks and one up quark. As stated earlier, the proton is defined as being composed of two up quarks and one down quark. The up quark has a net electromagnetic charge of +2/3. The down quark has a net electromagnetic charge of -1/3. Adding the values of up and down quarks for each case, the result would be the expected +1 charge for a proton and a zero charge for a neutron. Putting it in yet another form: the neutron equals (2/3 (up) -1/3 (down) -1/3 (down) = 0; the proton equals 2/3 (up) + 2/3 (up) - 1/3 (down) = +1.

These values can be achieved algebraically by assigning u to that quality which causes a positive charge and d representing a unit (down quark) responsible for the negative charge. By this we get two algebraic equations: $2u+d=1$ and $2d+u=0$: First, we multiply the second equation by 2 and subtract it from the first equation. We get the following:

```
    (2)u  +   (1)d  =   +1        : First equation representing a proton
   -(2)u  +  -(4)d  =    0        : Subtracting twice the neutrons makeup
   -----------------------------
    (0)u  +  -(3)d  =   +1        : Note the u-value is zeroed out
             -(3)d  =   +1        : Remove extraneous zero value
                 d  =  -1/3       : Divide 1 by -3 gives d = -1/3
```

Next, plug this value into the first equation and solve:

```
    (2)u  +(1)(-1/3) =  +1         : Substitution for d-value
    (2)u  +  -1/3    =  +1         : Multiplied variable by constant
    (2)u             =  +1 +1/3    : Subtracting -1/3 (both sides)
    (2)u             =  +4/3       : Simplified into improper fraction
       u             =  +2/3       : Divided equation by 2
```

Therefore the up quark is assigned a value of 2/3 and a down quark the value -1/3

The strange and charm quarks are second generation hadronic quarks. They are marked by their larger masses than up and down quarks, and even more famously for their unusual decay pattern. They have decay times that are outside the normal range with stronger energy being radiated. They have the same electromagnetic charge properties as the up and down quarks: strange quarks have a charge of +2/3, and charm quarks have a -1/3 charge.

Chapter 10: Quarks

The third generation of hadronic quarks is labeled top and bottom. Just as the second-generation quarks more massive than the first generation quarks, the third generation is even more massive. As stated earlier, the top quark was the last to be discovered and was much more massive than anticipated. The top quark has a +2/3 charge, while the bottom has a -1/3 charge. They decay in the normal process, meaning they have no attributes of strangeness. However, only the assigned "first generation" quarks provide us with any stable formulation.

Antiquarks and Mesons

An antiquark will have the opposite attributes of a quark, primarily in the magnitude of magnetic charge. For an example, an anti-up quark will have a charge of -2/3. An anti-electron (**positron**) has a charge of +1. However, there are also antineutrinos as well. This makes 24 quarks in all. Despite their name, they are not responsible for the decay of submatter, as all second and third generation quarks decay. Hadronic antiquarks are perhaps the most famous as they are often found with "normal" quarks bringing the total magnetic charge to zero. Next is the positron, the anti-electron. In placing a positron around a muon, we assimilate the antimatter of Hydrogen-1. Antineutrino has no charge, but is defined by process within the formulation of a lepton antiquark. As stated earlier, mesons are composed by two hadrons. These hadrons are pairs of quarks and antiquarks. The net magnetic charge of this kind of pair will always going to be zero. However, there are some formulations of mesons that will have a positive or negative magnetic charge.

Strangeness

Our next phenomenon to analyze is the strangeness attribute. This attribute is found in the strange and charm hadronic quarks. Its half-life lasts much longer than that which is expected in normal gamma radiation. This demonstrates that the xyzenthium crystal has two neutral faces that behave differently. In this case, the layers are inverted via the collision and recollected formation of the second-generation quarks. The external layers collide together at the impact. The external surface of the layer then becomes the internal surface, and the internal layers become external. The daysod xyzenscapes containing the microstrong force are facing out and the microweak force of nightsod is facing inward. It, naturally, will take the microweak force longer to push off the xyzenthium crystals.

Within the third generation quarks, the inverted material inverts again eliminating the strangeness attribute. Each generation requires more mass or layers to accomplish the task to hold them in place long enough to magnetically crystallize. If it were possible for a fourth generation to exist, there would be another generation of even larger hadronic quarks containing the strangeness attribute.

Part 3: Creation of Neutrons

Isobaric Spin

The **isobaric spin** is the angular momentum that most subatomic particles have. This spin is the result of an imbalance relationship between the hypertips and hypotips within a given assembly. In general terms, the hypertips exert a greater pressure upon the miortex than the hypotips; this generates an acceleration of the miortex. As a result, the entire crystal acceleration is in the direction indicated by the miortex. This occurs while the miortex redistributes kinetic pressure throughout the xyzenthium crystal.

This imbalance occurs only in the two charged layers and in chexosod. If we examine the internal structure of these crystal alignments, the imbalance becomes evident. We shall examine a layer of electrosod. The external faces of positive charge are met with microstrong and microweak faces of adjacent crystals. The microstrong diflohexet forming the microstrong face of the xyzenthium crystal is projecting hypertips into the positive diflohexet, which normally receives hypotips. However, on the opposite side of this diflohexius, the positive diflohexet is receiving the expected hypotips. Plus, the microweak diflohexet is receiving the positive diflohexet's hypertips when it is expecting hypotips. This will become evident as we illustrate the phenomenon.

There is another lateral kinetic imbalance at a $90°$ angle to the previously described phenomenon. This kinetic push is generated by the same phenomenon: hypertip energy exists in directions that hypotips need to be in order to establish kinetic balance. It should also be noted: the resultant vector of these two forces is at a $45°$ angle to the crystalline alignment. This assumption is based upon the crystals forming $90°$ angles. As noted earlier, they do not; it is only in a virtual sense as the distortion of the crystal's linear alignment slightly alters this angle.

To illustrate the alignments of the different xyzenscape, we will symbolize the phenomenon. The diflohexet pair's xyzenthium crystal facial surface is represented by a "<" (less than character) and a ">" (greater than character). Their mathematical implications are ignored in this illustration. The exterior of the diflohexet pair is symbolized "<>." Note: that the two characters point away from each other. This shows that the pair's surface face in opposite directions. These arrowheads act like a shell to the pair. They also indicate a direction of flow as there are forces of energy being projected outward from the miortex toward the crystal's surface. Now, we can fill the shells with energy. The flow of hypertips and hypotips is visualized as ones and zeros respectively. Recall that the diflohexius has a two-way flow. That which flows away from the miortex "presses" against the crystal's surface hence is placed against the shell. On the same line of thinking, the energy flow toward the miortex located at the center of the crystal is symbolized in the innermost positions. The colon punctuation between the two number sets represents the miortex itself.

For example: <10:01> is the illustration of the positive charged diflohexet pair. Observe that the ones are next to the arrowheads. This represents the hypertips

Chapter 10: Quarks

pressing against the crystal's surface. The zeros that are next to the colon indicate the flow of hypotips within the diflohexet toward the miortex. The following demonstrates the lateral configuration of the xyzenscapes:

Nightsod:	...<01:10><10:01><01:10><10:01>...
Daysod:	...<01:10><10:01><01:10><10:01>...
Electrosod:	...<10:01><00:11><10:01><00:11>...
Protosod:	...<01:10><00:11><01:10><00;11>...
Chexosod P:	...<01:10><00:11><01:10><00;11>...
Chexosod E:	...<10:01><00:11><10:01><00:11>...

Secondly, we need to devise a symbolized illustration of the exchange of energy that occurs in the outer xergopath layers of the adjacent diflohexet. Let us extend a line under each colon mark. To represent the flow between the two diflohexets, move the external tavtip symbol of the adjacent crystal representation next to the line under the colon on the side from which it arrived. Then we need to determine the balance between the two sides. This is accomplished by comparing that number with the number generated by the opposite direction. If they are the same number, then there is no kinetic change. This means there is no isobaric pressure asserted within the layer. To represent this in our diagram, we will put an equal symbol under the extended line. If there is an imbalance, we will place an inequality symbol representing the direction of the imbalance. Notice also the difference between charged and neutral layers.

Nightsod

```
...   <01:10>   <10:01>   <01:10>   <10:01>   ...
...    1|1       0|0       1|1       0|0       ...
...     =         =         =         =        ...
```

The analysis of nightsod is that the "pressure" exerted by the adjacent crystal's diflohexet is the same as the crystal's own pressure upon its miortex. Looking at the numbers adjacent to the extended line, we see two patterns. We find hypotips pressuring hypotips and hypertips asserting pressure against hypertips holding the kinetic energy of the crystal in balance. We are ready to examine other xyzenscapes:

As expected, daysod provides no input to the isobaric spin, as it is an inverted nightsod layer. Even though there is no difference internally, externally they are as different as night and day within the xyzenthium crystal.

Daysod

```
...   <01:10>   <10:01>   <01:10>   <10:01>   ...
...    1|1       0|0       1|1       0|0       ...
...     =         =         =         =        ...
```

149

Part 3: Creation of Neutrons

Electrosod has two kinds of conditions that occur in the exchange of energy between crystals. It has both the balanced and imbalance conditions occurring by the exchange. However, the imbalance expresses itself at two different levels. The pressure exerted by the exchange and the release of pressure between crystals by the exchange. Notice under the neutral diflohexet pair, <00:11>. Even though the internal forces are different, they still exert equal pressure upon the internal miortex. This can be best illustrated mathematically. If we add two different sets of additive inversed numbers, such as -7 to +7 or -2 to +2; it will matter in our case, as the answer will always be zero.

Electrosod

```
... <10:01> <00:11> <10:01> <00:11> ...
...   1|0     1|1     1|0     1|1    ...
...    >       =       >       =     ...
```

Protosod behaves in a similar manner as electrosod, but for the inverse reasoning. It is the hypotips that give the imbalance to the formulation. However, the result is the same as the imbalance is equal in strength. This does not imply that the imbalance is running in opposite directions, as the orientation between layers is not restricted to dictate such an occurrence.

Protosod

```
... <01:10> <00:11> <01:10> <00:11> ...
...   1|0     0|0     1|0     0|0    ...
...    >       =       >       =     ...
```

Chexosod has two alignment patterns occurring within the xyzenscape layer. They are chexosod-P and chexosod-E. Chexosod-P has the identical format to protosod, but occupies only half of the phenomenon occurring internal to the xyzenscape. This can be seen below. This xyzenscape also promotes the isobaric spin observed by today's scientists.

Chexosod-P

```
... <01:10> <00:11> <01:10> <00:11> ...
...   1|0     0|0     1|0     0|0    ...
...    >       =       >       =     ...
```

Chexosod-E behaves identically to electrosod. By this, we observe the alternate pattern between active input and inactive input to the isobaric spin by the xyzenscape layers. The isobaric spin is the result of all the pressure variance vectors experienced within the charged layers of a given particle of submatter.

Chapter 10: Quarks

Chexosod-E

```
...  <10:01>   <00:11>   <10:01>   <00:11>  ...
...    1|0       1|1       1|0       1|1    ...
...     >         =         >         =     ...
```

There is another pattern of chexosod possible that generates an inactive layer toward promoting the isobaric spin. This requires that the neutral faces of the xyzenthium crystal to also alternate in alignment, however this adds more complexity to an already complex layer. This added complexity would serve no purpose except to weaken the spin to a value less than half the mass involvement. This would generate an obvious deviation to the measurements observed by scientists.

Bubble Chamber Experiment

The existence of a chexosod xyzenscape is hard to prove physically, as it is so unstable that it would never last long enough to observe. However; it is because of this instability, it provides some interesting phenomenon pointing to its existence.

This brings us to another observation made by scientists that a proton can be made to lose a positron and two electrons. A **positron** is like an electron, except it has a charge of +1. By asserting a Gamma Ray against the isolated proton (Hydrogen-1 Nucleus), scientists accomplish this disassembling task. The initial image is as an electron and positron pair off in opposite directions, via an external magnetic stimulus, as two opposite degenerating spirals.

Meanwhile, another electron shoots off at a speed far greater than the original pair. This electron behaves in the fashion of normal **protonization** (neutron breakup). The resulting visual image is like a fancy drawn arrow.

There is a reason for the "docile" electron existence. When the Gamma Ray penetrates the proton's surface layers, its puncture extends throughout the xyzenscape's buffer sequence between the protosod and electrosod xyzenscapes. This perforation caused the layers to contract laterally. The image is that of a pin piercing through a piece of thin rubber stretched over a rigid surface. The newly acquired hole immediately expands as the rubber pulls away from the hole's center, via imbalance tension. Unlike layers of stretched rubber, the layers interact with each other electromagnetically.

Three layers follow the protosod before encountering the next electrosod layer. Since all four layers were contracting simultaneously, the chexosod xyzenscape separated from the neutral charged layer under the protosod. The layers group into two sets of xyzenscapes. By examining the xyzenscape generated by the neutral layer, we find the separation between layer sets. Since nightsod is always facing away from the neutron's center, this gives the xyzenscape of daysod facing the neutron's core per neutral layer.

Part 3: Creation of Neutrons

This implies that the microstrong force is pushing the charged layer away. The exception is that of the electrosod. The electrosod expects the input to complete its cycle. However, the push is normally absorbed. Therefore, the expulsion of these layers is virtually non-existent. There is a slight push as the outer layers have more crystals associated with them. Moreover, as the layer contracts laterally, the process intensities the expulsion tendency beyond the electrosod's ability to absorb within the same surface area.

The visualization, again, is that of two pieces of stretched rubber. By drawing two sets of equal-distant parallel lines at right angles to each set upon each surface, we observe two pieces of stretched rubber with a grid of squares drawn on each. These squares are equal in length and width (congruent) on each surface. Now, release the tension of one piece of stretched rubber. Notice that the sizes of the squares have shrunk. Examining a square inch of each (assuming that the side of the original square is a half-inch or less), we find that the non-stretched piece has more squares per square inch. Imagine to each corner of the square is attached a miniature air blower pointed straight up unaffected by the shrinkage of distance between them. In other words, these blowers do not shrink with the squares. The number of blowers per square inch has increased; thereby, they exert a greater force per square inch due to the lateral shrinkage. For this reason are these layers expelled from the surface by an increase of force.

As the structurally unstable chexosod began to form its own spherical entity, the positive charge of the forming positron repelled the positive crystals of chexosod causing the negative charged face of the xyzenthium crystal to flip into its place. This forms the electrosod xyzenscape.

The newly formed layer of electrosod then pulls in the opposite direction by the magnetic field giving the symmetrical appearance. There is no violent expulsion from the nucleon surface, as the layers already exist free from the surface by the impact of the Gamma Ray. Moreover, the positron does not shoot off as the positive charge emits hypertips as does the daysod facing the opposite direction of the fourth layer underneath. By this, the expulsion factor is in balance. This gives the pair's mild trajectory velocities. However, before the two fully escape from the surface, the remaining layers of electrosod and nightsod interact normally and eject like a bullet. This leaves the nucleon in a proton state. The expelled electron bumps the two other particles into acceleration unnatural to its existence. The effect of this acceleration eventually diminishes to zero as the natural isobaric spin accelerates.

Cold Neutrons

Scientists have cooled neutrons to near absolute zero (-273.16°C or -459.69°F). Unusual properties manifest themselves near this temperature. Not only does this cold slow down the neutron, but splits the surface xyzenscape, namely nightsod, into two

Chapter 10: Quarks

parts. This contraction exposes the chexosod xyzenscape underneath at the equatorial region of the neutron. The central width is about $40°$ providing there are three equal surfaces, the "valley" can be smaller. The newly exposed surface is now free to interact with the external magnetic forces. The ramifications of this observation are unnoticed by scientists, as they believe that the neutron has three parts intrinsic to its nature.

Our first noticeable observation is the stated split into three regions. The reason for this occurrence originates in the nature of its diminishing surface and composition. We shall soon see that the present-day neutron has less mass than the primary neutron. This loss exists primarily within the three surface layers of the neutron. This loss reduces the amount of juttoria available to the surface to hold it together during the reduction of excess juttoria. The second principle involved is that the neutron was originally a cube. The corners of the cube are the weakest regions of the spherical form as they experience the greatest stress of the contortion into a sphere. As the amount of heat reduces, the surface crystals contract temporarily and attempt to return to a crystalline pattern as the core of the neutron retains more heat maintaining its volume. Since it is nearly impossible to remove heat from a surface evenly, there will always be a cooler side, no matter how slight. This creates a split of the surface. The crack moves along the coolest edge from the corner and eventual splits into two hemispheres. Unlike the quark particles having mass of about one-third of the total mass of the neutron, the surface of a neutron configures into two plates separated by a large valley of the chexosod surface. The nightsod edge pulls into the chexosod surface via magnetic interaction. This brings the nightsod surfaces in contact with the "valley" creating a change reaction flipping the exposed chexosod crystals into a proton-like surface. The incoming hypotips provide the expected response of the magnetic reception to the positive charge. In essence, the plates behave as if they had a negative charged surface. Because of this polarity between the components, they are able to make tiny neutron magnets. By asserting more magnetic manipulation, scientists move the central component to one side of the neutron assembly pulling the two end components together on the other side.

Notice also the arrangement of the magnetic axial energies. These are not as the magnetic poles found in the macrocosm as both ends radiate the same energy. They are the flow of energy leaving one surface to an adjacent surface as energy from the adjacent surface enters back into the given axis. This phenomenon is not limited to quark-like experiences. We shall observe more of this phenomenon, as we examine the nature of the bondverse. However, there is a unique nature found in the cold neutrons. The negative poles are half as strong as the positive pole, as both negative poles are feeding a single positive pole. These poles are actually three central planes radiating the charge interactions. A side observation, the magnetic value of the hypotips of the two neutral components provide equal and opposite charge of the "valley." Since the two components are equal in area as the central component, the opposite magnetic charge of each is half and opposite of the central component. Following the same logic in the

Part 3: Creation of Neutrons

primary formation of the meson particles, we find that the two quarks are equal and opposite in charge as there are only two components of nearly equal mass. By this, the original quark and antiquark pair formulate. Unlike our neutron, they are able to create their quark surfaces while they formulate. If the up and down quarks preexist, then the natural occurrence of mesons would be that of some composition of up and down quarks instead of the quark and antiquark pair with different charge values than presently assigned to up and down quarks.

During the evaporation of the surface nightsod, the movement of the chexosod crystals underneath converts into the nightsod xyzenscape. This starts a chain reaction converting the imitated proton surface into nightsod. The reason for this reaction is that the nightsod/ daysod xyzenscapes are the optimum alignment bringing all the magnetic faces of the crystals into contact with each other. Daysod aligns itself next to the nightsod xyzenscape underneath imitating magnetic energy.

Another observation is that the size of these neutrons is much larger than those sizes found in normal conditions. After the central component losses its initial heat, the cooling process continues. At this point, the connecting xergopaths to the adjacent crystals break away and rejoin with the xergopaths of its own crystal. As this occurs, the crystal has less force pulling the crystals together via contraction lessening interactions between the magnetic energy of any two adjacent crystals. As a result, the crystals begin to return to their original size before the contraction from a neutron cube into a spherical form. Within this form, scientists are able to deflect photonic energy. The result is the appearance of a polished surface deflecting energy in all directions. This is because the photons have more energy than the decompressed xyzenthium crystals can absorb. Normally, the crashing photons are completely absorbed, as the mass of the nucleons are denser. The excess magnetically deflects and the kinetic energy mirrors off the surface of deflection.

An interesting observation contradicting the concept of quarks as a natural building block of the universe is the measurement of light across the face of a proton. It is not the numerical value of the distance that matters. It is the lack of the need for a minor axial reading. If the quarks exist as three distinct particles, it would generate a major and minor axis to the proton. Considering that a proton has two up quarks and one down, we need to separate the two up quarks by the down quark via magnetic requirements. In the "hard" quark image, the three quarks would be lined up in a row giving three times the width in comparison to the height. In a "soft" quark image, the three quarks merge together giving gravity the final input to the proton's shape. However, the problem with this is that a proton is formed from a neutron, supposedly formed by two down quarks and one up quark. In this, the up quark separates the two down quarks for magnetic reasons and would also give gravity the ultimate input to deform the quarks into a sphere. Within the transformation from a neutron into a proton, one of these down quarks would have to transform into an up quark. In doing

Part 3: Creation of Neutrons   Chapter 10: Quarks

this, it would be ejected from the central up quark of the neutron via magnetic demands. This does not occur. Moreover, we know magnetic interactions are stronger than gravitational interactions. This means that the "soft" quarks are very unlikely to exist. This brings us back to a "hard" quark image, which lacks the dimensional properties to define their existence. Quarks of a proton exist only as resulting particles after smashing the nucleon.

Hinted at earlier, the apparent change in mass via temperature is the result in change in ratio between juttoria and jammeria. As juttoria decrease in magnitude, weight increases for the same volume, creating a denser volume, hence increase in mass. The weight increase is because the upward bending of jammeria decreases, via less heat, giving jammeria (gravity) a truer reading.

Chapter 10 Quiz

1. What is the view held about Quarks?
   A. They do not Exist
   B. They are all Manmade
   C. Most of them are Artificial
   D. They are Ugly

2. How many known Generations of Quarks exist:
   A. 1
   B. 2
   C. 3
   D. Many

3. What are Mesons?
   A. Set of Two Quarks
   B. A singular Quark
   C. Set of 3 Quarks
   D. An Antiquark with Two other Quarks

4. That is Strangeness?
   A. Inverted Radiation
   B. Radiation from a Strange Quark
   C. Result of Exposure of Daysod
   D. All the Above

5. The smallest Quark is:
   A. Third Generation Electron
   B. First Generation Neutrino
   C. Second Generation Hadron
   D. All are Equal in mass

6. (T/F) Hadrons form nucleus submatter.

7. (T/F) Up quarks have a charge of $+2/3$.

8. (T/F) Charm quarks have an attribute of strangeness.

9. (T/F) Third generation quarks are stable.

10. (T/F) Antiquark of the up quark has a charge of $-2/3$.

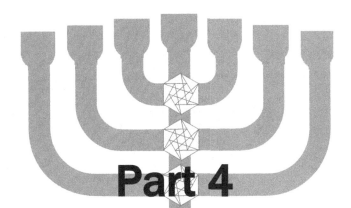

Part 4

Creation of Atoms

Chapter 11

Neutron Breakup

Nightsod Evaporation

The evaporation of nightsod is observed as gamma rays. They are photon-like particles containing large quantities of jammeria. Jammeria gives gamma radiation a greater destructive force as they impact submatter of other atoms. The rate in which nightsod evaporates from the neutron is determined by the impact of juttorial energy in comparison to the jammerial pressure exerted upon the surface by other nucleons. This ratio changes for two reasons. One, the external void surrounding the neutron releases pressure upon the xergopaths of the diflohexet formed by the xyzenscape surface. Two, the contact with other nucleons, which has the inverse effect, in that, it slows the process down. Initially, we will concern ourselves with the first process.

After the neutron shrinks into a sphere, a large void establishes itself externally to the sphere. This void generates an imbalance of pressure with the external diflohexet allowing the feeler xergopaths to expand outward from the xyzenscape surface. The appearance is like a multitude of swellings occurring upon the surface. This releases pressure upon the miortex in the direction facing out. Meanwhile, under the miortex is the juttorial force of the daysod xyzenscape pushing upward away from the center of the neutron. As the xergopaths upon the surface form larger loops, the miortex moves away from its original location. This is due to the set of xergopath loops giving way to the juttorial force underneath. Recall that the miortex formulation is from sets of irregular duocollisions occurring at its base rather than the septicollision observed in the kineverse. In this, the miortex begins to accelerate away from the xyzenscape below.

As this acceleration continues, the miortex accelerates to the speed of light away from the center taking the entire crystal with it. The lateral connected xergopaths first stretch as the crystal moves away from the neutron surface. This becomes short-lived as the xyzenthium crystal accelerates to the speed of light nearly simultaneously from its initial movement of its departure. Given enough time, the xergopaths would eventually rejoin into their loops via magnetic behavior. Usually, the ray is drawn into another nucleon by magnetic and gravitational energy. This escaped crystal is known as the gamma ray.

After this departure, the neutron loses mass, minute, but still a loss. The effect it has upon the surface is that a void is formed within the layer by the "evaporated" particle. This causes an imbalance of pressure within the layer. Adjacent crystals attempt to fill

Part 4: Creation of Atoms

in the gap resolve this imbalance. The result is that the imbalance spreads throughout the layer. This attempt to stabilize by the neutron is overtaken by the continued evaporation of particles from the surface. Eventually the entire layer disappears.

The particle that evaporates has photonic properties. As we shall observe in the anatomy of a photon later, the jammerial energy is the leading edge of the photon generated from the xyzenthium crystal. The juttorial force, naturally, becomes the trailing edge of the photon. In this, the space/time continuum is bent first downward and then upward forming the trough and crest of the wave. Their photonic properties cause the plasmaverse temperature to rise from near absolute zero temperature to the extreme heat of the plasmatic nature. Temperature is then defined as excess juttorial energy to magnetic processes. We stay near absolute zero during the absorption of xyzenthium crystals while forming a spherical neutron from a cubical neutron.

Lastly, before the surface layer finishes evaporating, the second layer begins the process. As pressure is released from the surface of the second layer by the outermost layer, the chexosod surface interacts forming another layer of nightsod. The evaporation process continues until the outermost three layers have completely vanished from the nucleon.

After the evaporation of the external three layers, another xyzenscape layer appears. This is the layer of electrosod. The reason that this layer does not evaporate exists by the nature of the internal structure. One, there is no juttorial force pushing outward. Two, the lateral structure exerts less magnetic "shrinking" tension. Recall that it is the internal magnetic pull per xyzenthium layer that causes the neutron to shrink. Within this layer, as with all the charged layers, there are neutral faces of the crystals being positioned laterally instead of vertically that weaken this function.

However, the electrosod layer still has a problem remaining upon the surface of the transforming neutron. The phenomenon that removes the electrosod from the surface exists in the reaction between the nightsod and the protosod layers directly underneath. Notice that the two xyzenscapes that are facing each other are both projecting hypertips, which contains juttorial energy. This increases the push between layers by the large juttorial presence. Being freed from the external three layers of nightsod, they begin to act aggressively to repel each other. This is called the **hyperpush**, as the hypertips are greater in number between these two layers than in any other two layers formulating the neutron.

Formulation of an Electron

To understand the formulation of an electron, we need to re-examine the neutron's formulation. Recall establishing the xergopath feelers exist projecting outward from each face of the xyzenthium crystal incorporated within a given xyzenscape generating a neutron, we realize that there are extra xergopaths moving toward the center of the neutron and its subsequent proton. Each xyzenscape layer contains fewer xyzenthium

Chapter 11: Neutron Breakup

crystals as we approach the core of the nucleon. We find another interesting pattern, when we examine the nature in which these extraneous xergopath feelers exist.

Let us analyze two large adjacent layers. Let the largest layer contain ten thousand crystals in a hundred by hundred square forming one face of the original neutron cube. The layer under it has 98 by 98 square giving a total count of 9,604 crystals. This means that there are 396 crystals from the upper layer that the layer underneath does not have. Moreover, the location of these crystals forms the edge of the cubical face.

Notice the ratio between the crystals in which the xergopath feelers that are connected to the crystals underneath to those that do not have any connection to the layer underneath. The ratio is the smallest at the surface and the nearest to 1:1 ratio at the core. Even so, the crystal count is greater at the surface than at the core. Along with this, each subsequent layer toward the nucleon core has a crystal count dimensionally two less than the layer above it. In this, the edge crystals will always be extraneous to the layer underneath. The feelers are extraneous to the layer underneath; they remain unassociated with any other crystal. They link between themselves. The image is that of forming a loop between the left and right side of the same diflohexet.

This process occurs for each of the six crystal faces. This anomaly between feelers gives the crystal to fold inward upon itself as the neutron forms. After the neutron forms, this unique formation establishes a weakness down to the nucleon core. However, they do not disrupt the lateral flow of energy per xyzenscape. The weakest formulation is at the corners as they are the only locations in which three crystals form together instead of four. Again, this formulation reaches down to the core of the nucleon. These weaker connections divide the nucleon into six equal quark-like sections. This is the underlying factor in the electron's ability to eject.

Returning: the hyperpush immediately separates the outermost remaining two layers, electrosod and nightsod, from the original neutron mass surface leaving the **proton** suspended centrally inside the shell via hypertip repelling force. From this point onward, we will call the main mass of the original neutron a nucleon. Actually both, the neutron and the resulting proton are both nucleons comprising the nucleus. The nucleon has an isobaric spin to its nature, which we described in the chapter dealing with quarks. At this point, we need to remember that the nucleon was originally a cube in form, as such, it has eight corners. Six of the corners are held in a spherical pattern by the spin and they push outward. The remaining two polar corners to the spin attempt to return to their cubical form by the push of the hypertips. The centrifugal stress upon the structure tends to push the edges jointing the six interior corners outward pulling in on the polar corners. The edges connecting the polar corners to the interior corners stretch the furthest away from the central nucleon. The greatest lateral tension is at their midpoints. The reason is the geometric pressure from the spherical transformation from a cube. This gives the hypertips freedom to push outward farther at the polar regions. For our purposes, we will exaggerate the image into a football-like

Part 4: Creation of Atoms

object. The image becomes a form similar to a crystalline football that is spinning on its side around its major axis. The rotational force is half as strong as the repelling force between the two layers as there is twice the hypertip activity.

One side of the "football" gives out first. The reason this occurs is that the net force external to the nucleon is locally put at a state of imbalance by the evaporation process. The result is that one side of the football-like surface extends further out from the nucleon than the other side. The stress tears the internal layer of the two because the structure has greater lateral tension than the outer electrosod layer. This tear is symmetrical in respect to that corner. As we shall observe, it engulfs about 1/40th of the mass of this layer. This piece of mass is the neutrino as it has a neutral charge. Even though it is internal to the structure, it is the first to break away.

As the neutrino pushes away from the nucleon it tears the electrosod layer. The electrosod layer tears at the corner and along the edge of the cubical form. The reason for this kind of separation within the electrosod is that the corner is the weakest point being that the internal facial connections are weaker than that of the neutrino. The same applies to the three edges forming the corner. After the neutrino escapes, there exists an imbalance of push between the nucleon. This formulates an electron in the opposite direction. Both the internal layer and the outer layer of the electron tear along these edges as the electron escape from the nucleon leaving behind a protosod xyzenscape that formulates the proton. The electron shoots off with an extreme force; it even pushes the proton away slightly. These tears along the edges play an important role in the experience of an electron as matter develops. Along with the neutron, the proton, and neutrino, these particles form all the natural submatter.

Accounting Mass

To understand the following information, we need to know the meaning of mass other than that it is matter. Some confuse mass with weight, there is a difference. Consider an object that weighs 90 pounds on earth at sea level. If that same object were to exist upon the surface of Mars, its weight would be approximately 30 pounds as the gravity of Mars is about 1/3 that of the earth. If gravity is a factor determining weight, there has to be another factor that is constant illustrating that the object has the same amount of matter. This quantity is known as mass.

Secondly, we must define the measurement system of subatomic particles. The "u" is used to symbolize the atomic mass unit. The atomic mass unit is defined as the average mass of a Carbon-12 nucleon, the nucleus of an atom contains six protons and six neutrons. The atomic mass of Carbon-12 is 12 u. As we observe, the actual weight of a proton and neutron can vary between isotopes or unique configurations of the nucleus forming an atom. As we shall see later: in some instances, the protons appear heavier than the neutrons. The mass of the atomic mass unit in kilograms is about 1.66E-27

Chapter 11: Neutron Breakup

(1.66 divided by ten 27 times). The masses given in our analysis is a study of a single "new" neutron as it breakup into atom of Hydrogen-1.

The first item to explore is the difference between the masses of a neutron and proton. The mass of the neutron is about 1.008664904(14) u. The "(14)" is tentative continuation of the decimal. The mass of the resulting proton is 1.007276470(12) u. The mass of an electron is 0.000548579903(13) u. The electron mass is symbolized within our equations by the letter E. From this point on, the tentative extensions are dropped; this was shown to give the most complete answer available at this time. Our first calculated mass is the subtraction of the mass of a proton from that of a neutron symbolized by the letters P and N respectively. The subtraction then is symbolized by N-P representing the value of 0.001388434 u.

Our next objective is to illustrate the mathematical premise for a five-layer count in the transformation from a neutron to a proton. Our approach is that the ratio will be somewhat elementary, meaning that it will not have fractional layers, nor be of a large set of numbers. For an example, we would not have 5.3 layers or have 5,000 layers when 5 layers are sufficient. The reason for this simplicity is that we do not have electrons that are of various sizes in mass. If we take the mass difference of a proton from a neutron and divide it into the mass of the electron, we get the following: E/ (N-P) = 0.00548579903 u / 0.001388434 u = 0.395092707. We also know that there is an unknown amount of neutrino that also escapes from the nucleon. This brings us closer to the value of 0.4 or 2/5. By this, it is easy to see that there are five layers expelled, three are evaporated in photonic-like "evaporation" and the other two belong primarily to the electron.

Now we need to examine the data for the size of the neutrino. For this objective we need to find the value of a single layer and multiply it by two 2*(N-P)/5 = 2 * 0.001388434 / 5 = 0.0005553736 u. We take the mass of an electron and subtract it from this value 2/5*(N-P)-E = 0.0005553736 u - 0.00548579903 u = 0.0000067937 u. For our ratio, we divide a single layer by our final answer. This is attained by the following equation: (N-P)/ (0.0000067937*5) = about 40.87. This gives the maximum mass of a neutrino is about 1/41 of a layer. This answer is not entirely accurate as we are assuming that each layer has the same mass. Actually, each layer differs in mass depending on its depth within the nucleon by crystal count. Even so, the difference between two layers of near exact equal thickness near the surface will not generate much variation in mass.

The Neutrino Experience

For simplicity sake, we will view the ratio between the mass present in a single layer of the nucleon and the mass of the forming neutrino as 40 to 1. Secondly, we will give a value of xyzenthium crystals existing at the layer as 2,400, which is extremely small to

Part 4: Creation of Atoms

the actual count. At 2,400 crystals upon the surface, we know that 1/6 of them forms one face of the cubical form giving 400 crystals per cubical side. We also know that the crystals are arranged in a square format giving sides of 20 crystals per edge. Next, we look at the neutrino. The neutrino breaks away at one corner; this geometrically affects three cubical faces, not six. The amount of crystals distributed among the three cubical faces is 2,400 divided by 40 giving 60 crystals. After dividing them between the three sides, we have 20 crystals per affected face. The twenty crystals do not form a rectangle of 4x5 crystals as they attempt to disperse the pressure evenly at the edge of the tear. The image is a rough triangular edge as the pressure is exerted at a diagonal between two opposite corners pushing the neutrino and ultimately the entire electron on the opposite corner.

The neutrino tears from the nightsod layer associated with it. This frees the neutrino to propel away from the protosod surface without be held back by the lateral magnetic energy within the nightsod xyzenscape. After breaking free, the neutrino pushes harder into the corner of the electrosod xyzenscape, which is already distorted as stated earlier. This distortion weakens the structural integrity of the layer. It breaks through the corner starting a chain reaction of tearing within the joining edges to the corner within electrosod and with the underlying nightsod layer. After passing through the layer, the neutrino continues to be propelled by the daysod xyzenscape. At this point, the required observation is that the neutrino is a singular layered object having two diametrically different sides. The initial shape is that of an open based pyramid.

Eventually, the magnetic energy existing upon the lateral edges of the tear fracture of the neutrino attracts nearby exposed magnetic crystals edges within the structure closing up the base gap. The internal daysod continues to repel internally as the neutrino forms into a spherical-like shape. Externally, the external void allows the nightsod surface to evaporate again. The evaporation continues until the neutrino no longer exists.

Volumass

By the preceding information we can derive the average volume (via mass) of a xyzenthium crystal and an approximate number of layers involved in forming a neutron. Since this information is in units of mass, we need to devise a method to translate mass into units of volume. This can be accomplished, as the density of the neutron is nearly identical to that of a proton. Therefore, we will create a volume measurement that is equal in scale defined by the scale of magnitude assigned to mass measurement. This can be algebraically reasoned by using the formula: density equals volume divided by mass.

Consider dividing the density of a proton by the density of a neutron. Since both are virtually equal, the answer is one. If the two densities are equal, then the proton's volume divided by the proton's mass is equal to neutron's volume divided by its mass.

Chapter 11: Neutron Breakup

This does not mean that the volume and mass of the neutron and proton are equal, but their ratios are equal. This is equivalent to saying 5/8 is equal to 10/16 as the latter will reduce to the first. Or the first could be multiplied by 2/2 and get 10/16. Their ratio is a scale between mass and volume. Even though the scale at this point is undefined, we can create a variable that will be usable to this scale at a later time. Note: the expression of 5/8=5/8 is like saying 1=1. Measuring 100 units of 5/8ths of an inch we will have 62.5 inches. Labeling 5/8ths of an inch as 1 pinch, then 100 pinches equals 62.5 inches. In the same way, we are labeling the ratio between volume and mass of a nucleon as being a single unit of measure. Even though this ratio is undefined at this point, we can establish this measure of volume to be equal to the scale of mass with a conversion factor that will convert it to the metric or English measuring system of volume. The establishing factor is the actual ratio between mass and volume by scale. To maintain a grasp of the ratio, the unit is called a cubic **volumass** (vol-u'mas) a combination of the words volume and mass (symbol: **vm**^{3}). Therefore: it can be said that a neutron has a volume of 1.008664904 cubic volumass (vm^3) and a mass of 1.08664904 u. This allows us to manipulate the mass data as volume data. The measurement is cubical in nature as it is volume. Since the measurement has a cubical nature, it also provides us with a linear component.

The next step is to use the equation of a sphere and solve for the radius. The result is that the radius is equal to the cube root of the quotient of 3 times the volume divided by 4 times pi (approximately 3.14159265). By plugging in our volume for the neutron, we get 0.622137099 vm for the radius of the neutron. Doing the same for the proton, we get the radius of 0.621851510 vm. By subtracting the radius of a proton from the radius of the neutron, we get the radius segment that is lost in the transformation of 0.000285589 vm.

Take the depth lost in radius by the neutron (0.000285589 vm) and divide it by 5, this gives a depth of one layer of xyzenthium crystals. The depth is 0.000057118 vm. By dividing this depth into the radius of the neutron, we will get the number of layers possible in the neutron. The raw answer becomes 10,892.18 layers.

𝕹eutron 𝕻rime

Obviously, there is no fractional layer. The primary problem is not with the data. The data is correct. The problem exists within the nature of the neutrons themselves. We truly do not have the prime neutron fresh from creation. We have a decayed neutrons preserved from becoming a proton by attachment to at least one proton. The problem is that all neutrons were evaporating before the some of them transformed from being a neutron into a proton. This is not some idle claim. If we examine data from other atoms, as we shall later, we will find that the mass of a neutron and the mass of a proton will vary in size. Even more peculiar, the average mass of the nucleons fluctuate between isotopes. However, we must return to our situation.

Part 4: Creation of Atoms

We must remember that Elohim is the creator of the neutron. HE implements a particular pattern into HIS Creation. For a moment, let us look at the problem from another angle. From our information so far, we have a six-layer sequence going through the entire neutron. Let us temporarily convert this number to 10,890. When we divide by six, we get 1,815 sets of layers. Next, let us remove the 15 giving 1,800 layers. Now, observe the results as we work back to the single crystal. First, we multiply 1,800 by six and get 10,800 crystals from the surface to the center. Now, we are going to calculate the edge of the neutron cube, which is twice that of going down to the core. This gives us the number of 21,600 crystals per edge. The number 21,600 equals 360 times 60. This brings us back to the pattern of a circle having 360 degrees and sixty minutes per a degree. Then if this pattern were to continue, the width of the xentrix is 1/60 the width of a xyzenthium crystal for the "60 seconds." This would make the xentrix 1/216,000 of the size of a xyzenthium crystal in its original form. Another interesting aspect is that in numbering system of base 60, the number of crystals measuring the cubical edge would be 60; as opposed to the hexadecimal, decimal, octal or binary numbering systems. We also see the 360-day year before the destruction caused by the "dividing the earth." Even though any of these by themselves may seem extraneous, but there seems to be an accent on the base 60 numbering system in creation. Our measurement of time (seconds to minutes to hours) comes from this ancient science, which formulates our present day science.

Estimating the mass of neutron prime (N') is relatively easy. We use the volumass measurements derived earlier. We use the volumass measurement of the proton's radius (.621851510 vm) and divide it by the number of layers left to the nucleon (10,795). The result of this division is 5.760551274E-5. Then multiply this value by 5 giving 0.0002880275637 vm. Add this value to the radius of the proton in volumass gives 0.622139537. Then we reapply the volume formula for the sphere, $V=(4/3)\pi r^3$. Then we substitute back to the atomic mass units, one for each volumass cubed units. We get the value of 1.008676762 u for the mass of N'. Compare this to the mass of a neutron, 1.008664970 u. The mass difference is a little over 1/100,000 more. If this seems like too much of a variation, we should look at the variation of sizes in neutrons and in protons within the isotopes of elements. This we sill do very shortly.

After looking at the problem at the viewpoint of creating submatter, we can also observe that there is some kind of decay other than the **protonization** process occurring in the universe. The model neutron value expressed in mass today is less than that found at the beginning of creation. There is the evaporation of mass within the subatomic particles themselves. The decaying process of the particles is also decaying. The process in which mass is lost is the photonic phenomenon. The reason for this loss is that not all the photons hit the nucleons completing the cycle; some are temporarily lost in the void of space as it disperses to reach the far ends of the universe. The decaying of this process is observed in the slowing of the speed of light. The loss in mass is not evenly distributed throughout the material formulating subatomic

Chapter 11: Neutron Breakup

particles. The reason is that the surface layers supply energy to the electron to formulate the photon, leaving the core submatter virtually untouched, but not totally.

The loss of mass will continue until Elohim intervenes and recreate the universe. This intervention is about 1,000 years after HE returns to rule from Israel. This event occurs after the final war is waged upon the earth. Even so, the decaying process will not have even come close to reaching its limit. The limit is half of the mass missing from each xyzenthium crystal giving half the mass of a neutron per neutron. At this point, only the feelers between crystals will continue to exist. These would then in turn attempt to reform into the "padded" crystal format. Again, this phenomenon occurs by design and not chance.

The loss of matter does attempt to reach within the neutron structure. Recall the replacing phenomenon that occurs as the surface nightsod evaporates. This process draws "mass energy" from the core as well as the lateral surface. However, there is "magnetic friction" that slows down the reach from the core. The reason for this friction is the lateral magnetic conditions of the layers. The tavtips within the xergopath attempt to hold on to their pattern via distortion conservation. This means that the bend of the space/time continuum by the tavtips would rather continue attempting to join with the adjacent tavtip rather than leaving it for another unless the tavtip moves closer in. The result is that the surface layers loose more mass than the internal layers, and the internal layers do lose some mass. For this reason we are unable to gather accurate information concerning the original state of matter from pure observation of today's matter by physical observation.

Here are three specific examples of the inconsistent masses of neutrons and protons. If the neutrons and protons were constant in nature, we should be able to derive the same numerical values for them no matter which set of nuclei we compare. However, this is not the case. We will examine those existing in Hydrogen-3 and Helium-3, Lithium-7 and Beryllium-9, plus those found in Magnesium-26 and Aluminum-27. These examples are not "handpicked," we could choose any set of isotopes and discover the same phenomenon.

Let **n** equal the mass of a neutron.

Let **p** equal the mass of a proton.

Let **a** equal the average mass associated with the two nuclei.

Hydrogen-3 & Helium-3

$2n + p = 3.01604927$ $n = 1.00535641$
$n + 2p = 3.01602931$ $p = 1.00533645$
Average $a = 1.00534643$

Part 4: Creation of Atoms

Lithium-7- & Beryllium-9

5n + 4p = 9.0121822	n = 1.0274654
4n + 3p = 7.0160030	p = 0.9687138
Average	a = 0.9980896

Magnesium-26 & Aluminum-27

14n + 12p = 25.9825937	n = 0.9996611
14n + 13p = 26.9815386	p = 0.9989449
Average	a = 0.9993030

Notice that the nucleons of magnesium and aluminum are slightly larger in mass than that of lithium and beryllium. However, both sets are lighter than those of hydrogen and helium. All of the sets nucleon averages are lighter than the single neutron and proton. As we shall observe later, these are only averages. If we look at the average neutron size of the nucleons formulating Lithium-7 and Beryllium-9, we will find that the mass of the neutron much larger than that of our neutron prime. By this, we see that in smaller isotopes that the radiation of photons can actually increase mass upon the remaining neutrons while protons continue to lose mass in their formulation during the protonization process.

Xyzenthium Crystal Count

Using the prime neutron as the model, we can establish an exact count of xyzenthium crystals existing within a neutron. Recall that the original number of crystals per edge is one more than the resulting number existing at the edge. Since the resulting number is 21,600 crystals, the original count is 21,601. Next, we recall that the neutron was originally a cube. By this, we cube 21,601 giving 10,079,095,744,801 crystals. Then, we have to subtract the ones removed to form the sphere. This number is actually 21,601 again, despite the diagonal direction of the removal. The result is 10,079,095,723,200 crystals in all. We could say, for simplicity sake, slightly over ten trillion crystals. However; to differentiate the neutron's count against the proton 10.079 trillion or 1.0079E13, we need to view the digits beyond their approximate values.

Computing the number of crystals existing within a proton exists similarly to that of the neutron. Since it is missing five layers we can add all five layers and subtract it from the neutron count or a couple of easier ways. We could subtract five from 10,800 giving 10,795, and multiply by two giving 21,590; add one giving an answer of 21,591 and cube the answer and subtract the number we cubed, or subtract 10 from 21,601 giving 21,591; cube this number and subtract 21,591. The reason this works is that the cubical process defining the crystals in the neutron is manifested in all the layers within the neutron. The first solution is more complicated, but it shows why we subtract 10, which

Chapter 11: Neutron Breakup

equals the multiplication of 2x5) as opposed to any other number. This number comes out to be 10,065,104,106,480 crystals.

Computing the crystal count of an electron and its associated neutrino is a little more difficult. While, the raw answer is not exact because of significant figure limitations, we can get very close. We take mass of an electron and divide it by the difference in mass between a neutron and proton and multiply that ratio to the number of crystals existing in the top five layers of the neutron, which is the difference. The crystal count of the top five layers can easily be calculated. We subtract the number of crystals in a proton from the number calculated for the neutron. This gives 13,991,616,720 xyzenthium crystals multiplied by the ratio of 0.395106936 we get about 5,528,184,800 crystals for an electron leaving 66,907,000 for the neutrino. Because we have only five significant figures defined by the neutron, we can only say that there are approximately 5,528,200,000 crystals in an electron and about 66,900,000 crystals in the neutrino. By this, we see that the mass of an electron-neutrino is much larger than that we might think. It is much larger in comparison to a single xyzenthium crystal.

Our next objective is to calculate the mass and volume of the xyzenthium crystal itself on the face of a neutron. As noted earlier, the xyzenthium crystals are not entirely equal in mass or size. However, we can give an average size and volume of a crystal for a hydrogen atom for a general answer gathered from our given data. We take the given mass of a neutron and divide it by the crystal count of the neutron: 1.008664904 u / 10,079,095,723,200 crystals. The answer is 1.0007494042E-13 u. For the measurement in kilograms (kg) using the conversion factor of 1.6605402E-27 kilograms equals a single atomic mass unit (u), we multiply the two numbers and get 1.66178462E-40 kg. To find the mass by the energy expressed within the xyzenthium crystal, we multiply 1.0007474042E-13 by 931.43432 MeV/c^2 (million electron volts divided by the speed of light squared) and get 9.321305E-11 MeV/c^2. For the average volume, we divide the volume of the given proton of Hydrogen-1 (4.93E-46 meters cubed) by 10,065,104,106,480 giving 4.9E-59 meters cubed. We are actually limited to two significant figures by the measurement of the proton is given in two significant figures as we shall calculate later.

Below is the information derived from our calculations.

Xyzenthium Crystal Count Table

Neutron	10,079,095,723,200
Proton	10,065,104,106,480
Electron*	5,528,184,800
Neutrino*	66,907,000

* Calculation depends upon Neutrino Mass

Part 4: Creation of Atoms

Average Xyzenthium Crystal Table

Mass	1.6617846E-40 kg
	1.0007494042E-13 u
	9.32132E-11 MeV/c²
Volume	4.9E-59 meters³

Cubical Earth Illustration

When we look at the mass of subatomic particles, we focus in their extreme microscopic relationships to the space that we are accustomed to perceive with our senses. The extreme density of these particles seems meaningless to us no matter how these figures look mathematically unless we can relate it to familiar objects. The enormous amount of space between subatomic particles obscures this density. For this cause we are going to take out all the space between the subatomic particles of the earth and attain the size of mass that forms the weightiness of the earth. Note: we are not deleting the spaces between bethtips within any xyzenthium crystal, just between particles formed by the crystals.

From observations within nature, we know that it takes light about 3.3 multiplied by 10 raised to the -24 power in seconds to cross over a proton. This amount of time is extremely short. If one thousandth is a number with only two leading zero and one thousandth of a thousandth (a millionth) has five leading zeroes, imagine how small this number is with 23 leading zeroes with a 33 at the end. Since the exponent is negative, we are dividing 3.3 by ten 24 times. We take this number and multiply by the speed of light, which is in a vacuum 299,792,458 meters per second exactly or approximately 2.998E8 meters per second. This number is read as 2.998 times 10 raised to the eighth power meters per second (approximately 186,280.1 miles per second) multiplied by 3.3E-24 seconds gives the actual distance traveled. This gives the diameter of a proton of 9.8934E-16 meters. This is still a very small number dividing 9.8934 meters (32.444 feet) by 10 sixteen times, even though 32 feet is somewhat large.

Divide this number by 2 to give us the radius of the proton 4.9462E-16. However, we are limited to two significant figures, so we need to round to 4.9E-16 meters. We are assuming that the proton has a spherical appearance. This is relatively safe as we are looking at a Hydrogen-1 (lone proton) atom. The required observation is that the cubical form is lost in overall appearance only. This formation has remembrance within the layers themselves by the crystals and their alignment.

Our next objective is to find the volume of the proton. After establishing the radius of a proton, this is now possible by utilizing the mathematical formula for a sphere: $V=(4/3) \, r^3$. We find the proton's volume is 4.93E-46 meters cubed. Unfortunately,

Chapter 11: Neutron Breakup

the margin of error is somewhat sever. If we use all the significant numbers in the speed of light with the total fractional values in our sub-step calculations, we get approximately 5.068782E-46 cubic meters. Part of the problem is that we are cubing a number that has a root factor rounded to the nearest two digits. Another part is the rounding that we do of other significant figures to reflect the general absence of significant figures in other factors in the equation.

Next, we need the mass of the earth. The mass of the earth is 5.98E24 kilograms. If we were able to weigh the earth on a scale at sea level, it would weigh approximately 6.578E21 tons. We take this value and divide it by the mass of the proton: 1.672623E-27 kilograms. We get the number of protons it would take to equal the mass of the earth. This number is about 3.575E51 protons. This is the number of hydrogen protons required to equal the mass of the earth. Imagine 3.575 times a trillion times a trillion times a trillion times a trillion, we will still need to multiply this number by a thousand to equal 3.575E51.

We now are able to acquire the volume of subatomic material needed to attain equal volume in size of the earth. Multiplying the number of protons needed to equal the mass of the earth (3.575E51) by the volume of each proton (4.93 to 5.068782E-46 cubic meters) gives a volume of about 1.76 million to 1.81 million cubic meters. While this sounds like a large number, it is actually not very big.

If we were to make a cubical form of this material, it would have edges of only 120.7 to 121.9 meters. Multiply these answers to the ratio between feet and meters (3.281) we acquire an answer at the low end 396.3 feet and at the high end 400.01 feet. However, when we round the answer to the required two significant figures, we get 400 feet per edge of a cube both times.

In order to convert this figure into cubic inches, we cube the sides and the inches per foot: (400)(400)(400)(12)(12)(12)=1.10592E11 cubic inches. We divide this number into the number of tons measured at sea level 6.578E21; a cubic inch of submatter would weigh about 59.5 million tons (using a 2,000 pound ton measurement). By this, we observe that the resulting size of the neutron is extremely small in comparison to the electron shell levels surrounding the atom. Another observation is that despite the extreme density of submatter, the quarquid has a fluid-like quality inside the xyzenthium crystals. This, in turn, makes the crystal flexible to external submatter influences.

Pluto-Electron Illustration

Now, let us look at an electron's orbit around a hydrogen atom and compare it to the orbit of Pluto for another illustration of submatter density. We are making three assumptions here: One, the hydrogen atom is of the isotope, Hydrogen-1. Hydrogen-1 is a solitary proton as opposed to other heavier forms of hydrogen containing neutrons as well as the single proton. Our second assumption is that the electron moves at the

Part 4: Creation of Atoms

outer edge of the shell. Lastly, we are assuming that the electron moves in a manner as a planet. The s-level orbital electrons move the most-like a planetary orbit most of the time. In our case, we have only one electron giving us a better chance, at least in a spherical sense.

The shell size of a hydrogen atom is 0.32 angstroms. An angstrom is 0.0000000001 or 1.0E-10 meters. Divide by two for the radius. The hydrogen shell radius expressed in exponential notation is 1.6E-11 meters. Recall from our example concerning the volume of our planet that the radius of a proton is approximately 4.9E-16 meters. The ratio between the shell edge and the size of the proton is about 32,653 to 1 using the raw data in our calculations and answer. Within the limits of significant digits, we must say 33,000 to 1, as our base measurements have only two numerical figures.

The ratio between the average radius of the sun (432,000 miles or 6.95E8 meters) and the average distance Pluto is from the sun (3.671 billion miles or 5.91E12 meters) is almost 8,498 to 1. This time we can be accurate to three digits and say 8,500 to 1 with zero being the last significant figure.

In comparing the two ratios describing the orbit radius to the radius of the central object, we will divide the smaller number into the larger. As we can observe mathematically, the ratio of the electron's orbit is the larger number. We find that the electron exists about 3.8 times as far from the proton as Pluto does from the sun. This means that if the proton were the size of the sun, the electron would be about 14 billion miles away. Even if the electron exists only halfway from the shell edge, it still would be little under two times the distance from the sun as Pluto's average orbital path in ratio comparison. The movement of the electron also becomes unbelievable, when we consider its orbital frequency per second.

If the earth were far enough away from the sun to view the orbit of Pluto as a $10°$ arc around the sun, the electron orbit would form about a $34°$ arc. The half value view of the electron's orbit would be a $19°$ arc. These values are easily found by taking the tangent of $10°$ and multiplying it by the calculated ratio and taking the arctangent of the results. However, to see the orbit of Pluto as a $10°$ arc, the earth would have to be almost 21 billion miles away. This can be found by taking the cotangent of $10°$ and multiplying that by the orbit of Pluto, which has an average orbit measuring about 3.7 billion miles.

Another observation to ponder upon is that the distance existing between the electron and the proton surface is due to the microstrong force generated by juttoria within the hypertips. If the interaction were defined purely by the magnetic polarity existing between the electron and the proton, the electron would crash into the proton and with no hope of ever reforming.

Chapter 11 Quiz

1. Nightsod evaporate from the Neutron Surface because:
   A. Daysod pushes it off
   B. There is an External Void
   C. Juttorial Energy accelerates the Crystal
   D. All the Above

2. How many Layers are lost by the Neutron via Evaporation?
   A. 1
   B. 2
   C. 3
   D. 4

3. What Xyzenscape remains after the Evaporation?
   A. Electrosod
   B. Protosod
   C. Daysod
   D. Nightsod

4. What occurs after the Evaporation?
   A. Hyperpush removes two more Layers
   B. Electron Forms
   C. Proton Remains
   D. All the Above

5. Approximately how many Xyzenthium Crystals exist in a Neutron?
   A. 10 trillion
   B. 7 billion
   C. 5 thousand
   D. 22 septillion

6. (T/F) Four layers evaporate from the surface of a neutron during breakup.

7. (T/F) Immediately, a proton forms after evaporation.

8. (T/F) When an electron forms, no mass is lost from the nucleus.

9. (T/F) Neutrino is the byproduct of a forming neutron.

10. (T/F) Electron's rayring forms a miortex in the forming photon.

Chapter 12

Plasmaverse
Atomic Nuclei

Within the plasmaverse, we begin to observe two different types of phenomenon occurring within finite space. The first is the continuation of the development of subatomic particles. The second concerns itself with the development of galaxies and all its various mass formations. Each has its own set of dynamics that shape the outcome of this phaseverse. However, we are limiting the scope of this particular chapter to subatomic particle development. Again, Elohim intervenes to move creation to a functional design. In the Torah, this is the beginning of the Second Day of creation. The process in which the neutrons collect is called the firmament. If it were not for HIS involvement, the entire universe of isometric neutrons would turn into hydrogen gas within about fifteen minutes with a multitude of neutrino "evaporating." The firmament will be described within this book in chapter 14, as we examine the formation of the galaxies on downward to the formation of the earth, for which the firmament was created. Within the scope of this chapter, elements develop from the collecting neutrons from a formally isometric universe.

In a nucleus, a proton can exist by itself, but a neutron cannot. As noted earlier, a neutron will break up into a proton, an electron, and an electron-neutrino. Two protons cannot exist in a nucleus without any neutrons. This is because the protons repel each other as both have a positive electromagnetic charge. The number of protons within the nucleus determines the element of that particular atom. The number of neutrons can very within a given element determining the isotope of the element. The isotope number is determined by adding the number of protons and neutrons together. Each isotope (nucleus configuration) is given the name of the element with the number of nucleons (protons and neutrons) after a hyphen. Hydrogen-3 is a "heavy" isotope of the element of Hydrogen. Since Hydrogen has only one proton, the number of neutrons is two; one proton plus two neutrons are components of three nucleons.

The number of neutrons and protons are nearly equal in number. However, in general, the number of neutrons will outnumber the number of protons per nucleus. There are exceptions. For an example, hydrogen normally has only a proton. Then there is a form of helium that also is an exception to this rule. Another general rule, no two

Part 4: Creation of Atoms

protons touch nor do any neutrons touch. Again, there are exceptions concerning the neutrons.

Two neutrons can touch each other easier than two protons as the neutrons are neutral in electromagnetic charge. When they touch, one of them will transform into proton by the phenomenon previously presented. Adding a neutron to the nucleus will alter an isotope. This happens either by man's intervention or some other natural cause. While the nucleus maintains its element composure, it gains atomic weight (the added neutron) and becomes a heavier isotope of the element. It will cause the element to become radioactive when two neutrons come into contact as a result of this increase. After the "protonization process" is complete the element becomes the next element higher via the proton count.

Even though it is nearly impossible for two protons to touch, it does occur in nature. In the resulting universe, this phenomenon occurs primarily within the large radioactive nuclei. This occurs when the protonization process finds conflict within the nucleus structure of neutrons during the initial protonization process. The reason this occurs is that a transforming neutron is touching a proton. After the transformation, two protons become next to each other. These two protons then repel each other. At this point, there is an ejection of one proton or another with great force. Usually, the proton closest to the surface within the structure of nucleons is expelled. As seen in elements today, the fragment catapulted from the assembly of nucleons form the configuration of a Helium-4 nucleus. However, in the formation of the original elemental isotopes, the division can be much greater or even less.

There are few natural occurring isotopes of elements that have an equal amount of protons and neutrons: Hydrogen-2, Helium-4, Boron-10, Carbon-12, Nitrogen-14, Oxygen-16, Neon-20, Magnesium-24, Silicon-28, Sulfur-32, Argon-36, and Calcium-40. With the exception of hydrogen, boron and argon, these atomic nuclei comprise the majority of the natural isotopes occurring per element. Hydrogen-1 and Helium-3 are the only natural occurring isotopes that the protons dominate in number within the nucleus. Since 14 of the 301 known nuclei combinations have special numerical relationships between protons and neutrons, the majority of the elements have a greater number of neutrons than protons. It should also be noted: the ratio of neutron dominance over protons increases as the isotope configuration increases in the number of nucleons.

Subatomic Forces

A pattern is necessary for the nucleus to be stable (nonradioactive) in the various elements. This formulation is called a **nucleon flake**. The image is that of a flat snowflake. Unlike the snowflake, the nucleon flake spins by its own internal torque. Another attribute that occurs differently is the flakes will stack on each other as the

Chapter 12: Plasmaverse

number of composing nucleons increase. Stacking occurs as the nucleons respond to gravity and magnetic energy collectively, as well as individually. Recall that the actual format of a nucleon is a cube, giving structurally weaker areas located at the six faces of the original cube.

The hierarchy of influences in the formulation of a nucleon flake then appears as the following: magnetic energy, centrifugal force (via isobaric spin), and lastly gravitational force. Despite this hierarchy of influences upon the formulation, the initial force is gravity. The reason is that the neutron is the initial state of all submatter. Since neutrons are magnetically neutral, the only force of attraction available is gravity (spatial Jammeria). This force brings the neutrons into their initial nucleon clump.

After acquiring contact, the neutrons secure a unified rotational momentum which establishes a flattened disc-like pattern. The isobaric momentum is reached during the protonization process mentioned earlier. The protonization process introduces interactions that will alter the clump arrangement via electromagnetic charges.

During the protonization process: we have the acceleration of angular velocity and the magnetic crystallization of the nucleus. The interactions that occur between a proton and a neutron stabilizes the neutron (prevents it from becoming a proton). There is another aspect displayed in the protonization process. The process releases the neutron's "stress" in the fewest number of protons possible. However, a given flake within a nucleus may have more protons than neutrons if warranted by the presence of an external flake or resulting neutron conflicts.

Another observation exists by examining the atomic weights of the isotopes. This topic will be discussed in further detail after discussing the photons. However, the required observation is that the neutrons absorb energy from other neutrons in the clump during protonization. The result is that there is a fluctuation in nucleon sizes.

To examine all the isotopes that naturally occur in nature is enough to fill an entire book in itself. However, in examining the first 50 nucleon combinations, we will establish sufficient data concerning the nature of protonization process. We will also examine some of the unique nucleon crystalline patterns beyond the 50 nucleon combinations.

Hydrogen and Helium Formations

There are three natural occurring isotopes of Hydrogen. In Hydrogen-1, the lone proton spins by its own rotational momentum. This is the primary manifestation of a point-flake formation. Outside Hydrogen-1, it is very rare to find a point-flake to be a proton. The reason being that protonization formulates the fewest protons possible. **Hydrogen-1** is the most prevalent within our universe. **Hydrogen-2** is comprised of a proton and a neutron. A special name is given to this isotope by scientists. It is called deuterium. They spin around each other forming a very small nucleon flake. Even

Part 4: Creation of Atoms

though it is nonradioactive it has proven to formulate heavy water. **Hydrogen-3** has two neutrons and one proton as described earlier. Unlike the two other isotopes Hydrogen-3 is unstable. Another name for this isotope is tritium. This isotope has a half-life of 12.5 years. Within the Hydrogen-3 nucleus, gravity has a dominate role as the neutrons have a neutral charge. It forms a triangular shaped flake that spins around a point central to the three nucleons.

Consider for a moment that by a process, which will be discussed later, three neutrons come together. During the protonization process they begin to spin. Assume for a moment that they were able to form a pillar (possible, but not probable). After the central neutron transforms into a proton, the angular velocity accelerates by the proton. The change in motion will destroy the balance because none of the particles are exact spheres. Gravity takes over and draws the two neutrons together again. The general course of Hydrogen-3 formation is a circuit of three nucleons with an isobaric spin. The one that usually breaks into the proton form first is the central neutron during the grouping period before the circuit forms. It is conceivable that two would collect first and one later. The neutron in which the latter one contacted first would become a proton. If both were contacted simultaneously, we still get two protons with one neutron.

The protonization process continues, the two touching neutrons in Hydrogen-3 interacts. One of the neutrons will become a proton. At this point, the protons will repel each other. Sill being bound to the neutron, they seek the farthest distance from each other. They end up on the opposite side of the neutron in respect to each other's position. The image, now, is that of a pillar formed by three spherical like objects rotating around the central unit of the assembly. This newly formed element is **Helium-3**. From this, we observe the crystalline force of magnetic energy that generates a pillar-like formation rotating about its center.

Helium-4 shows that protons can exist at 90° angles to each other provided there is a neutron to hold it in place. The circuit spin manifested here is in its simplest stable form. This formation appears leaving radioactive elements as explained earlier as the alpha particle.

There is another formation seemingly possible. Imagine a clump of neutrons coming together and forming a single line of four neutrons. After protonization, they would form an alternating pattern of protons and neutrons. Note: the pattern requires that a proton to exist at one end of the pillar and the neutron at the opposite end; there is an isolated proton within the pillar also. Since the isobaric torque upon the proton is slightly different than that of a neutron, there is an imbalance of energy exerted upon the assembly of nucleons. This causes a rolling action to occur which would result in the previously described formulation. However, the pillar form is highly unlikely to occur as it requires the isobaric spin of each to be in unison to the direction that they collect. Otherwise, gravity would cause the pillar to collapse into a four-nucleon circuit.

Chapter 12: Plasmaverse

Lithium the Isolated Isotopes

Scientists have not yet found either Helium-5 or Lithium-5 isotope configuration in the naturally formed universe. Helium-5 would require three neutrons to exist with two protons. Lithium-5 requires the converse configuration. Imagine five neutrons coming together to form a single clump of neutrons (**Neutron Clump**) via gravity. They would form a clump that most resembled a sphere. The image appears as three neutrons forming a triangular formation (a neutron at each corner) with a neutron centered upon the gap formed by the three joined neutrons on both sides of the triangular face. These two external neutrons apply pressure to the triangular formation, as gravity demands them to come closer together. This conflict is further enhanced as the assembly begins to spin generating centrifugal force.

The problem is structural in nature. Despite the neat symmetrical formation of the neutron clump, the entire formation becomes destroyed as the assembly moves into the protonization process. The two "polar" neutrons within the formation transform first leaving the central three as neutrons, but the neutrons are still touching each other. The interaction between the three neutrons requires that one of them to transform into a proton. At this point, there are three protons touching resulting in a fragmentation process of the nucleus. Another possible scenario is the five-circuit arrangement of the neutron clump generated by centrifugal force. Generally, the pattern still starts as our original geometric configuration as a clump formulating via gravity. However, this is not necessarily the case. The arrangement of the neutrons by their spatial potions may be two rows of neutrons with three in one row and two in the other. There are even others. Even so, the result is the same. Centrifugal force takes control and flattens the assembly into the five-nucleon circuit pattern. Naturally, two of the neutrons transform into protons. The resulting arrangement of the nucleons now has two protons separated by one neutron on one side and by two on the other side of the ring. The two adjacent neutrons are still interacting. One turns into a proton. The newly formed proton is ejected out of the assembly by the adjacent proton while the electron shoots off from its surface. The resulting formulation is one Hydrogen-1 and one Helium-4 isotope. The only possible pattern in which the five nucleon isotope could exist in "peace" is that it formed a pillar of alternating neutrons and protons. In order for this to occur, they would have to initially formulate as a pillar of neutrons spinning around on the minor axis before the protonization process begins. Within our part of the universe, this simply is not the case as indicated by the gap in the table of naturally occurring isotopes.

Lithium-6 is composed of three neutrons and three protons. The process is much like the Helium-4 process. When gravity brings six neutrons together, it forms two sets of triangles. These two sets are placed at a 60° angle (or flipped 180°, the result appears the same). It gives the appearance of being interlocked. One set becomes protons and

Part 4: Creation of Atoms

force the neutrons into its geometric plane as spacers between protons as they push outward from each other. The centrifugal force forces the heavier neutrons outward generating a larger gap in the central region. It depends upon the protonization process, meaning the sequence in which the protons form, whether or not these neutrons remain together as their hold is weaken by the enlarged gap.

In the formation of **Lithium-7**, we find the neutrons clump together into a somewhat peculiar pattern. The initial pattern can be visualized by using the three-dimensional axis. We will center the first neutron upon the origin. The remaining six neutrons we will center upon the six axial directions tangent to the central neutron. From this formulation, we can easily see that it has a potential to be a natural development. It would formulate more abundantly. Naturally, the first neutron to transform into a proton is the central-most neutron. As this occurs, between the gravitational pull between neutrons and the acceleration of the isobaric spin, the other neutrons "roll" into each other and interact. When the second neutron forms into a proton, it is repelled by the first neutron via their positive charge. At this point, the two protons are on opposite sides of the neutron clump. While the protons repel, the dynamics of the geometric structure alters. A neutron is pulled between the two protons forming a line; the other neutrons become rearranged into a pattern more suitable for the centrifugal force generated by the isobaric spin and then to gravity. The result is that of a single flake in a 2-3-2 formation causing the central neutron to turn into a proton.

The resulting flake can be viewed as a three-by-three square of nucleons that has two corners diagonal form each other missing. The protons are then seen as occupying the remaining two corners plus the central location. The remaining four outer locations are occupied by neutrons.

The next gap in the table of naturally occurring isotopes ends the lithium configurations. There is neither Lithium-8 nor Beryllium-8. While we can imagine two Helium-4 configurations coming together to form one, it does not happen in the natural phenomenon as the neutron clump forms. The primary reason is the torque generated in the isobaric spin unifies. Centrifugal force eventually compels the nucleons into a single flake. From the mathematical perspective of geometry, we observe six spheres surrounding one touching each other in the ring-like formulation, and this leaves the eighth one external to the formulation. In this scenario: As the seven-nucleon flake forms, one is left outside in an offset location and generates an imbalance. The result is that the assembly of seven forms a Lithium-7 nucleus and the solo ejected neutron becomes a proton forming Hydrogen-1.

The second scenario: The offset of forces moves one neutron off via centrifugal force or brings the eighth one into the ring pushing the central one out for balance. The flake of nucleons visualized in the three-by-three square of nine positions leaves the central position vacant. By leaving any other position vacant, the crystalline assembly of nucleons becomes structurally imbalance for rotation. For this reason the four nucleons

Chapter 12: Plasmaverse

position between the corner nucleons move away from the center. This causes the circuit to widen. The diameter of the central gap becomes much greater than the diameter of a single neutron. The entire structure is weaken by the increased distance and breaks into two pieces forming usually two Helium-4 nuclei. This usually occurs before the proton phenomenon manifests. The primary reason is that the gravitational energy between two nucleons is weaker than the isobaric spin of the assembly.

Of the two scenarios, the first is the most likely to occur. That is not to say that there are not occurrences of the second scenario. The Beryllium-8 nuclei could also occur under certain conditions. These holes, in the natural formation of isotopes that isolates lithium, point to the fact that the nucleons all started as neutrons gathering into clumps that the isobaric spin reforms before the protonization occurs. Moreover, the newly formed protons within these clumps often form irresolvable conflicts as a single clump.

Beryllium and Boron Isotopes

Beryllium-9 is then the next natural occurring isotope. There are a couple items of interest here: This particular configuration composes 100 percent of the element. It is the last common single flake isotope. The Beryllium-9 flake becomes a three-by-three square of nucleons. The center and corner nucleons are neutrons. By this, we have a completion of the simple square flake. Notice: This pattern gives the fewest possible protons (four) and prevents the other five neutrons from touching.

Boron-10 has five neutrons and five protons. Unlike the eight-nucleon pattern, when the seven neutrons join together to form a hexagonal flake with a center, there remains three nucleons to ride upon the primary flake. The central nucleon remains a neutron as the end ones become protons. The reason that the central one remains a neutron is that it is sitting upon a proton and the upper edge neutrons are between this neutron and their respective neutrons underneath. Thereby, the edge neutrons become protons.

The two opposite protons within the top flake are drawn into their respective corners by the neutron "valleys" generated by the seven-nucleon flake scenario. This leaves a lone neutron for the top nucleon flake. This forms the point flake. The point flake consisting of a single neutron is held in position by a few factors: the flatness between itself and the central proton, the magnetic interaction between the two nucleons, being concentric to the spin generated by the nucleus. If the neutron moves away from the center then the pattern partially spins apart creating other elements.

Another phenomenon manifests; the nucleons will flatten. This is not so unusual when we consider their original form to be cubical in nature. However, this does not account totally for this behavior. The interaction of these forces is extreme in nature. Recall that the magnitude would cause steel spheres to behave as flimsy as a rubber foam ball in comparison by the forces expressed at the subatomic level. Although a cubical alignment is favored over other positions provided, the structure allows other types of

183

Part 4: Creation of Atoms

movement. The hexagonal prism form is favored by the primary geometrical sphere-like form in the isobaric spin. The formulation becomes like two separate entities held together by the magnetic interactions with the central proton. Note that the central axis of each clump is protons. This is because the central neutrons contact more neutrons.

Boron-11 can have the same basic configuration as Boron-10. The difference is within the format with four neutrons within the top circuit instead of three. Let us return to the original clump of eleven neutrons. During its initial formation before rotational momentum builds, gravity attempted to formulate them into a spherical format. There is another acceleration involved, which is beyond the topic in general, but does have an effect in the formulation. There is a linear acceleration occurring as these atoms formulate their nuclei. The effect is the imbalance in flake sizes when there are partial flakes involved, which is anything under the central flake count. The heavier flake is always in the direction of the acceleration. This acceleration will be covered in the second chapter on the plasmaverse, discussing the forces formulating galaxies.

Let us return to the development of Boron-11. By the above influence, the clump initial configuration is 7-3-1, in the hexagonal form. A set of three central linear neutrons in the bottom layer transforms first. The second is the top lone neutron. These four protons move to reach the farthest location away from each other. The image is four corners on a square. Since two of these were initially central in the nucleon flakes, the center becomes vacant. This reformation is unstable as another proton forms as the neutrons realign. This occurs because the neutrons are pulled together by their gravitational forces, causing them to come into contact with each other. The fifth proton forms within the second major flake leaving a neutron on opposite sides giving a total of three. The bottom flake reforms into the three by three square formation with the central nucleon missing. The neutrons occupy the corners of this squarish flake as they hold two adjacent protons into the flake at right angles.

The result is a formulation that is quite unique. The protons from the bottom view control the central vacant region exposed to the external universe. This gives the structure the ability to attract and keep neutrons in atomic reactors. These lone neutrons are literally sucked into the microweak vacuum almost magnetic in nature. The neutron ends up in the proton "pocket."

Carbon and Nitrogen Isotopes

Carbon-12 is a configuration of six neutrons and six protons. Now, we are firmly into the two-nucleon flake realm. The nucleus configurations consist of two flakes greater than point flakes. The image of Carbon-12 is that of two Lithium-6 nuclei stacked upon each other. The six neutrons on the bottom half of the clump formulates protons in the opposite pattern as the six above. In this, we have two six-nucleon-circuit flakes spinning in unison.

Chapter 12: Plasmaverse

An important observation to acknowledge is the nature of the circuit. Recall that the neutrons are generally heavier than the protons especially true in nuclei with relatively few nucleons. This divides the circuit into two separate rings of nucleon centers. These rings are imaginary circles defined by the actual positions of the nucleon's geometrical center. The outermost ring is comprised of proton centers as centrifugal force and the repelling force between protons pushes them further out as both protons and neutrons are traveling at the same angular velocity. This is true for both flakes of Carbon-12. This isotope forms nearly 99 percent of the carbon element.

Carbon-13 can be visualized as a neutron "plugging" one of the flakes "hole". However, more generally the neutron appears as being in a central position caught between the two flakes via gravity. Our immediate observation is that this configuration is very delicate. It would not take much variation to upset the proton balance causing the entire assembly to breakup into smaller element isotopes. It is of no surprise that this isotope is rare.

In **Carbon-14**, neutrons plug both holes of the two six-nucleon circuit flakes. The two central neutrons touch causing this isotope to be radioactive. Normally, when the protonization process is completed the isotope becomes Nitrogen-14. Again, it is the manner in which the neutrons are collected that determines the outcome. It is also possible that a Hydrogen-1 atom to be formed and the carbon element returns to the stable isotope of Carbon-13 or split into boron-10 and helium-4, both of which are stable.

Nitrogen-14 could also emerge without fist becoming a carbon isotope. The formation is the same, only that it was established during the initial protonization process. Unlike the carbon isotopes the two seven-plugged-circuit flakes are not identical. One is inverted; the proton ring is external as the central proton pushes away the protons and pulls in the neutrons. This makes the two flakes of the Nitrogen-14 nucleus workable in both nucleon flakes.

The **Nitrogen-15** nucleus is much less common than Nitrogen-14 isotope, at least within our sector of the universe. There are two flakes, one containing eight and the lighter one containing seven. The inverted flake gains another neutron. The central neutron moves externally holding the flake in centrifugal equilibrium. The resulting form is an eight-nucleon flake with neutron corners. This flake aligns its neutron corners with the top flake's proton dominated central axis.

Oxygen and Fluorine Isotopes

Oxygen-16 is formed by "impossible" eight-nucleon flakes. They are called impossible, as they cannot exist by themselves within the nature of local neutron clumping phenomenon. These flakes are aligned inversely above each other. As they rotate, there is a hole forming larger than the diameter of a neutron. The required observation is that

Part 4: Creation of Atoms

the neutron clump forms from the gravitational format leaving the missing component of the simple flake on an external corner; it is only after protonization that the hole becomes evident.

There are other alignments possible, even a single flake. The most common one maintains the vertical alignment. If the formation of the flakes were squares, the squares will lay directly in line with each other. The resulting pattern is that these nucleons alternate in each dimensional direction. This gives each flake four protons and four neutrons. This isotope forms over 99 percent of the oxygen found in nature.

Oxygen-17 formulation becomes a variation of the first kind of Oxygen-16 flake configuration. However, it is first of a series of double-pyramid neutron formations. This pattern is the most diminutive of the formulation. The central neutron flake has nine neutrons and the two external flake starts with four neutrons each. The initial normal protonization turns only seven neutrons into protons three in the central and two at each exterior flake. These repel each other and gravity pulls other neutrons to the center, one of which forms a proton. This proton is touching another and is immediately repelled. The result is that another neutron is drawn into the center, and we get our final formulation. For more about their nature, examine Oxygen-18. One flake has neutron corners giving place for a neutron core. This is a rare configuration indicating that these normally break into two or more clumps during protonization. This is the rarest form of oxygen. When we examine the clump of neutrons of Oxygen-18 we will uncover the reason.

The proton "boiling" process is called the **rejoinion resolve**. The word derives its formulation from the words **rejoin** and **ion**. However, the long "i" sound in ion is shortened as in the word, un**ion**. Ions are charged particles, whether subatomic or atomic and whether positive and negative. Rejoinion Resolve refers to a rearrangement from a gravity controlled environment between neutrons to the centrifugal force environment generated by the isobaric spin and lastly magnetic control taken by proton formulation. This transfer of preeminence of energy creates instability within the nucleus. The charged particles manipulate the nucleons moving their masses as individual objects within the nucleus. The primary objective of a proton is to get as far away from other protons as possible, and it matters little if a neutron is in its path. In the smaller nuclei, the results are identical per isotope. As the number of nucleons increase per isotope, the combinations increase. Even before the number of nucleons gets very large, a neutron clump can produce more than one element via number of protons produced. The result of the rejoinion resolve is that the clump of neutrons is transformed into a positive ion that will become the nucleus of an atom.

Oxygen-18 may seem unequivocally paradoxical to the phenomenon in that we would think that Fluorine-18 would be the natural solution. However, there is another factor dominating the formation other than just making sure that no two protons touch. This answer will show the reason that Florine-19 is the natural and only choice for fluorine.

Chapter 12: Plasmaverse

First, we need to start with the geometric structure generated by gravity in the neutron gathering. Starting with the largest layer, it is the standard three-by-three flake. On top of this layer are four more neutrons resting in the "valleys" formed between the nine in the larger flake. Then on top of this flake is another neutron seated in the single "pit" formed by the four neutrons underneath. This leaves four neutrons. These fit underneath the nine-neutron flake as the four-neutron flake exists upon the nine. If this pattern were to be completed, there would be one more neutron on bottom making a total of 19.

Our next task is to illustrate the protonization process forming Oxygen-18. If we were to cut into the figure with a plane that is vertical to the structure and at a 45-degree angle to the three-by-three neutron flake, the number of neutrons that would be sliced is eight. These would be the eight that become protons, which makes an isotope of oxygen. However, there is a slight variation to this by the nature of protonization itself. The two four-neutron flakes form protons at right angled to the pattern existing within the nine-neutron flake. This is because the protons do not form all at once.

The first proton to form exists, as expected, within the nine-neutron flake having contact with 12 neutrons in all. There are four in the flake of nine, and then both adjacent four-neutron flakes are also touching this central neutron of the nine-flake. The next set to transform into protons exists within the two four-flakes. Before the central proton formed, these were touching five neutrons, while the corners were only touching four. After the central proton formed, they are left touching four. The energy received by the central proton is weak as it is divided between 12 neutrons. Returning to the plane cutting into the assembly of neutrons, two neutrons from each of the four-neutron flakes intersect this plane. After these turn into protons, three corners transform. The two corners within the nine-flake opposite of the corners formed in the four-flakes, and finally the single neutron in the point flake. The point flake transforms because the protons underneath are touching five neutrons, while the two protons under the nine-flake are touching only four. By this, the top flake receives the least aid in sustaining its surface and transforms into a proton.

The previous process occurs with relative ease in comparison to the rejoinion resolve of the protons after they are formed. The protons within the two four-flakes are the first to move. The central proton in the nine-flake is repelling them. As they move away, they are drawn to the neutrons located centrally to them via gravity and surface interactions. The result is that the top proton then is drawn down between the two neutrons. As this occurs, the central proton is pushed by the proton and also is draw in between the two neutrons underneath it. This occurs also due to the centrifugal force by the isobaric spin upon the neutrons within the four-flakes pulling them away from the center. The result is that the original protons formed within the four-flakes are repelled by the entry of the proton, and rotate around their central neutron to get away from the central protons entering into their respective flakes. The two sets rotate in the

Part 4: Creation of Atoms

opposite directions in effort to get away and move 45-degrees in their respective rotation forming right angles to each other as a result. They are drawn into contact with another neutron stabilizing their condition. The result is a somewhat exotic form. However, they are magnetically stable. After this point, even more manipulations are possible leading to different solutions. Perhaps, the most stable are two nine-flake nucleons that are neutron dominated. This may seem easy enough, but they are also offset so that no two protons are touching each other generating spinning issues.

Fluorine-19 is the only isotope of fluorine that naturally occurs. As stated earlier, the initial clump of neutrons completes the dual pyramid form of 19 neutrons. Again, we can cut a plane through the center vertically at a 45-degree angle and get the exact number of protons. This isotope consists of 9 protons and 10 neutrons. It is a composite of three flakes. It has two simple square flakes one of an inverted polarity and a point neutron flake. The two square flakes are joined squarely. The neutron point flake covers the central proton of the inverted flake in the final form.

Neon, Sodium, Magnesium, and Aluminum Nuclei

Neon-20 needs 10 protons to define the neon element; therefore, we find that the isotope of Neon-20 has an equal number of protons and neutrons. This isotope forms about 90 percent of the isotopes composing this element. The primary configuration is a composition of three flakes: a nine-nucleon bottom flake, an eight-nucleon central flake and a three-nucleon top flake. Again, we observe an imbalance along the vertical axis, with the largest on the bottom and the smallest upon top illustrating an external acceleration upon the assembly. Another possible solution is two ten-nucleon flakes with inverted polarity. However this configuration requires heavy rearranging of nucleons during the protonization process.

Neon-21 is the rarest of the three forms of neon, despite the ease of assigning three seven-nucleon flakes to its formulation. Notice that the central nucleon is a neutron and not a proton. This requires much reconfiguration in the formulation process as a proton naturally fills this location. Secondly, the seven- nucleon flakes have the two week sides allowing unwanted neutron movement. Since it is the rarest form, the reconfiguration process often results in a breakup of the clump into smaller nuclei.

Neon-22 has three flakes. The central flake has eight nucleons while the top and bottom flakes have seven each. The interesting factor involved with this combination is that there is no central nucleon. The reason being all the central protons repel each other moving the central proton away from its original position. This isotope forms nearly 10 percent of the element. From looking at the pattern, it appears as if another isotope of neon could form by inserting a neutron within the central location. The problem is that the formulation of this isotope requires that the neutron to be drawn into the middle of the entire assembly. This is unlikely to happen. Despite the lack of

Chapter 12: Plasmaverse

probability, somewhere in the universe this kind of phenomenon could happen through some prearrangement.

Sodium-23 requires 11 protons and 12 neutrons. This is the only natural isotope of sodium. The flake pattern is 7-8-8. This means that there are seven nucleons upon the top flake and the remaining two have eight nucleons each. The two eight-nucleon flakes have inverted polarity to each other and the top flake has protons for the central three nucleons. The next isotope is very unlikely to ever develop as it requires a central neutron.

Magnesium-24 needs 12 protons for the element and 12 protons for this particular isotope. These nuclei can be seen as an isotope consisting of three eight-nucleon-circuit flakes. This only requires that the central flake be in the inverse pattern of the external flake. Note: there are two ways to accomplish this task since the number of protons in each flake is an even number. The central theme is that each flake forms a nine-flake pattern with the central nucleon missing.

Magnesium-25 is an extension of the same flake configuration. This configuration requires that one of the flakes acquires a central neutron. Normally, we view the neutron being locked into either the top or the bottom flake. However, the neutron could be drawn into the central flake by gravity aided by the magnetic pull of the central protons.

Magnesium-26 appears as both end flakes being the simple square flakes with the central flake being hollow inverse flake. This is accomplished as the external flakes unifying before the internal protonization process was complete (a boiling effect occurs as protons develop).

Aluminum-27 is another lone isotope that occurs in nature. This is the completion process described in the magnesium isotope sequence. The central proton fills the central hole of the central flake making it a complete inverted nine-nucleon flake. Naturally, this flake has a normal 3x3 square flake on each side. This gives the 13 protons necessary for aluminum to exist forming a perfect 3x3x3 cube.

Silicon, Phosphorous, and Sulfur Isotopes

Silicon-28 has 14 protons and 14 neutrons. Perhaps, the easiest solution is to add another point flake to one of the outer flake surfaces. While it would work, there is another answer that is more centrifugal in approach. In this the neutron is added to the central flake. It tries in most cases to emulate a five-count flake. The flake does this by separating, leaving a hole for the five-count nucleon process. This appears in the central flake of ten nucleons. The central nucleon is missing with two neutrons external to the flake. The resulting equation is 9-1+2=10. The pattern of the flake is actually three nucleons on each side with four in the middle, but the middle separates in halves

Part 4: Creation of Atoms

and spread out to act as a five count. The result in flake order is 9-10-9. Note: other solutions exist in considerably less abundance.

Silicon-29 adds one neutron to the Silicon-28 formulation. This neutron fills the central most location within the central flake. While the central neutron is not the natural location, within this structure it is the most logical resulting location. Even if the central neutron started at the edge flake, it would be drawn into the center by gravity and the proton interaction. The result is a 9-11-9 flake sequence.

Silicon-30 is the rarest of the three silicon forms. It also follows the pattern set by Silicon-28. Instead of filling the central location, these neutrons find themselves external to the formulation. Even so, they are drawn to the central flake and are drawn to the protons. The energy that holds them in place is centrifugal force. The flake pattern is 9-12-9. However, a point-flake pattern is possible this would give a flake pattern of 1-9-11-9.

Phosphorous-31 formulates 100 percent of the natural occurring phosphorous. This nucleus has 15 protons and 16 neutrons. The general configuration is 9-13-9 favored by the original neutron clump. There are other combinations that work using three flakes such as the 8-15-8 and the 10-11-10 sequence. As inviting as the 10-11-10 sequence appears on the surface, it requires seven protons with only four neutrons in the central flake, this is a little extreme. The 8-15-8 flakes have central nucleons missing in the external flakes, which is not a cause for alarm, but does give the central flake about twice the nucleons as the external flakes. Despite all these objectionable qualities of these latter flake sequences, they may all exist given the proper conditions. Often, we will find that many have several solutions. Even so, there are those few that have nearly one. The multitude of solutions is why we limited ourselves to the first 50 isotopes.

Sulfur-32 is the first of the naturally occurring isotopes of sulfur; it has 16 protons and 16 neutrons. It composes about 95 percent of the isotopes found in sulfur. There are many possible configurations. However, there are the more likely combinations. Two of these may dominate the isotope configuration. Our objective is to present the most common answer, which is not the only good answer. The only real unlikely formulation is four flakes of eight nucleons. While it seems reasonable abstractly, when we consider the isobaric spin of the assembly, we find that the dimensions are contradictory. The assembly will not be taller than it is wide. The 8-8-8-8 flake sequence is four high but only three wide. Within this scope, we find the first appearance of the 16-nucleon flake. It is interesting to note that this flake has two forms the even flake form 4x4 and the 5x5 form with missing nucleons. Strangely enough it is the latter formulation that is favored. However, there are nuclei that use the even flake format. The odd flake divides the 16-nucleon flake into fourths forming 2x2 squares and slides then away from the center until it aligns itself with the next nucleon. Our flake sequence is 8-16-8. Another good flake pattern is the 11-11-10 configuration. However, we find that the central flake has an extreme proton/neutron ratio.

Chapter 12: Plasmaverse

Sulfur-33 is the extension of the Sulfur-32 isotope. It is the second rarest of the four common sulfur forms with a flake sequence of 9-16-8. The neutron merely fills in one of the edge flakes centers. The reason for the rarity is that this neutron usually becomes a proton as the neutron moves to the center and contacts other neutrons. However, it destroys the isotope through protonization and the violent result of the newly formed proton touching other protons breaks the assembly into smaller isotopes.

Sulfur-34 is the most favored trace isotope of sulfur of nearly 5 percent of the entire element. Again we return to the Sulfur-32 isotope. In this sequence, 9-16-9, both ends are closed. This closes the central hole enough that the weak magnetic hold of the protons upon the neutron is more likely to keep the neutrons in place. The last isotope is Sulfur-36. It will be covered with the Chlorine as an isotope of chlorine occurs next.

Chlorine, Argon, and Potassium Nuclei

Chlorine-35 has 17 protons and 18 neutrons defining the isotope. This isotope composes about three fourths of the element. The flake sequence continues filling the central axis of the three flakes. The formulation becomes 9-17-9. The nine nucleon pattern is the standard neutron dominated flake. The seventeen nucleon flake pattern is somewhat complex, but has symmetry in respect to the center. Imagine the central flake as a spinning three by three nucleon flake; we add two nucleons to the leading edge of each edge. The resulting image is a modified plus sign. The central most nucleon has to be a proton by demand of the external flakes. This establishes the checkered pattern throughout the flake. Notice that the additional central neutron added to the Sulfer-34 clump could only become a proton to finish the prevailing pattern formulating this element. This illustrates that gravity and the isobaric spin of the nucleus both still play a major role in the development of the nucleus.

Sulfur-36 is next in the sequence even though it defines an isotope of the previous element. When we add one nucleon, we actually subtract one proton and add two neutrons to the outcome. Statistically speaking, this is the rarest form of sulfur. Another interesting side note is that the arrangement of the neutron-proton-neutron extension occurring at the edge of the central flake could also align itself at a right angle to the flake giving 11-14-11 flake sequence instead of the 9-18-9 which seems to be favored. For this to occur, gravity has to play a greater role than centrifugal force during the rejoinion resolve.

Chlorine-37 is the continuation of the Sulfur-36 flake pattern. The central hole within the isotope composition fills with a proton as it is surrounded by neutrons. Again, the extended end formulations of the central flake could formulate at a right angle to our illustration via gravity playing a greater role in the rejoinion resolve. It is a matter of timing within the process. The gives the result of 9-19-9 instead of the 11-15-11 flake sequence as might be expected.

Part 4: Creation of Atoms

Argon-36 has 18 protons and 18 neutrons. The three-flake sequence is 12-12-12. Despite its evenness, it forms less than one percent of the isotope. The twelve-nucleon flakes are even flakes, meaning that the crystalline axis exist between nucleons in one of its dimensional formation. All the Argon isotopes have this uniqueness. The tendency is for an odd dimensional count of nucleons per flake until the nuclei becomes large. Note: A 36-nucleon flake or two 18-nucleon flakes are possible; however, they would require special conditions. This is not to say that these conditions do not exist.

Argon-38 is the next isotope adding two more neutrons. These are added to the central flake filling in the neutron deficient corners. This formation is the rarest form of the argon element. The flake sequence then becomes 12-14-12. This isotope is about five times as rare as Argon-36.

Argon-40 provides yet another variation using 18 protons and 22 neutrons. There are many possible structural shapes that can formulate during the rejoinion resolve. The easiest resolve provides us with the 14-14-12 flake sequence. However, the primary stable format is the 8-24-8 nucleon pattern, and the second forms a 9-22-9 pattern. As unreasonable it may seem that the central flake to contain three times the nucleons as a given external flake, it must be remembered that 25 is five squared and nine is three squared. Moreover, the difference in flake count from the center of the flake is only one nucleon per lateral dimensional direction (two directions per dimension).

Before continuing, we need to look at the two isotopes of Sulfur-36 and Argon-36. This is the first example of the same number of nucleons generating two different isotopes. Both of these isotopes provide rare forms of their respective elements. Even so, it shows that the rejoinion resolve process works with more variables than just the number of neutrons. Other factors of variance are the vectors in which the neutrons collect into neutron clumps and their arrival times.

Potassium-39 has 19 protons defining the element and 20 neutrons defining this particular isotope. About 93 percent of the element is this isotope. A more stable flake sequence fills is an 8-23-8 flake configuration. The central flake in the 5x5 format with the two opposite corner protons missing, and the external eight-nucleon flakes have the standard central nucleon missing. In this instance they are neutrons. Another somewhat stable sequence shows a 13-17-9 flake sequence.

Potassium-40 is a slightly imbalanced flake pattern. The central proton is filled in the central flake and a central neutron in one of the external flakes. This is the rarest form of potassium. This is to be expected primarily because most of the time Argon-40 or Calcium-40 will be the rejoinion resolve. This gives the flake sequence of 9-23-8 with 19 protons and 21 neutrons.

Potassium-41 is the completion of the potassium pattern. There are nearly 700 times more of this isotope than that of Potassium-40. The flake pattern is 9-23-9 giving 19 protons and 22 neutrons.

Chapter 12: Plasmaverse

Calcium to Vanadium

Calcium-40 has 20 protons signifying the element giving 20 neutrons for this isotope. This configuration provides about 97 percent of the isotopes found in calcium. Our illustration shows a 12-16-12 flake configuration. While the twelve-nucleon flake is rectangular in nature, it is still concentric sound for the composite isobaric spins of the nucleons. A variation of this pattern that is more equally distributed within the given flakes is an 8-16-16 flake pattern. For this alteration, remove the extraneous four nucleons on the top layer and place them on the bottom layer. The easiest flake pattern to conceptualize is the two flakes of 20-nucleons. Both are variations of the five by five-square flakes; both have all their corners missing, as well as, their centers. The flakes are an inverse of each other. This gives one flake that has 12 neutrons and 8 protons while the other has 8 neutrons and 12 protons. The side with the twelve protons provides room for further isotopes of calcium. However, this isotope formulation is unlikely in Calcium-40 (nucleon ratio). It would occur if the neutrons arrived in a pattern that made it difficult for the rejoinion resolve to achieve this particular stability.

Calcium-42 is an extension of the Calcium-40 format. This is the second most popular trace isotope of calcium composing a little more than a half of one percent of the element. The two new neutrons arrive late into the assembly and are place either upon the outer edge of the central flake or become the central neutrons in the two external flakes. Our illustration shows the first scenario of the 12-18-12 flake sequence, there is also a 9-24-9 sequence exists when the neutron arrives earlier in the rejoinion resolve. It is also possible for the clump to form from a different collection configuration.

Let us backup a moment to the 40-nucleon isotopes. The first set of elements formulated by the same number of nucleons was trace isotopes. Now, we are observing even the dominate form of the elements becoming a matter of the initial neutron clump formulation rather than an automatic protonization response within the rejoinion resolve. As we have observed, as the nuclei become larger, the combinations increase. By the time we reach the initial configuration for nickel, consisting of 60 nucleons, we will be "swimming" in combinations. It is no longer a matter of finding the accurate account of the formulation of an isotope, but it is finding the configuration that occurs the most common within a given isotope. Examples of different configurations are as follows: Argon-40 flake having a 14-14-12 sequence, a Potassim-40 composes within a 13-17-10 sequence and a Calcium-40 having a 16-16-8 flake pattern.

Calcium-43 requires that the neutron and proton pattern to be inverted in order for the single neutron to be inserted into the assembly without becoming radioactive. This neutron locates itself in the central position of the central flake. The required observation is that the first isotope of calcium could also be inverted and keeps the same flake sequence. Our flake pattern for Calcium-43 is 12-19-12.

Part 4: Creation of Atoms

Calium-44 is the most favored among the trace isotopes of calcium forming about two present of the element. The reason it is favored is that its initial clump forms a complete geometric structure. This structure forms a 4x4 flake in the central location. The two flakes just above and below it are 3x3 forming nine. Above and below them are two 2x2 flakes and two single flakes above them. Despite the neatness of this pattern, its formulation is not well favored. It requires a stacking of three layers vertically on either side of the central flake. While this is not bad in itself, the problem is that the central flake measures only two layers horizontally and any direction from the center. In essence it requires the vertical direction to be longer than the assembly is wide. This is difficult considering the centrifugal formulation model. However, it does happen. Within our model of the rejoinion result, it removes the central neutron and inverts the arrangement of protons and neutrons within each flake and adds two neutrons. It either fills the central location of the exterior flakes or adds to the edge of the central flake. Our illustration is that of larger exterior flakes, but this may not be the primary arrangement. Another possible factor is how it relates in the galactic schema.

Scandium-45 is another unique isotope forming 100 percent of the element. Once again, the gathering neutrons have a solid geometrical shape to follow. Another double pyramid forms. The central flake is 5x5 giving 25 neutrons and the two flakes forming above and below are 3x3 giving nine each making 18 total. Then we have the outermost flakes consisting of one each. This gives the grand total of 25+18+2 which is 45. However, it does not maintain this structure through the rejoinion resolve. Providing there is not much rejoinion movement, the two exterior single neutron flakes roll down the slopes and find their place upon the edge forming the 9-27-9 flake sequence. The next possible solution is that of a 13-19-13 flake, making the nuclei as an ellipsoid rotating on its minor axis. Our flake sequence shown is 17-19-9, giving other forces a factor. However, in general, the forces promote the large central 5x5-flake formulation above evenness in nucleon count between flakes.

Calcium-46 is the rarest isotope of the six found in calcium, only about one out of 25,000 isotopes examined will be of this structure. One example of this isotope is the 16-20-10 sequence. **Calcium-48** is much like the Calcium-46 except a little more balanced centrifugally and gravitationally. Continuing with the same pattern, we arrive with a 16-20-12 flake pattern. Both patterns have the central nucleon missing in each flake. Generally, this collection of neutrons will form an isotope of titanium.

Titanium has 22 protons defining its element. Titanium seems to undergo the same process as calcium. It gathers neutrons even after protonization begins. The most unique and yet most common form for titanium is Titanium-48. This neutron clump starts out as a complete pattern formed by the 4x4x3 neutron flakes. However, it does not remain the rectangular form during the rejoinion resolve. The primary reason is that some of the flakes are even and centrifugal force desires odd numerical counts for its dimensions. Because this crystalline structure of this formulation requires the flakes

Chapter 12: Plasmaverse

to have an even count, it is harder for gravity to generate this formulation giving way to other formulations creating other isotopes of other elements with the same amount of nucleons. The isotopes that occur in titanium are continuous from 46 to 50. The formulation of a normal set of 50 nucleons will form Chromium-50. Occasionally, the 50-nucleon process will form Vanadium-50.

Lastly, notice again the collection of neutrons increasing with each isotope. Our selected flake pattern of the titanium series provides the accent upon the neutron collection within this element. The even flake pattern is 16-16-14 for **Titanium-46**. This is the only titanium isotope that remains in the 4x4 flake pattern. Each increase of the following isotopes attempts to imitate the 5x5 pattern. **Titanium-47** then becomes a flake pattern of 16-23-8 vacant in the central neutron position of the top flake. It follows that **Titanium-48** has a 16-23-9 flake pattern inserting the central neutron in the top flake. This isotope is the prevailing pattern for titanium. **Titanium-49** adds two neutrons symmetrically outside the 3x3 top flake and vacates the central neutron position giving a 16-23-10 flake pattern. Again, the mathematical logic is 9+2-1=11. Finally, the isotope, **Titanium-50**, refills the vacant central neutron position in the top flake giving a flake pattern of 16-23-11.

Vanadium has two isotopes utilizing 23 protons to define its element. As noted earlier, **Vanadium-50** is a trace isotope of this element because of the original neutron clump arrangement. Our flake pattern of 17-19-14 reflects the gravity dominated formulation during the rejoinion resolve. The fourteen-nucleon-flake pattern within the top flake has the central nucleon missing. **Vanadium-51** fills the central neutron within the top flake giving the flake pattern of 17-19-15. Again, the reason for the lop-sided counts of nucleons is due to an acceleration of neutrons, as we shall observe shortly. The larger count occurs in the top flake, as it is pushing into the massive neutron matrix.

Few More Single Isotope Elements

Manganese-55 is the only isotope of manganese, and it contains 25 protons and 30 neutrons. The primary reason that there are no other isotopes of this element is not the lack of neutron variations, but because of the rejoinion resolve. The initial geometric pattern of the neutron clump is also peculiar. The base flake is the 5x5 with the four corners missing giving 21 nucleons, the flakes above and below are 4x4 flakes with its corners missing giving 12 nucleons each. The flakes outside of them are the 3x3 flakes with its corners missing giving five neutrons each. The pattern cannot continue, as there are no "pits" formed between neutrons underneath because the corner neutrons are missing. The total is 21+12+12+5+5=55. The flake sequence that can formulate is 17-21-17. This flake pattern has room for other nucleons; however, neutrons occupy the adjacent locations. The collection of more neutrons then can only produce more protons to the existing formulation. This pattern also appears as the only natural

Part 4: Creation of Atoms

resolve for this number of nucleons, other resolves would allow more neutrons to collect generating more isotopes that are natural. The reason for the missing corners of these flakes is that gravity gains control over the initial flake formulation as the number of nucleons increase. The exception exists in Arsenic-75 having an initial pattern of 5x5x3 giving three sets of twenty-five nucleons. However, this formulation satisfies the centrifugal ellipsoid constraints. In the formulation of cobalt, we shall see that the centrifugal energy of the isobaric spin holds control over the neutron clumping within the rejoinion resolve.

Cobalt-59 is another element formed by one isotope. This isotope has a similar double pyramid formulation as seen in Manganese-55 except the central flake has all of its nucleons for the total of 25. The count then becomes 5+12+25+12+5=59. These geometric figures promote uniformity within the rejoinion resolve as the gathered neutrons are "messaged" into the familiar formation via gravity and centrifugal force before protonization. The result of the rejoinion resolve is an 18-23-18 flake sequence. The external flakes both have their central neutron nucleon missing. This illustrates that protonization can alter the count established by the isobaric spin. The two protons, one each, moved vertically outward to their respective flakes through the center causing the protons to shift in these flakes leaving a hole in both.

Gold-197 is our last isotope to examine. The reasons for choosing this element are that it forms 100 percent of the element, and has many more neutrons than protons. Gold requires 79 protons to define the element, this leaves 118 neutrons giving 39 neutrons beyond the one to one ratio between neutrons and protons. Since this is an element composed of one isotope, the chances of having only one solution is much greater. Another interesting attribute is that being a singular isotope element means that there is only one initial configuration defining the clump pattern of neutrons. The pattern found for gold is the central flake is a 7x7 full flake; the flakes adjacent, above and underneath, are both 6x6 full flakes; the ones external to these two flakes are 5x5 full flakes. At this point the tendency toward the spherical formulation removes the corner neutrons of the 4x4 flakes external to the 5x5 flakes making them a 12 neutron flake each. The phenomenon worsens the condition for gathering neutrons on the external most layers. Instead of leaving five out of nine neutrons, it allows only one neutron for each flake resting in the flake's only "pit." This gives the total of 197 neutrons.

The rejoinion resolve usually does not leave a mixture of even and odd orientated flakes within the same nucleus as neutrons will touch causing the nucleus to become unstable. This reformation in itself will create major shifting within the process. The central flake loses five nucleons in the process and the end single flakes become absorbed into the flakes supporting them. Two of the nucleons were drawn into the adjacent flakes from the central flake, one each, while the other three other nucleons "boiled up" to the top. The term "boiled up" represents that there was an upside and downside during the rejoinion resolve because of the kinetic acceleration experienced by the assembly of

Chapter 12: Plasmaverse

neutrons. The reasoning is a little beyond the scope of this chapter as it concerns itself with the creation of galaxies. The resulting flake sequence is:

13-25-37-44-37-25-13-3

The unique attribute about this flake sequence is that each all the flakes except the top one has absolutely no more room for another neutron. The next nucleon within these flakes must form a proton in order to be stable. Even on the top flake, it would require two neutrons to hold the balance of the flake. By adding one just neutron, it would have to formulate another proton in order to formulate a balance in all the flakes. Subtracting one neutron would also throw the flake balance off; again, another proton has to form to maintain flake balance. In both cases the proton fills the central flake hole formulating the element of mercury. Therefore, gold has only one isotope. Another observation along the same line is that the "broken" edge of the flakes exposes only neutrons.

From the examination of the isotopes in calcium and titanium we see the formulation of a phenomenon occurring. The process of increasing the ratio of neutrons continues as the elements increase in nucleon count. This process is responsible for most of the trace isotopes found naturally occurring within the universe. The primary reason finds its roots within the reason that larger nuclei takes longer for the protonization to complete beyond that which can be attributed linearly by the increase of mass. The resistance of formulating protons increases as more protons form. As the nuclei sizes increase there is a tendency for them to develop a neutron shell.

A post collection of neutrons adds to the existing assembly of neutrons. This collection is called **neutron snow** for the apparent nature of the collection. The image is similar to that of snow upon the earth's surface where the neutron snow is responding to the collective plane of positive ions (plasmatic isotopes- nuclei). The "storm" of neutrons generates a leap in the isotope table in the heavier elements. The isotopes jump from Asatine-210 to **Radon**-222 with nothing in between forming naturally. In different regions of the universe, we expect this gap to be different. Another interesting item is that the effect is the sudden and short in duration of this phenomenon, as we do not observe the gradual shift to a gap of 12.

Lastly, looking at largest set of elemental isotopes found upon the earth, we can inscribe each of them within a 7x7x7 nucleon cube (343 nucleons) and still have room for more nucleons. In fact until recently, we could have inscribed the isotopes within five flakes of 7x7, now we will have to add another 22 nucleons to these flakes for 267 nucleons. From the standpoint of being in the center of a given flake, the nucleons are stacked at most three above the central location in any direction. The concept is that the stacks of nucleons above the central nucleon do not need to be very high to engulf large numbers of nucleons. This is not to say there are not extremely large nuclei within the universe because there are. A neutron star is a prime example of this phenomenon.

Part 4: Creation of Atoms

Neutron snow exists in response to the limit of the normal increase in nuclei or clumps of neutrons. The neutron shell that was being developed through the isotopes was actually forming a cubical structure. The elements beyond lead, without the neutron snow, would have been stable elements. We could theoretically complete the neutron cubical shell as Astatine-231 and still be stable. However, this was not to be; for these elements to be stable, the neutron snow needs to occur later forming radioactive isotopes with larger nuclei. If we were to examine the core of Saturn, we would not find any radioactive material as it contains no heavy core material. This is true for the other outer planets as they contain only light core material. However; within the sun, if we could examine its elements, we would find heavier isotopes than within the earth, as the sun is the center of our plasmatic storm system even though its net density is under half that of the earth because of the ratio magnitude of light elements upon its surface and the juttorial heat of plasma. Working within the earth's element system, scientists have concluded that all radioactive material will eventually breakdown to lead.

Mass Shrinkage

As noted earlier, there is a shrinkage factor involved in the protonization process. For example: in examining the mass of each isotope in relationship to the number of nucleons, we find the average mass per nucleon decreases as the number of nucleons per an isotope increases (average = mass / number of nucleons). Examine these three samples: Hydrogen-1 has a mass of about 1.0078 u (atomic mass units) giving an average of the same. Boron-10 has a mass of approximately 10.0129 u giving an average of 1.00129 u. Ruthenium-100 has a mass of about 99.9042 giving an average of 0.999042 u. These are averages and not the actual measurement of each nucleon.

Another observation is that the protonization process is not a singular function. The average nucleon size shrinks until we reach nickel. Nickel has 28 protons with approximately 30 neutrons. However, it is a particular isotope within the nickel element in which the shrinkage process "bottoms out." Being consistent with the numerical design, the isotope is **Nickel-60**. The shrinkage process then seems to gradually recede but it never makes it back to the sizes found in the first few isotopes. Another interesting item is the jump in the average sizes of nucleons past the **Astatine**-210 and Radon-222 void of isotopes. All nucleon averages of the isotopes approaching this void are slightly less than 1u; conversely, all the isotope nucleon averages, after the void, are slightly greater than 1u. The reasons for the two phenomena are one; it is the increase of neutrons not converted into protons. If we look at Plutonium-244 we observe 94 protons with 150 neutrons. Within the Nickel-60 isotope there are about 1.14 times as many neutrons as there are protons. Compare this to Plutonium-244 having nearly 1.6 times as many neutrons as protons. However, the core material still shrinks; the counter balance is that the external neutrons forming the neutron shell around the core of the nucleus.

Chapter 12: Plasmaverse

The apparent shrinkage of the core is actually an illusion. The reason is that we assume that the space/time continuum bends in one direction. The general image we are given is that of an elastic plane that has a funnel-like depression caused by gravity. A graphical cross-section of the funnel gives us a v-shaped figure that curves outward flat at the shape's top in both directions. Let us draw a horizontal line slightly above the flatten ends of the v-shape representing the asymptote or limit of the function in which the bend of the continuum is zero. Our focus is at the bottommost point of the v-shape. The depth that the funnel submerges below this asymptote via gravity provides us our measurement of mass. We will also make a vertical line that bisects our v-shape at the base focus. This marks the center of gravity generated by the mass. Note: centers graphically equate to a zero measurement. Now, we are going to make a slight modification at the base of the v-shape. Instead of the tip made by the v-shape pointing downward, it will point upward forming a little rounded w-shape. This slight flip represents the effect of juttoria. Recall that juttoria is responsible for the microstrong force which dominates as we approach zero. This moves our reference point above the "normal" projection of the v-shape causing our measurement of mass to be less.

Nuclei Analysis

Within analyzing the average nucleon size per isotope, we find three different phenomena at work within the graph. The first phenomenon, the first notable characteristic is the increased juttorial exposure. Its signature is the steep initial incline of the general slope that flattens out nearly parallel to the horizontal axis. This occurs from Hydrogen-1 to Nickel-60. As previously stated, it is interesting that it is the isotope 60, not 59 or 61 nucleons that produces the turning point within the graph. The second phenomenon, we will call the neutron shell counter shrinkage. This phenomenon becomes evident at Nickel-60 and beyond. There is actually another sub-phenomenon that occurs within the last few elements by the involvement of neutron snow that elevates the average size of the nucleon beyond that of the neutron shell counter shrinkage. The third phenomenon provides us with a "vibration" of the line throughout the graph. The vibration refers to the tiny "peaks" and "valleys" that move with the graph. This phenomenon, we will call the nucleon geometrical resolve.

The rate of the apparent "nucleon shrinkage" decreases as the number of nucleons increase. The reason for the increase is the exposure of new protons as the nucleon count increases. This effect continues to dominate the graph until the building of the neutron shell initiates. As previously examined, this begins after Nickel-60. The increased juttorial exposure still continues after Nickel-60, it continues at a very shallow, almost zero slope. The nature of the curve is similar to $1/x$. The initial steep slope in the isotope curve is indicative of the slope observed in the graph of $1/x$. Unlike the $1/x$ graph, the mass does not start at infinity. This is because we are starting at the count of one nucleon or Hydrogen-1 instead of zero. This is the upper limit of the function.

Part 4: Creation of Atoms

The lower limit of the function is harder to find. We can visually observe the bottom of the shrinkage of the average nucleon at Nickel-60; however, while Nickel-60 is very close, it is not the bottom limit. The bottom limit is estimated to be about 0.9985 u. We will call this the heavy limit. Heavy implies a nucleus with an extreme number of nucleons. This limit would sink slightly over a large period of time as more protons are formed in the heavier nuclei, if we had that much time left within this physical universe.

Even though we can readily observe that the initial part of the curve resembles 1/x, there are other curves that also resemble 1/x, but have steeper slopes. We also know that there is a constant k that 1/x is going to function within giving k/x. Our k value exists by subtracting the initial value from Hydrogen-1 from the base limit. This gives approximately 1.0078u - 0.9985u giving 0.0093u. Then we observe that the 1/x stretches over the graph by a factor m giving k/mx. This factor turns out to be about 2/5, which resembles the amount that an electron removes from the neutron in relationship to the total mass removed. However, this equation still has a problem, it "blows up" as the nucleon count approaches zero. The reason for this problem is that there is only one nucleon count from the center outward in Hydrogen-1 to Lithium-6. By this, the equation has a limit. We then can establish another 1/x equation from Hydrogen-1 to Lithium-6 giving a constant of 0.0053u (1.0078u - 1.0025u). Actually, the problem will persist down through the graph as the flake dimensions increase in nucleons per dimensional direction. In other words, we gain neutrons to the flake exterior. The reason that we do not readily witness this phenomenon is that this part of the equation is very slight, and it is buried in the third (primarily) and the second phenomenon. By this, we see that a unified equation becomes nearly impossible. However, we can make some general mathematical observations.

Our second process is the neutron shell growth. After Nickel-60 the neutron shell becomes more evident or consistently growing. For a prime example, look at the solitary isotope of gold; the isotope is Gold-197. We will find the primary composition of the surface as neutrons. The surface has 102 nucleons, 79 of them are neutrons leaving 23 protons on the surface giving a surface of about 77% neutrons. By the time we get into the "neutron snow" series existing at Radon-222 and beyond, the condition becomes more severe. Within this process, we are looking for an upper limit. It would be a mistake to assume that the upper limit is our given mass of a neutron. There are a couple of reasons for this statement.

Working within the isotope table itself within this region, we can subtract adjacent isotopes from the same element, or even, of an isotope by an adjacent element. We will get the increase of mass by an additional neutron. The problem with this mass increase does not measure all the increase by the neutron itself. There is also the juttorial energy generated by heat. This also lessens the measurement of mass per nucleon. It should be noted that this alteration is minute in comparison to that generated by an additional exposure of a proton within the assembly. For this reason, we will not find a neutron

Chapter 12: Plasmaverse

measuring 1.008664204 u in mass within any isotope. This, however, does not negate the reading provided to us by scientists.

Scientists have been able to isolate a neutron from the nucleus. However, examining a moving neutron is extremely difficult, if not impossible. Therefore, the neutron is slowed down. We are examining a "still" neutron. The result is an appearance of a slight mass increase, not because of volume but by weight. This is an illusion, as there is no gain in material. The reason this appears is that "heat" (juttoria) bends space in the opposite direction than gravity. To slow down the neutron is to remove heat, as defined by the first law of thermal dynamics, from the neutron. To remove heat from a neutron is to remove juttorial energy. The result is that the jammerial bend is given more freedom to express itself by this unnatural imbalance. By this, the neutron weighs more. This is the reason that a free "still" neutron weighs slightly more than those found in almost all atoms. However, we are not without recourse. For our constant, we will examine is the Helium-4 nuclei, as it points to the limit as we shall soon observe in alpha radiation. This does not negate the attempt for a uniform measurement in respect to a proton as the proton is also measured at the same temperature.

Alpha radiation occurs in nuclei that are unstable. While the protonization process of a single neutron does not produce alpha radiation, it does exist in large nuclei. The attributes of alpha particles is that it has atomic mass of 4 and has a positive charge of +2. The atomic nuclei that discharge the particle loose a charge by +2 and become an isotope of an element that requires two less protons. Helium has the mass of 4 and has two protons. From these facts, it is easily to visualize the equivalence. After our examination of the nuclei shrinkage, we see that Helium nuclei are much smaller per nucleon than the isotopes of similar size. The reason for this particular variation is that much of the element derives its existence from the alpha particles. It should also be observed that the size of the Helium-4 isotope is closer to the size of the heavier elements, but still larger than the average nucleon from which it was ejected.

The average mass of a nucleon within the Helium-4 isotope is 1.0006581 u. This could be considered the upper limit of the neutron shell building function. While there are neutrons that measure heavier, their masses are gotten from the initial protonization process, see the Lithium-7 and Beryillium-9 isotope comparison. There are others, but not as dramatic. Despite of all of this, their nucleon averages are less than that of Hydrogen-1. Working within the range of the neutron shell development via nucleon increase within a given nuclei, Helium-4 nucleon average is greater than any of their averages, even including the "neutron snow" elements. Even so, the growth limit rises about only 1/5 from the initial increase of juttorial exposure.

Not all Helium-4 isotopes form from other elements. There are some Helium-4 nuclei created at the time of creation that are heavier. However, the Helium-4 generated at Creation was a onetime event. Helium-4 is constantly being produced by heavy radioactive atoms as alpha particles throughout creation since the Beginning of Creation

Part 4: Creation of Atoms

to now. The result is that they far outnumber the original Helium-4 isotopes. The ones formed at creation should be slightly heavier than those formed as alpha radiation as less time involved by protonization via fewer neutrons involved.

The reason, for the larger measurement of Heilium-4 than the average nucleon from which the alpha particle radiation source exist, is the juttorial saturation limit for four nucleons as opposed to the multi-nucleon assembly of 200+. Notice again, the dip in nucleon average between Helium-3 and Lithium-6, that Helium-4 forms. This shows that the natural juttorial input per proton is less than that which the assembly can hold. In other words, the alpha particle lost juttoria input as it ejected from the large nucleus, yet maintained more juttoria than it would have acquired from the formation of its two protons.

The last variable to examine is the geometrical resolve. For this, we need to examine the result of the rejoinion resolve, which involves the original configuration of the neutron clump and the resulting geometric configuration of the isotope nucleus. This function behaves as a step function as there are no partial nucleons. An example is that there is no such thing as 2.37 nucleons; it will be either two or three nucleons with masses near 1.0 u each. The other variable is whether or not the additional nucleon formulates into a proton or remains a neutron.

All three phenomena can be expressed as one. It is the geometric properties of the protonization process itself. The basic concept is that with a set of three linier nucleons, the middle nucleon will become a proton. The remaining outer neutrons are just as important as the fact that the central nucleon transformed into a proton. This is the process that forms the neutron shell in our larger nuclei. This checkerboard process between protons and neutrons determines whether the isotope formulated by the added neutron forms into a proton. In this, the last phenomenon becomes evident.

The final image of the phenomenon is that of a juttorial envelope surrounding the nucleus of an atom. Its magnitude is affected by two properties of the atom: the number of protons in the assembly and the amount of heat. Some may think that with the given parameters that adding heat would attract more electrons. The problem with this conclusion is that the assumption that juttoria itself attracts the electrons, but it does not. It is the count of xergopath configurations (established only by protons) that contain juttoria that gathers the electrons. The density of juttoria per xergopath determines the heat expressed by an atom. It is because the juttorial energy expressed external to the nucleus that any mass distortion is present. The image is as looking at a fish in water and trying to establish its exact position. Most of the juttoria within the nucleus is immeasurable as distance to the surface diminishes its value to near zero.

Chapter 12 Quiz

1. Can two Neutrons touch each other without reacting?:
   A. No
   B. Yes
   C. Depends upon the Temperature of the Plasma
   D. Depends upon the Direction they Face

2. The Nucleus of an Atom is:
   A. Spherical in Nature
   B. Randomly Shaped
   C. Crystalline in Nature
   D. None of the Above

3. The Rejoinion Resolve occurs so that:
   A. No two Neutrons touch
   B. No two Protons touch
   C. No two Atoms touch
   D. Electrons can Escape the Nucleus

4. Nickel-60 is the:
   A. Only isotope of Nickel
   B. The Largest Isotope
   C. Radioactive Isotope
   D. The Smallest Isotope

5. Isotopes are:
   A. Unique Elements
   B. Radioactive
   C. Unique Nuclei Configurations
   D. All the Above

6. (T/F) Fluorine and aluminum both have singular natural occurring isotopes.

7. (T/F) Touching protons generate gamma radiation.

8. (T/F) Rejoinion Resolve forms magnetically crystalline nuclei.

9. (T/F) Helium-3 is the outcome of Helium-4.

10. (T/F) The average size of neutron increases until the neutron shell process.

Chapter 13

Bondverse
Shell of the Atom

Within the plasmaverse, we observed the formation of nuclei of many elements without regard to the electrons emitted from the assembly. We were able to omit their activity, as they existed entirely outside the realm of the nuclei structure. The reason for their exclusion from the atomic assembly is that the intense heat from the initial formation of the universe excited the electrons beyond the ability of the atomic nuclei's magnetic hold. Within this phaseverse, matter loses this heat within the various constraints established by the creation of natural laws governing the natural physical universe. These forces are the topic of the next chapter. The primary scope of this chapter is to describe the environment and anatomy of the electrons and their photons.

Initially, our analysis of the bondverse involves the magnetic interaction between the magnetic surfaces of the protons and electrons. This brings us back to the xyzenthium crystals that compose the two surfaces. The electron surface composition comprises only of the electrosod xyzenscape alignment. The proton surface contains only the protosod crystal alignment. Both alignments are only half "true" magnetically charged, as two crystal sides supply polarized neutral charged diflohexets within both surfaces.

Neither surface emits pure hypertips or hypotips. Actually, each emit hypertip/ hypotip ratios the reciprocal of each other. This ratio is 1:3. Recall that the atom has two hemispheres attracting electrons. From the electrosod, xyzenscape the ratio translate to two hypertips to six hypotip emitting xergopaths for the total of eight xergopaths. Recall that the one xergopath nearest to the center is the "feeler" from any set of xergopath loops. As "feelers," they are the first to interact with external xergopath loops. This number of four xergopaths will play an important role as we examine both the collection of electrons and the bonding of the elements.

Magnetic Lines

After the xergopaths emitting from the proton stabilizes the surrounding neutrons, except for Hydorgen-1 consisting only one solitary proton, the xergopaths continue to engulf the surface of the nucleus. After saturation, these xergopaths begin to emit outward away from the nucleus as the void draws them. While this occurs, the entire nucleus usually spins. Recall the image of the xyzenthium crystal as the loops revolve

Part 4: Creation of Atoms

around the dimensional axis. The same pattern attempts to develop with the xergopaths forming the magnetic lines. The difference is in the nature of the xergopaths are controlled by the protosod xyzenscape and the entire assembly is spinning. The basic concept is that tavtip types (hypertips and hypotips) begin to behave magnetically as seen in xyzenthium crystals. The observed tavtip flow identified with positive charge energy moves along the rotational axis, and the negative charged xergopaths along the equator. Ultimately, both hypertip and hypotip energy moves along both xergopaths only in opposite directions. However, there is an imbalance of energy. The "polar" moving tavtips are three times as many as the horizontal xergopaths. This aligns with the observed count in the protosod xyzenscape "feelers."

There is one final twist to this scenario. The vertical emitting xergopaths rotate somewhat as they return to the nucleus. The reason for this twist is the isobaric spinning by the nucleus and the extreme distances away from the nucleus that these tavtips are moving comparatively speaking in relationship to the size of a nucleon.

The xergopath loops upon the electron experience a similar phenomenon. The major differences are the direction of flow and the reciprocal ratio of hypertips and hypotips. These xergopath loops of both particles come into contact and exchange energy. The energies are mutually receptive to the flow of energy within both xergopath loops. The hypertips are moving against the hypotips as needed for the exchange. As the two loops approach each other, the attraction pulls the tavtips "off course" to join into the xergopath loop of the external xergopath loop. After the loop passes beyond the external loop, the exchange path is broken and rejoined to its regular course. However, this reunion does not last long, as another external xergopath loop approaches our xergopath loop. This continues throughout the progress of its spin. Our analysis of this phenomenon takes much longer than the process itself, for it is nearly instantaneous.

Shell Levels

The shell of electrons exists in **orbitals** over the nucleus; most are quite dissimilar to a planetary orbit, these orbitals exist within defined regions of the layers called electron shell sublevels. The electron orbital sublevel patterns compose seven layers called electron shell levels or levels by scientists. Scientists alphabetically label these layers from K to Q. K is the closest level of electrons toward the nucleus. Another method of reference is relationally to each other a lower level is closer to the nucleus. Each level further away from the nucleus contains more electrons than the one under it.

Each level divides into **sublevels**. The number of sublevels that exists within each level increases with each level. The first level (K) has only one sublevel. The second level (L) has two sublevels. This progression continues; at Level Q or Level 7, there are seven sublevels. Each sublevel increases in number of electrons, but maintain the same number of electrons at all levels. Example: The second sublevel (p) has six electrons,

Chapter 13: Bondverse

which is more than the first sublevel (s) containing only two electrons. However, at any level, L through Q, there are only six electrons possible within the second sublevel (p). Another interesting note: While there may be more shell levels in extremely large nuclei, seven is the number found on earth.

The image is that of building a square with square tiles. When we lay down the first tile, we have a square already of one tile unit. However, when we want to make a larger square using the least amount of tiles, we need to use three square tiles forming a square two tile units high and two tile units wide. Notice that if we multiply each of these squares by two we get the number of electrons in a sublevel 1*2=2 for the s-sublevel and 3*2=6 for the next sublevel p. This process continues throughout the sublevels sequence s, p, d, f, g, h and i. We can calculate the number of tiles that we will need to build the next square one tile larger. Notice that our second larger square needs four tiles to formulate its square, but we only added three. Actually, we had to subtract the first square from the second square to get our count. This is represented by $x^2-(x-1)^2$. Doing a little algebra, this equation will reduce down. The value of $(x-1)^2$ gives the value x^2-2x+1 and we subtract that from x^2: $x^2-(x^2-2x+1)=x^2-x^2+2x-1$ and $x^2-x^2=0$ leaving $2x-1$. This answer, of course, is no surprise geometrically. Let us continue our progression: for the next square of three by three tiles, we know by the equation that we need five tiles. We set x to 3, we get 2*3-1=6-1=5. However, for our electron count we need to multiply by two.

$$2(2x - 1) = 4x - 2$$

Multiplying the answer by two seems a little extraneous to the process other than that is the actual count. However, when we examine the nature a little more closely, we observe that it is necessarily true. Let us return to the tile example once more. Using the single unit tile 1x1 in dimension, we are going to change its properties. Imagine being able to pump air into the middle of the tile. As it expands, we can observe that the tile has both a bottom and a top side. If we were able to pump enough air into the tile, it would become spherical in nature just as we observe around the nucleus of an atom. However, we know that there are two sides to the sphere. Even without using the tile there is a side facing us and a side away from our view equal in size (without getting technical- our line of vision causes us to see slightly less than half). Going back to the tile, if we were to divide the tile into four units, we would also have to divide the bottom into four units doubling the number.

Along the same lines, two times the total square units is the number of electrons within the entire level. The lowest level (K Level) has the maximum of two electrons. The next level (L Level) has the maximum of eight. Electron levels operate in a type of geometric progression. The sequence computation is expressed in the following equation: let L equal the count of levels away from the nucleus (K Level = 1); let T

Part 4: Creation of Atoms

equal the total. The equation then is: $T=2*L^2$. This means the count is multiplied by the count and two multiplies this result. Here is another example: Q Level (7th level) = 2 x 7 x 7 = 2 * 49 = 98 electrons possible for this level. The total number of electrons collected by all the sublevels within the seven levels is 280. This figure derives its value from adding 2+8+18+32+50+72+98 from L to Q level respectively.

Ks (1s) Sublevel

The first sublevel of s is different from all the other s-sublevels. The electrons are gathered perpendicular to the plane of the normal s-sublevel electrons. In respect to the isobaric spin of the nucleus, they are polar orientated. However, in hydrogen the electron is free to move around the K shell as there is no other electron to contend for the space. When the second electron exists, the two polarize into two separate hemispheres. Even though they are polarized into stationary locations, these electrons are not still. But, for a moment let as view them in stillness. The strongest influence of the electron is upon its surface, as we move out away from the electron, its influence weakens by one divided by the distance squared multiplied by some constant expressing the magnitude of the magnetic charge. This constant is about 1.602 Coulombs. However, this constant does not help us in examining the nature of the magnetic field within an atom. Nevertheless, it is important that this constant is the same for both electrons and their respective protons.

Returning back to the examination of the weakening magnetic field by distance from the electron: As we move around the nucleus at the same altitude as the electron, we observe their weakest point is halfway in between the two electrons. The actual size of the electron is extremely small in comparison to its giant magnetic field. If we were to measure the distance from the nucleus and the distance from the midpoint between electrons, we will find that the midpoint is a larger distance. The reason is that a diagonal line measurement to a plane is always greater than a perpendicular line measured to the same plane. If the magnetic field were completely spherical in nature, there would be room for four more electrons within the same plane. This can be observed by taking seven coins of equal size and placing the six around the one chosen to be the center coin; they all have room to fit around and touch the central coin. This is not the case with the electron. The two fields engulf the entire atom leaving no room for another electron. If there were room for another electron then helium would bond with another atom. However, it does not; it is one of the noble gases. By definition, they do not bond with other atoms. This is getting a little ahead of our present topic.

Despite the engulfment of the entire shell level, there is a weakness along the aforementioned midpoint between the electrons. Using the distance that the electron is from the nucleus as the radius, we can return to the coin scenario. The radius is half the distance of the sphere of influence along the dimensional plane. Recall that we were able to insert two coins between two opposite coins. The midpoint between the two

Chapter 13: Bondverse

electrons then becomes defined by the point in which the adjacent coins touch. The ratio to the radius of the field reaching the nucleus is 3:1 the actual ratio is pi or 3.14159265... because of the curved surface. However, we will keep it somewhat simple. Recall that distance from the electron dilutes the strength of the field by the square of the distance. Therefore, the distance to the opposite electron is three times that of the distance to the nucleus. This gives the strength at the midpoint to be 1/9 of the strength existing between the nucleus and the electron examined. This weakness is compounded by the attraction to the nucleus by the protons. The energy is drawn into the nucleus pulling the magnetic lines inward giving a near zero field at the midpoint. The result is a "magnetic valley" in the negative charge at the midpoint between the two electrons. Unlike our two-dimensional coin example, the magnetic field of the electrons is three-dimensional. This generates an equatorial circle defining the midpoint between the two electrons. The "magnetic valley" then is a circular crevasse between the two electrons, like a stripe on a billiard ball.

The orbital of an electron is quite different than that of a planet orbiting the sun. The sun's hold upon the planet is gravity, and the force that prevents the planet from crashing into the sun is the centrifugal force. Centrifugal force generated by the kinetic energy by the planet is perpendicular movement to the mutual pull of gravity between the sun and the planet. Gravity has very little to do with the pull upon the electron by the nucleus. As noted earlier, the proton existing within the nucleus of the atom pulls the electron by magnetic energy. But, the electron is also stopped from crashing into the nucleus. The force that stops this from occurring is called the microstrong force generated by the emitted hypertips that attract the electron. The juttorial force emitted by the hypertips functions like heat and excites the electron causing greater movement. As the electron approaches the nucleus, the amount of hypertips encountered increases, once again, by the square of the distance. Eventually the increase of juttorial energy "outweighs" the incoming jammerial force emitted by the electron in the magnetic interaction. When this occurs, the nucleus repels the electron.

An analogy of this phenomenon is that of flying a kite. The kite is the electron; the string is magnetic energy. The wind is the microstrong energy. Instead of the wind blowing horizontally across the surface, the wind is blowing upward toward the sky. Just as the kite swirls around in the sky, the electron swirls around in a tight region being held taunt by the string, magnetic energy. Much of the spinning process is isobaric within the electron itself. As more juttorial energy is encountered, the isobaric spin of the electron increases. Since the magnetic lines are not continuous in their radiation from the nucleus, the input into the electron is somewhat uneven as it crosses through the magnetic lines by the spinning of the nucleus. The result is that the electron wobbles in its spin as the kite does by the puffs of wind. This is how most electrons exist within their orbitals; the exception is the s orbitals outside the K level. These electrons appear to move primarily in a circular fashion. However, even they will osculate toward the nucleus.

Part 4: Creation of Atoms

Ls (2s) Sublevel

Moving up to the next shell level, we find another s-sublevel. The next electrons moving in upon the nucleus must work with the existing electrons in the level below them. The reason is that the magnetic fields generated by them create favorable and unfavorable positions. The unfavorable positions are those, in which they encounter the negative charge of the electrons before reaching the positive charge of the nucleus, they need a direct connection to the nucleus. Recall the "magnetic valley created by the two electrons within the first s-sublevel, this is the location in which these electrons are allowed to find "rest." The term "rest" is used very loosely, as we shall observe. The orbitals of these two electrons are confined to the magnetic valley. When we examine the results, we find that the valley generated by the first two electrons becomes a ridge created by the following two electrons.

Electrons existing in the Ls or 2s-sublevel form right angles to the Ks or 1s electrons. If the four electrons exist at the same distance from the nucleus, their magnetic fields approaching the nucleus would form right angles offset by 45 degrees. The reason for the offset is that the strongest energy is that of the most direct path existing between the nucleus and the electron. This line crosses the center of the electron and the center of the nucleus, which bisects the right angle giving $45°$. However, this is not the case; the L level s-sublevel electrons exist at a farther distance from the nucleus than the K level s-sublevel electrons. The result is that the angle favors the Ks electrons, meaning that the angle of control by the Ks electrons is greater than that of the Ls electrons. With a little cheating by examining data of the next sublevel, we find that this angle is $35.3°$ beyond the "equator" of the "magnetic valley." This angle will be explored in more detail later. However, it should be observed that two more valleys have been formulated between the two s-sublevels at $35.3°$ on either side of the central ridge with a mound rising in the middle upon both sides of the atom by the Ks electrons accenting the valleys, or more precisely, "magnetic basins."

Movement by the s-sublevel electrons within the L level is anything but stationary. They actually rotate in the direction of the nucleus. They connect with the horizontal magnetic lines leaving the nucleus, as described earlier. The rotation of these electrons causes the electrons to move further away from the nucleus via centrifugal force. By this the magnetic field of the Ls (2s) electrons extends beyond that of the next sublevel. Referring back to the kite scenario, the wind is blowing nearly vertically. Another interesting observation is that these electrons engulf the entire sublevel leaving no room for another electron. Unlike the Ks (1s) electrons, their ability to ward off other electrons is not entirely dependent upon their own strength. Recall that their weakness is at the midpoint between them. The distances between these midpoints are greater as they are farther from the nucleus generating a greater weakness. This weakness creates another valley within the ridge. However this valley is moving fast. Even without the

Chapter 13: Bondverse

movement, other electrons coming into the atom by the increased number of protons, find that the central region of the polar regions more inviting as the electron energy level is low. The only electron energy existing at the polar regions are the Ks (1s) electrons. The swirling effect seen in the Ks electron also exists within the Ls electrons. The primary difference is that the downward movement becomes restricted by the centrifugal force of its rotation. The movement observed within the Ls (2s) electrons resembles a spiral or a twisted slinky surrounding the nucleus.

Other s-sublevels

Examining other sublevels, we observe that the phenomenon repeats itself. Each of the sequential s-sublevels rotate like the Ls (2s) sublevel. Recall the magnetic valley that rotates with the electrons. This is the location that the next set of s level electrons moves. The result is that these electrons spin around the nucleus at right angles to the electrons underneath. Ms or 3s moves precisely in that manner as long as there are no external s-sublevel electrons. Only the Ks (1s) electrons congregate at a right angle to the s-plane.

Entrance of more s-sublevel electrons requires the nucleus to contain more protons to become present. However, we can cover more s-sublevel electrons without discussing the other sublevels as this region of the atom is untouched by the other electrons. One item to point out is that even though the electrons exist further away from the nucleus, their magnetic charge does not increase in magnitude. The result is that the valleys formed by these electrons become steeper as they formulate further away from the electron. The reason is that the shell circumference increases in size while the magnetic fields of the two electrons remain the same size within each succeeding s-sublevel.

The next s-sublevel set beyond the M shell level attempt to continue the right angle routine to the set of electrons underneath. However, there is a slight "bump" in the middle of the valley by the L level electrons under them. The new arriving electrons move to one side of the bump. As they do so, they affect the electron orbital pattern underneath. They are forced to move somewhat in the opposite direction maintaining equilibrium between electrons. The result is that the N level electrons create a valley offset from the right angle ideal. This causes the next set of s-sublevel electrons to move into their locations at a near right angle. However, the phenomenon balances itself allowing the right angle relationship to continue somewhat. The process continues through the known isotopes of atoms consisting of seven levels.

Next, we need to examine the relationship that distance exerts between levels within the equation of s-sublevel electrons occupying the atom. Recall that the angle between the K and L sublevel electrons did not form right angles because of the difference in distance by the two sets of electrons. The same is true with the additional sets of s-sublevel electrons. For instance, the M sublevel does not require that the electrons

Part 4: Creation of Atoms

underneath to comply with the 60 degree interval. It only requires that room be made to keep the balance of energy between them. Even though the resulting angles are not right angles, each new set attempts to form right angles to the previous set. In other words, the next set of electrons does not accumulate continuously in a clockwise or counter clockwise manner, but in an alternating pattern. From this point onward, we will use the transitioned standard notation of the sublevels. Example: for Ls to Qs, we will use the symbols 1s through 7s. The primary reason for the convention is that a number and letter together is less confusing than using two letters even though we were using capital letters for the level.

Sublevel p

While s is used to represent the word spherical, p is used to represent polar electrons. These electron orbitals are called polar because their positions can be ascribed to occupy both sides of a set of x, y and z axis that are right angles to each other as the dimensional axis. These are labeled by scientists as sets of p_x, p_y and p_z orbitals. However, when we examine them more closely, we observe that these orbitals actually exist at $120°$ angles to each other.

If we were to take a perfect cubical object and balanced it on one of its corners upon a perfectly flat surface and looked straight down upon the cube, we would see only three edges of the cube pointing exactly $120°$ apart from each other. If we were to measure the corners at the end of these edges from a point directly underneath them, we will find that these distances are equal. We will also find that the opposite set of corners offset by $60°$, which perfectly bisects the angle of $120°$ between the edges viewed. These corners are also equal distant from the table, but at a smaller distance than those previously observed. We will go into more detail later, but the understanding of the $120°$ angle is important in analyzing the relationship between the electrons existing within the p-sublevel. In this, the three electrons repel each other equally within a spherical domain. This formation creates different magnetic valleys and mounds. Interesting enough, the p-sublevel electrons on the opposite side of the atom exists at $60°$ angles to those on the side analyzed. The greatest valleys formed are at the edge between the electrons. This pattern is quite different from the s-sublevels.

The 2s sublevel acts as a basket holding the three p-sublevel electrons. This is true for both sides of the atom. It controls the elevation angle of the p-sublevel electrons as they repel each other within the "basket." Recall the angle that exists between the two sublevels is $35.3°$. Now, we need to illustrate the derivative of the $35.3°$ angle.

We can calculate the angle as it exists between the flat surface and one of its edges of the cube. For measurement purposes, we will give our cube a measurement of 1x1x1 units (the size of the units does not matter as they result with the same angles). We will initially place the cube flat upon the plane. Our first modification is that we are going to

Chapter 13: Bondverse

lift one side of the cube up until it is balanced upon its edge. If we draw a diagonal line from one corner touching the "floor" to the opposite corner, this line forms a $45°$ angle to the cubical edges of that cubical face. Now, we lift the back opposite corner up until the cube stands upon one corner in perfect balance. Imagine a line from the bottommost corner to the uppermost corner. By these two lines, we can find the angle.

Using the Pythagorean Theorem, we add the lengths of the cubical edges inscribing our first line and take the square root of the answer. Naturally, we end up with the square root of two (1.141421 units), as 1+1=2. The next line length is found by taking the square root of the addition of the squares of the previous line and the edge inscribing the line. Here, we end up with 1+2=3, and then we take the square root. The square root of three is about 1.73205 units. If we divide this number into one, we get the sine of the angle, which is about 0.57735. Then we can take the arcsine of this decimal to attain the angle which is $35.264°$ rounded to $35.3°$. Now, we need to relate this angle to the surface. Thus far, we have established that if we lift and balance a cube on one edge and draw a line upon its surface at its point of contact perpendicular to the "floor" and then lift the opposite bottom corner to stand it in balance upon that corner, the angle between these two lines is $35.3°$. This means that we lift the cube's bottom-back corner the same angle. So, when we pivot up lines $120°$ apart $35.3°$, they form $90°$. Using this information in conjunction with the observed data, we know that the electron orbitals are elevated $35.3°$ to the nucleus. This angle lessens with each succeeding s sublevel. And after the 7s sublevel is filled, the higher "energy" sublevels bury the phenomenon.

Our next observation concerns itself with the relationship in size of the 2p sublevel orbitals to the 2s sublevel orbital. The p-orbitals are slightly smaller than the s orbital. The reason for this difference is that the 2s orbitals are rotating around the nucleus at a rapid speed giving centrifugal force enough energy to move the electron further away from the nucleus. Another interesting observation is that the p-sublevel "dives" closer to the nucleus than all the upper s sublevels, which gives it the tear shape. Actually, the p-sublevel attempts to act as a level 1s electron. The primary reason is that it gathers energy from the polar xergopaths emitting from the nucleus. However, the void external to the electron is greater than the domain of the magnetic field generated by the electron, and the electron "slides down" to the edge of the "basket."

One last observation of the p-sublevel, we observe from the data gathered by scientists that the p-sublevels are somewhat concentric in nature. There is a variation to the concentric nature of these orbitals. Imagine that we had a set of concentric pipes that fit loosely within each other, and that we laid this set of pipes on the ground horizontally. If we were to look at the ends of the pipes, we would observe the pipes resting at the bottom side toward gravity. Similarly these orbitals meet with resistance near the nucleus like the ground is to the pipes. By this, we see that each p-sublevel orbital one level above engulf the space a p-sublevel orbital underneath it. The reason for this returns us to the nature of the xyzenthium crystals forming the energy pattern.

Part 4: Creation of Atoms

Recall that the "feelers" were of a ratio 1:3 per side of the atom. Multiplying this by two gives the two horizontal s sublevel magnetic pulls and the six, three each side magnetic pulling line sets. The ratio sets the stage for the optimum electron relationship, called the octet rule within chemistry between atoms, as we shall observe shortly. The exterior most p-sublevel reaches closer to the nucleus because of the long acceleration time in the approach to the nucleus giving it slightly more momentum.

Sublevel d

Remembering d as a symbol of this sublevel can be achieved by relating it to the word decimal meaning pertains to tens or tenths. There are ten electrons composing this sublevel. Again, the sublevel is divided into two hemispheres giving five electrons to each side. These electrons form at right angles to each other giving four with a central electron making five. While the pattern between sides are somewhat symmetrical, the two sides exists independent from each other as the p-sublevel orbitals separates them.

The nature of the d-sublevel orbitals resemble the p-sublevel orbitals in that their magnetic source stem from the polar magnetic energy coming from the nucleus. Like the p-sublevel, these electrons attempt to imitate the 1s polar electrons existing at the K shell level and exist stationary within the shell. As observed with nature by scientists, the 4s electrons fill before the 3d electrons.

When the electrons come to fill the 3d level, without filling the 4s sublevel they slide down the 3p valley off to the edge. At the edge, there exists a deeper valley wherein the first two arrival electrons are drawn into and stay. The next three d-sublevel electrons move into the p-valleys forming the first 3d sublevel orbitals, but not at $90°$ angles, but $120°$ as demanded by the electron fields of the p-sublevel below and kept in place by the 4s electron orbitals. By the data collected upon this matter, it seems that one side develops and then the opposite hemisphere develops. When the next electron enters, it shifts the orbitals into $90°$ angles moving them out of synchronization with the p-sublevel underneath. In this, only the first electron that enter the d-sublevel routine is lost to the 4s sublevel increasing the energy level just enough to make it higher than the central location between the four electrons wherein the last electron is drawn for a total of five. This process repeats itself for the opposite side after losing the first electron back to the 4s sublevel.

The question arises pertaining to the filling configuration of the electrons. In reaching the d-sublevel within the third shell, a deviation occurs. Not only does it fill the 4s sublevel first; it also removes electrons from this sublevel to fill the d-sublevel in certain intervals. This occurs when the fourth and ninth 3d electron is needed. It gives the feeling that one side fills before the other giving a favored side and an unfavorable side. However, other possibilities exist. More observation data would be necessary to assess which phenomenon prevails. However, the reasons for the 4s electrons to fill first and

Chapter 13: Bondverse

the irregular 3d filling are both caused by the geometric shape of the electron field generated by the configuration generated below the last gathered electron. Example, the 4s filling locations prevails over the 3d filling location because of the size of the gap between the 3s electrons is greater than the gap between the 3p electrons. Along with this, the five electron configuration does not fit well over the three electron formation.

The next d-sublevel is the 4d orbitals. There is a slight variation due to the extended room generated by being further from the nucleus. The 5s fills first as expected. But the desired right angles are easier to achieve because of the gain looseness provided by the added distance from the nucleus. By this, the sublevel is able to leap to the four pattern before losing the second electron to the 5s sublevel. Actually, there is a sinking effect upon the sublevels, as we require more electrons further from the nucleus, by the sheer fact of distance. The attribute that makes distance a factor is that the electron's magnetic influence does not increase because of its distance; they are uniformly sized.

Sublevel f and Beyond

There is no sublevel e; sublevel f has fourteen electrons within its sublevel giving seven for each side. The symbol f can be remembered as it is the final observable sublevel within the natural elements found upon the earth starting and the sublevel starts on the fourth shell level. The atomic shell surface at this sublevel is very complex making it harder to follow. Only the s and p-sublevels in the higher shell levels are easily discernible and the d somewhat.

The next sublevels are sequentially assigned g, h and i. The patterns are somewhat dubious because we have no real pattern to observe at ground state. However, it is interesting that the last sublevel forms a Star of David with a central point per side. Some scientist believe that after a certain point, electrons just collect about the atom in a somewhat random manner by the dictates of the number of protons within the nucleus. Then there is another problem: What if a nucleus contains more than 280 protons? Naturally there would be another level, but it may become meaningless as the surface of the atomic shell becomes even more integrated into a single cloud of electrons as the sublevels increase. Again, this is beyond the elements found on earth.

Orbital Limits

We are now going to examine in more detail the nature of the orbitals at ground state. The basic concept is that there is a spherical shell-like groove surrounding the nucleus caused by the bend in the space/time continuum by the two primary forces jammeria and juttoria. The basic equation is the addition of two functions: **$y=g(x)+h(x)$**, where x equals the distance from the centroid of a nucleus of an atom. However, in examining the two functions or factors, the formula becomes quite complex. There are a few

Part 4: Creation of Atoms

variables involved. The first two variables are the mass and heat in modifying the jammerial curve and the juttorial curve respectively via distance. In essence, the equation defines one fiber of the space/time continuum from the centroid outward.

Reysh Factor. This is the g(x) function of the formula. The g is chosen, as jammeria is spatially expressed as gravity. The magnetic discernible manifestation reaches further outward than the microstrong force as expected. Jammeria initially pulls the fibers downward to the groove, the curvature of this pull resembles the graph for the square root function or $y^2=x$ and x is greater or equal to zero, or more precisely, looking down a funnel that fans out near flat on top. A better description is $y=-1/x^2$, when x is greater or equal to zero. The -1 flips the graphic results about the x-axis. The curve flattens out below the positive x-axis to touch the x-axis at infinity, and drops near parallel to the negative y-axis. Rotate this line around the y-axis to generate the funnel-like shape, and we have a vertex at negative infinity. There are alterations, mass is not a constant because of the different isotopes of nuclei and the effect temperature has upon a nucleon. Even so, this does not destroy the general image, only the slope of its curvature. The size of nucleus in relationship to the graph is as a tiny dot covered by the intersection by the lines drawn for the two axes. Imagine an electron sitting somewhere upon the near flat surface of the funnel. Eventually, it would roll toward the center; the incline reaches a point where the fall is greater than the distance directly to the center (x=1). At this point, the electron is trapped by the nuclei. This would be the image if attraction was the only energy involved, with a minor alteration by the multiplication of the jammerial constant. However, juttoria also has an input.

Quuf Factor. This is the h(x) function of the formula. The h is chosen, as juttoria is spatially expressed as heat. Quuf is the nineteenth letter of the Hebrew alphabet; it demonstrates the least within the picture language. While the Reysh Factor generates a large funnel, the Quuf Factor generates a central small mountain at ground state of an atom. Heat is only the excess of juttoria brought in by the constant impact of photons upon a given nucleus. On the opposite side, absolute zero is not the absence of juttoria but, the inability of juttoria to produce a photon from any of the electrons associated to the particular nucleus because of the lack of excess. By this, we find that this factor varies the greatest, while heat alters the Reysh Factor by a very small fraction. The image of this function is that of two flattish curves rising to a sharp peak in the center bending the space/time continuum upward at the center, this is after adding it to the negative vertex of jammeria, showing the inward functions (0<x<1) of jammeria and juttoria as illustrated in the microweak and microstrong comparison on page 78-9. If the two function we just added together, there would be a noticeable "crease" in the valley formed, but the tension between the two makes the valley smooth. The final image looks similar to a cross section lengthwise of a semi-deep-welled spoon.

The electron is pulled by magnetic energy down the jammerial slope. As this occurs the electron gains momentum. After reach the bottom of the groove, the electron starts to

Chapter 13: Bondverse

climb up the cliff via its accumulated momentum. Then it reaches a point in which all the kinetic momentum becomes exhausted. The potential energy by the height the electron "climbs up on the cliff" is transformed into kinetic energy as the electron falls back into the groove gaining momentum and climbing the jammerial slopes again. The electron continues this rocking back and forth between the two forces. The height of the "climbs" doesn't change as long as the atom remains in the ground state or without excessive heat or excessive lack of heat. There is more that happens to the electron during this period that causes the vibratory movement as we shall soon observe.

Now let's examine another isotope containing more protons than the previous example requiring an electron to be drawn into the same region. The first electron to enter the groove alters the landscape of the valley by its own magnetic repelling force upon the second electron as it moves toward the valley. There is now a bump that was not there when the first electron arrived giving it a "higher ground" or shell level. When the next electron enters the scenario, it seeks the lowest ground possible. This may not always the lowest point in the shell as there may be higher pit-like valleys. However, within the initial fill from the plasmatic state, the lowest possible point is equal to the lowest point with the shell.

Atomic Radii

Previously, our view of the shell levels and sublevels has been static in electron field sizes. However, electron shell levels and sublevels are not equally sized between atoms. In essences the magnetic fields of the electrons becomes somewhat flattened by the energy exerted from the nucleus. Gravity is not the real issue, but does have a small effect aiding the process. The primary reason is that the magnetic pull increases with each proton added to the nucleus, but the shell level and sublevels do not increase in level with every proton. The result is that a given electron level and subsequently its sublevels pulls inward with each additional proton. This process has a counter process constituting the requirement by electrons of adding another shell level. If we were to examine the periodical chart for the elements, we will find that the atomic radii of an atom shrinks as we move right to left within the chart and grows as we move from top to bottom within the chart.

Shrinkage of the radii from left to right is not linear in nature excluding the first two that appears linear because there are only two points of reference. Looking at the second line down or the second period, we observe a severe difference in sizes. The line connecting their values form a steep slope downward as we move to the right of the table and then curves outward to a near level angle. This pattern occurs for each period underneath. However, the lines are less severe and of larger sizes. Scientists measure the radii of the atoms in angstroms, which is a unit of measure equaling $1.0E-10$ centimeters or $1.0E-12$ meters or trillionths of a meter. We will examine the first ten atomic radii. Hydrogen is the smallest atom with helium being a close second

Part 4: Creation of Atoms

measuring 0.31 and 0.32 angstroms respectively. The second period consists of lithium through neon. The first and largest atom of the period (row) is lithium at 1.23 angstroms, which is about 3.97 times larger in radius of hydrogen or about 62.5 times the volume. The following radii are 0.89 (beryllium), 0.82 (boron), 0.77 (carbon), 0.75 (nitrogen), 0.73 (oxygen), 0.72 for fluorine and ending with 0.71 angstroms for neon. As readily observed, there is not much variation between the last five elements in radii. Lithium is large in comparison to the first ten elements. However, there are larger atoms. The largest atom radii is francium at 2.7 angstroms holding 87 electrons while uranium holding 92 electrons has a radius size of only 1.42 angstroms. Francium is slightly over twice that of lithium giving a volume about 10 times greater. This makes the volume of helium slightly over 660 times smaller than that of francium. The volume of oxygen will fit slightly over 50 times into this atom.

Bonding of Atoms

If magnetic energy were just the matter of the balance between positive and negative charges between protons and electrons, bonding would not occur because these charges already exist in balance at ground state. Secondly, there would be no change in the radius of the atom having the same exterior shell level save the slight change by gravity. The magnetic lines connecting the electron to the proton's energy are composed of hypertips and hypotips which contain juttorial and jammerial xevim (energy angles). These radiate energy fields similar to a gravitational field. It bends space without regard to the magnetic connections that exist between the electrons and protons involved. The term scientists give to this nature is electronegativity from the words electron and negativity. In essence, it is the positive threshold level that exists upon the atoms surface shell level. This system shows the relationship between covalent and ionic bonding. With values from zero to four, values of two and greater are considered to be ionic.

Imagine a surface that is only able to pull one electron to itself. For our first example, we will use hydrogen. It can only attract and hold one electron, but has room for two. The juttorial force does not selectively bend space on one part of space and ignore the other. The bend in space continues all around the nucleus. The magnetic lines are primarily involved with that one electron, but the energy formulating these lines are not dedicated only to this electron. Consequently, there is a side of the atom which has a positive charge, and the opposite side of the atom the electron emits a negative charge far beyond the boundaries of the nucleus. This gives the hydrogen atom the oxidation number of +1. It means that it is more apt to give its electron to another atom leaving it with a charge of +1. When two hydrogen atoms come within vicinity of each other, they both grab the others electron to fill in their hole in the negative charge field. The result is an image similar to the helium nucleus as a circuit of alternating particles; the exceptions are that the particles are separated and the electron has a negative charge.

Chapter 13: Bondverse

Continuing the scenario, in the second shell level lithium behaves much like hydrogen in bonding. However, within the second shell level there is room for eight electrons total. As the nucleus is able to gain more electrons within the second level, the positive radiation strengthens by the gain of protons within the nucleus, which also gatherers electrons. The strength increases into fluorine as the maximum which has the strongest electronegativity. Actually there is one stronger, neon; however, the electron slot is filled and becomes a noble gas (bonding with no atom).

This process brings into being- the octet rule. This is the tendency for atoms to imitate noble gases. The noble gases have eight electrons within its outermost shell. The octet has some exceptions that seem illogical until we examine the phenomenon more closely. Primarily, the octet rule applies to the first 18 elements, hydrogen through argon. After that the octet rule fades in and out as we pass through the transition elements, called that by their bonding natures and reflect other sublevels like d and f. However, well over 95 percent of the elements existing upon the surface of the earth as solid, liquid or vapor, with the exception of iron are of the first 18 elements. Iron exists to be about five percent of the crust. By this, the octet rule's influence is magnified in observations.

Some exceptions can be explained. Boron trifluoride (BF_3) only provides only six electrons to the boron atom's outer shell. Since the boron atom has only three electrons in its outer shell, we could assign it an oxidation number of 3+ (giving three electrons to bond with other atom(s) being in Group IIIA) and fluorine 1- (taking one electron from the bonding atom being in Group VIIA) and be satisfied. From this assignment, we observe 3 fluorine atoms required to equal 3-, which balances the 3+ in ionic bonding. This would leave the s sublevel as the outer layer being full with two electrons. However; this is not the observation, the three fluorine atoms gains one electron each from the outer layer of boron. It should also be observed, this gives a total of eight electrons to each fluorine atom. The boron atom has a complete shell of two electrons. This abstractly maintains the octet rule. Nonetheless, there is another principle functioning that inscribes the ionic bonding and the octet rule and the apparent exceptions; it is the interaction between the positive charge generated by the structural holes in the electron configuration and the strength of the positive charge.

Another case of apparent violation of the octet rule is arsenic pentachloride ($AsCl_5$). However, in further examination, we will discover that it's actually maintaining the rule's dictate. Arsenic has 33 protons holding 33 electrons in its various shell patterns. The filling pattern is the following (inside the parentheses displays the actual electron count): 1s (2), 2s (2), 2p(6), 3s(2), 3p(6), 4s(2), 3d(10) and 4p(3). Notice the total count of electrons upon the outer shell is five.

This scenario is quite similar to that of boron trifluoride. Chorine has a high electronegativity attribute. Despite the break in filling of the third and fourth shell levels, the 4s and 4p electrons exist upon the outermost shell and are picked off by chlorine to be shared with the arsenic atom. Arsenic behaves similarly to that of the

Part 4: Creation of Atoms

boron atom. The remaining shell of the boron was complete within itself, so is the arsenic atom complete with the remaining third level shell. It seems somewhat like a coordinate bonding, in which the electron is "given over" to the exterior elements.

Obviously, there is much more that could be said about bonding. There is double and triple covalent boding readily observed in the oxygen and nitrogen molecule respectively. Coordinate covalent bonding, ionic bonding could also be discussed in more detail, but we would be reinventing the wheel. The primary chord that unites all the bonding issues is the magnitude of the positive threshold level existing within the surface shell level of two atoms as defined by electronegativity and the number of electrons existing upon each atom's outer shell level. This controls whether both atoms are sharing electrons, or just one, or even if the electron is just taken from another atom.

Fluorine has the strongest electronegativity of 4.0. The noble gases have the least at zero because they do not readily interact with other elements. Oxygen has the strongest electronegativity within the schema of abundance within our planet's composition. Electronegativity increases as we move from left to right in the element table and decreases as we move though each period within the table. These values are not absolute. For an example, Silver will be replaced by Mercury even though they have equal electronegativity of 1.9. There is not the expected equilibrium indicated by their limited two digit numerical values.

Electron's Anatomy

The anatomy of the electron is our next focus. As stated earlier, there are two layers to the electron: the surface electrosod layer and the core layer of nightsod. Under the nightsod layer is a large "empty" chamber called the photon furnace. Recall that there is a hole about 1/41 of total area of the nightsod layer. Attached to this hole are three crystalline fractures. Each fracture in the fabric of under layer leads from the hole to the closest corner following the original "neutron cube" edge. This generates three trapezoid forms by the breech. These trapezoid structures have composition of the neutral xyzenscapes, giving their internal structure magnetic pull. This is the same pull that caused the neutron to shrink into a sphere in the onset. The result is that these trapezoid forms shrink from their three free sides, via the slits, as this alleviates lateral tension holding them apart. The hole never reaches the equatorial division between front and back. If it did, this layer would shrink into nothingness. There are three reasons that this kind of shrinkage does not happen. One, the crystals of the hole's edge would have to expand to pass through the electron's equatorial belt. Two, the gravitational influence between the two layers holds the interior layer in place as both electrosod and nightsod both have hypotip activity upon their surfaces between them. Three, is the repelling force of daysod interior. The result is that the internal hole is a latitudinal circle about 20 degrees above the electron equator. The internal hole creates

Chapter 13: Bondverse

an imbalance in mass between the two sides of the electron equator. By this, the internal hole is held away from the nucleus.

The external layer of the electron covers the nightsod hole. The ceiling of this chamber is the exposed surface layer of electrosod, the larger floor of the chamber is daysod. The two hemispheres of the electron are different. One hemisphere has an uninterrupted surface, while the other has fractured surface. Unlike fractures seen in ordinary rock, these are quite crystalline in nature forming straight and nearly isometric lines. These fractures formulated their pattern during the electron's departure from the "evaporated neutron surface." Recall the lateral formation of electrosod. It has three faces of the four faces projecting hypertips making for a weak interconnection. This coupled with the shrinking away via division causes the fractures to be maintained. These three fractures upon the exterior surface of the electron are aligned with the internal crystalline fractures of the interior layer. The nightsod hole formed is called the photon passage and the electrosod fractures forms the photon gate. The ring of concentrated daysod defines the **rayring** (ray-ring).

Anatomy of an Electron

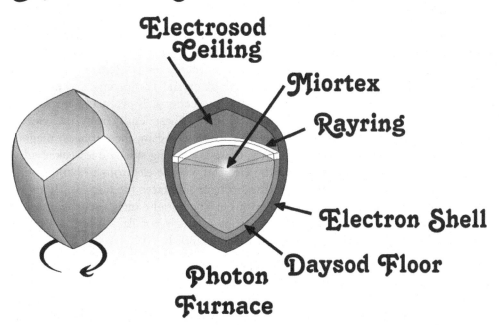

The electron's fractures, by which the photon escapes, remain as fractures. The reason is that the xergopaths of the lateral faces forming the fracture join to the adjacent crystals within the newly formed surface. The process in which these fractures maintain their independence from the opposite crystal is found in the nature of the xyzenthium

Part 4: Creation of Atoms

crystal itself. The lateral faces contain two faces of positive charge, one weak neutral face and one strong neutral face. These exist within a uniform pattern. When the fracture occurs, two different alignments appear on their fracture edges. On one side we see an alternate pattern between microstrong and positive faces, and on the other an alternation of microweak and positive faces. Of the two neutral faces, microstrong and microweak, it is the microweak faces that interacts with the magnetically charged positive faces adjacent to them. By this, half of the xergopaths become adjacently connected, while the other half reconnects their original interface. This weakens the strength of the magnetic hold within the fracture. When a strong contracting tension is applied, the fracture separates.

The interior of the electron is basically a large cavity in relationship to the dimensions of electron's exterior. This interior region is called the photon furnace. The rayring is a catalyst for formulating the miortex. The miortex behaves like a seed in which the photon develops. The interior xergopaths stacks the miortex collective. The electrosod ceiling is the interior of the surface shell provides hypotips needed for the development of the photon's components. The hypertip flow from the daysod floor determines the wavelength of the photon. The daysod floor serves also as the launching pad for the departure of the photon as the surface contracts from exposure to the external coolness beyond its orbital.

The magnetic energy radiating from the electron behaves in a similar manner as the collective proton energy radiating from the nucleus. We have four groups of energy lines per side forming eight lines total. These energy lines stem from the four diflohexets forming the faces of each xyzenthium crystal composing the electrosod xyzenscape of the surface. One-fourth, of the projections, is from the microstrong diflohexet emitting hypertips instead of hypotips. These emit from the electron horizontally, just as the hypotips emit from the nucleus. Again, the strongest magnetic force emits from the polar regions. Similarly, there is a strong side and a weaker side as the polar magnetic energy winds around the electron, via the electrons isobaric spin, on its course to the central plane. Naturally, the stronger side faces the nucleus. It appears that the s-sublevels from 2s through 7s electrons exist at a right angle exchanging energy from their central planes propelling the electrons on their circular "orbit" around the nucleus via photon-like activity that will be explained shortly.

The size of an electron varies greatly; it depends upon the heat that it holds. If the xyzenthium crystals were to remain the same size as when it left the nucleon, the electron would be the same size as a neutron. However, the crystals shrink as there is less energy holding them from the center and have the same lateral magnetic energy pulling together. On the other hand, there is less neutral layer energy present to shrink the electron. The shrinkage is stopped by the daysod interior repelling itself. As more heat enters into the electron, the size of the photon furnace grows from the pressure of the growing photon within. This stretches both layers of the electron.

Chapter 13: Bondverse

Surface appearance of the electron has also altered by the arrangement and nature of the two layers. The original cubical shape of the electrosod layer tends to regain control over the shape, as there is less internal energy pushing outward. If it were only this process involved, the electron would become a cubical form with rounded edges and corners. However, there is a partial internal layer with a different nature. While it also has the same original cubical nature, the magnetic mechanics existing internal to the layer desires to shrink the assembly, but as stated earlier, the daysod xyzenscape provides resistance to prevent this from occurring. Recall that the electron shot off the main mass with extreme force. This is similar to the energy holding the electron out from the center as the daysod in 3/4 of the electron is facing itself. The force is actually stronger than the hyperpush. The proton surface partially has hypotip activity, but the daysod has zero hypotip activity. The repelling force also takes a semi-crystalline appearance because of the lack of internal matter pushing at its walls. However, 1/8 of the shape has either been destroyed or shrank creating a line near 20 degrees above the electron equator. The exposed electrosod layer to the interior and also has a negative charge as the surface. This generates a lack of repelling by the "ceiling of the proton furnace to 1/8 of the daysod floor for the total of 7/8 (1/8+3/4). This part of the electron collapses somewhat flattening the top eighth of the form. The top eighth is a corner of the crystalline form of the electron. The image is that of a cube standing on one of its corners with the top corner flatten. The opposite corner from the flatten corner also experiences some flattening, but not as noticeable. The reason is that there are two layers holding the integrity of the corner instead of one.

Next, we have the isobaric spin of the electron. This causes the electron to spin like a top using our "flattop" corner-standing cube image. The effect of the spin has upon the cubical form is that it tends to pull the crystalline pattern into a circular pattern about the spinning axis. The resulting image is that the spinning top has a slight crystalline appearance about the spinning axis and lengthwise along the spinning axis a squarish form with rounded corners and a flatten top. This form will alter as the photon forms within the furnace. The whole assembly becomes more spherical in nature as the photon pushes outward upon the electron's walls via juttorial energy.

We will start at the point that the electron draws closer to the nucleus. As stated earlier, the electron is drawn into the nucleus, the juttorial energy gets stronger. The variation via distance becomes more pronounced. This appears as we examine the difference between the ratios by the square of two different pairs of sequential numbers: 2, 3 and 4, 5. The squares are 4, 9, 16, and 25 respectively. The ratio of the first set is 4 to 9 generating 1:2.25. The ratio of the second set is 16 to 25 giving a ratio about 1:1.56. The same holds true with $1/x^2$ for the inverse reason as x approaches zero.

Imagine the radiation of positive charge radiating like the sun upon the face of the electron. The central regions get more "sunlight" than the edges. Unlike the sunlight, the night-side of the electron receives energy as well. Even so, the radiation is weaker

Part 4: Creation of Atoms

via length of the energy line. So, we observe at the point closest to the nucleus receiving more energy than another point further away. Juttoria energizes the electron causing it to spin more wildly. This is observed by the increase in the electron's wobble as it moves away from the electron. The wobble continues to grow and decreases in growth up to the maximum wobble at the **microstrong limit**.

The number of hypertips entering into the electron outnumbers the hypotips leaving the electron as the electron moves closer to the nucleus. The reason for this imbalance is the density of mass. The electron has become far less dense in mass and radiating the same energy. In essence hypertips are being shot into the electron like a high-speed machinegun while the hypotips are being shot into the proton like a semiautomatic weapon. When the extra hypertips enter into the surface electrosod xyzenthium crystal, they initially behave as part of the xergopath stream. However, when they reach the miortex, there is no counter tavtip-2 to deflect it. Eventually, it just passes on through into the core (photon furnace) of the electron.

Now we can concern ourselves with the observations describing the formulation of the miortex within the photon chamber. Recall that the initial form of the layers is cubical in nature and are bent into a spherical format. The spherical format does not destroy the fact that there are six faces involved. Each face interacts with their opposite cubical face. The bent nature of the wall causes the crystals' xergopaths to cross the center of the chamber. The physical finite distance between the inner walls of the photon chamber has no bearing upon the formulation of the miortex. Within infinitesimal space, the miortex reformulates regardless of the finite distance the tavtip travels to reach this finite location.

Initially the tavtips within the photon chamber just cross on through. Eventually, they become more intense until they intersect simultaneously. The tavtips move away from their linear path to join with perpendicular tavtips. Recall that they do not collide in the normal duocollision manner. They are drawn to the electron's centroid creating a tavtip-2, which continues the kinetic direction by the combined vectors of the original two tavtips. These collide with other sets of formed tavtip-2 bethtips and establish the miortex within the central region of the of the photon chamber. This will, in turn, strengthen the distortion of the space/time continuum within the photon furnace that promotes xergopath stacking to the core centroid in the miortex collective pattern.

Remember that the internal layer had three trapezoid-like structures that shrank down to form the rayring near 20° above the electron equator. The shrunken crystals of the rayring produce the same amount of daysod energy per crystal. These crystals lose volume and not mass. Because of this shrinkage, the emitting energy of the rayring is more intense. Because of their structure and angle, their energy reaches a focal point at the center of the electron. This acts like a barrier for the xergopaths moving vertically and semi-vertically and deflecting hypotip energy coming from the electrosod ceiling. The exposed electrosod inserts some jammerial force into the chamber giving it the

Chapter 13: Bondverse

counter force. This force is much weaker in this scenario than in the xyzenthium crystal as the exposed electrosod is 1/8 at maximum instead of half of the influence. This makes photons from electrons less dangerous than those originating from the xyzenthium crystals. Ultimately, it is the ratio between the jammerial and juttorial forces within the photon chamber that determines the resulting photon's wavelength.

Formulation of the photon occurs as the electron continues to get charged by the incoming energy into the electron structure. The return of the depleted hypertip energy to the electron, as it approaches the proton, energizes the shell. In the process, the proton pushes the electron away as the juttorial energy influences the miortex centers of the xyzenthium crystals upon the shell surface. Recall also that hypotips also project from the protosod xyzenscape (surface of the proton). By the radiating intensity differences between the nucleus and a given electron, even the hypotip depletion of the electron replenishes with hypotips. By this, we see the hypertip phenomenon feeding the photon furnace as very intense.

After being pushed away from the nucleus and reaching the **microstrong limit**, the phenomenon of the increased "heat" stops, and the wobble of the electron deteriorates. The amount of "heat" reaching the electron becomes less than needed to maintain its "warmth" and cools. However, the surface cools faster than the internal layer surrounding the photon furnace. The internal photon becomes independent of the electron as no more energy from the shell feeds it. Similar to the electron's isobaric spin, the juttorial force pushes upon the miortex for the linear escape. Unlike the Electron, this push does not promote a spinning force. The twisted magnetic lines within the photon begin to straighten out seeking the shortest possible distance. They become as the xyzenthium crystal in form. In this internal form, they still rotate.

After, the momentum carries the electron beyond the range that is "comfortable". The electron moves beyond the distance in which the heat from the nucleus can maintain the shell. The electrosod shell becomes "cold" and contracts. This opens the "gate" of the electron (the fractures separate via shell contraction). The photon then shoots out of the electron as a bullet via release of pressure. Moreover, there is a decompression of the photon. As the photon escapes, the photon gate seems stationary to the decompression. The gate holds the magnetic lines from rotating as the photon expands. The decompression of the photon is limited by the amount of juttoria present. The escaping photon pushes the electron back toward the nucleus with the exception of the 2s - 7s s-sublevel electrons. This process propels their electron in its circular orbit.

In normal cases at ground state, the photon is a heat photon unseen by the naked eye. It does contribute to the general feel of warmth above absolute zero. Note: even one degree above absolute zero is considered warmth despite that it is below the freezing point of all elements except helium which freezes just a fraction under a degree above absolute zero. In the excited state of an atom, it releases a photon that usually occurs within wavelengths of visible light. The electron at ground state behaves similar in

Part 4: Creation of Atoms

nature as an electron in the excited state. The only differences are that they stay within their orbital and emit waves longer than visible light. Unlike the photons from excited atoms, these electrons release the energy in a less explosive manner. The reasoning is that they are at near equilibrium magnetically; hence, they are near heat balance.

After the photon departs, initially, the internal layer remains expanded beyond the limits needed for the cool external layer to close. In other words, the electron's photon passage remains unobstructed by the photon gate. This allows hypotips to be attracted into the internal surface like a magnet restoring lost hypotips from the magnetic interaction with the protons. As this process becomes complete, the interior surface shrinks via hypotip's jammerial influence allowing the gate to close and then starts again.

Anatomy of a Photon

The term, photon, comes from Quantum Mechanics, and is used to describe the energy packet formulating the nature of an electromagnetic wave. As known in science, the electromagnetic wave has two wave components these components exist in right angles from each other. These waves are viewed as transverse waves that have the same wavelength. These paths can be inscribed on two perpendicular planes. The wave moves in direction "K" which is inscribed by the intersection of the two planes. The vertical plane is labeled E for being the Electric Force Plane. The Magnetic Force Plane is labeled B. Scientists also note that electromagnetic waves behave as both a wave and a particle.

The anatomy of the photon as seen in xyzenthium terms infers the previous model. The xyzenthium photon incorporates the two planes as a central representation of the volume of energy involved. The shape of the photon in general is similar to a long cylinder-like ellipsoid. Unlike a planet having a wider horizontal axis, the photon has a much longer vertical axis. Within the photon a double teardrop image appears along the K-axis. The ends are rounded and the central region collapses at an acute angle. There are two other sets of teardrops forming perpendicular to the K-axis at right angles to each other. These forms distort the space/time continuum. The length of a photon varies with the size of the wavelength dictated by juttoria. Most photons are much longer than the diameter with the exception of those leaving the nucleons. All photons move in the streamline fashion as a missile.

A photon then appears as containing a single cycle of the wavelength. Unlike the wave model, which has the wave continuously repeating itself, it is self-contained. The internal image is like the symbol for infinity. The infinity symbol does not reflect the diflohexius formulating the photon, but of their **ergitude**s (er-gi-toods). An ergitude is the general bend in the fibers of the space-time continuum. Recall that the juttorial and jammerial forces bend these fibers via alephtips in opposite directions. As we observe the forces rotated at different angles it becomes difficult if not impossible to determine

Chapter 13: Bondverse

which way is up or down. The three dimensional image then appears as two tear drops joined at their extended points from being a sphere. However, in examining a single strand, we observe that the relevancy of the wave format. It has both a trough and a crest as depicted by the ergitude of the force. Juttoria is naturally responsible for the crest, and jammeria for the trough. As the photon passes through space as a particle, the warping factors move with it, giving the wave an appearance of motion.

The two magnetic responses exist at right angles to each other perpendicular and centered on the K-axis. Both of these charges contain both hypertips and hypotips in proportional amounts by both planes. This brings the bend of the space/time continuum back to zero for the node existing in the middle of the light wave. Both gather and disperse energy into the two neutrally charged diflohexets at right angle and in opposite direction to each other. The spatial warp is the collective effort of all hypertips and hypotips involved. The front view looks similar to a missile-like object with two sets of fins setting at right angles to each other.

As noted in science: visible light waves have a wavelength around 2,000 times that of gamma rays. Infrared (heat waves) are longer still while ultraviolet is shorter. However, in general the ratio is about 2,000 to 1 between visible light and gamma rays. This is similar to the ratio between the mass of a neutron to the mass of an electron. This illustrates that mass and density has a role in the photon's host during its creation. The various lengths in gamma rays occur by the various masses of atom nuclei. The shortest wavelength comes from the radioactivity of neutron stars that are extremely massive. The size of a photon can reach beyond the size of an atom as a heat photon. The longest of these come from electrons in plasmatic conditions. Even in sub-plasmatic states, the electrons release some heat photons beyond the range of visible light during their orbital movement.

This brings into focus the five expressions of photons. The different kinds of photons derive their significance by their source and the phenomenon forming them. The expressions of photons are: cosmic, gamma, strangeness, light and heat photons. These five expressions have two basic classifications, nucleophoton and electrophoton. Nucleophotons originate from the nucleons via ejecting surface xyzenthium crystals. Electrophotons are formulated and expelled from electrons.

Nucleophotons exist in the ultraviolet side there are cosmic and gamma rays. Strangeness from strange and charm quarks are gamma-like rays and are nucleophotons. Cosmic photons formulate from neutron stars. These never reach the earth. Because of their density, they are drawn and smash into atoms in space, and the byproduct of the collision reaches the earth. Like the gamma-photons, they formulate only from xyzenthium crystals. Gamma-photons are less dense than cosmic-photons; hence, cosmic photons are smaller than gamma photons. Both have an excess amount of hypotips. Their total bethtip count is much greater than the light-photons. Inversely, infrared heat waves have a smaller count.

Part 4: Creation of Atoms

Electrophotons overlap in the electromagnetic spectrum with the nucleophotons. In the ultraviolet end of the spectrum, we observe x-rays having the same wavelength as a gamma ray. Light-photons formulate when electrons are removed from their standard orbital. Heat-photons formulate as the electrons move within its standard orbital. Moreover, heat-photons are quite similar to light-photons as electrons form them both. The heat-photons are longer than light-photons. Unlike the light-photons that are ejected, the heat-photon streams out gradually (comparatively to a sudden ejection) as the electron moves toward the nucleus. Therefore, the length of the heat-photon is multiplied by time (very minuscule time-lapse) unlike the instantaneous light-photon.

The plasmatic photon is a cross between the heat-photon and the light-photon. It is classified as a heat-photon as it does not instantaneously discharge photons unless it comes into contact with cooler substances. Plasmatic electron's emission of visible light and heat are inhibited by extreme hypertip activity. The outer shell of the electron never cools enough to release all of its stored energy. Instead, the energy occasionally leaks out like a heat-photon via coolness, except its energy level is internally dense enough to emit light. This propels the electron through the plasma on its virtually aimless course. If the outer shell were able to cool enough, it would release all of its stored energy allowing itself to be drawn into a nearby nucleus. The result is a light-photon giving a greater measurement of energy, hence, more heat.

Contrary to popular belief by scientists, photons have mass. If they did not have mass, they would not have any physical attributes. Moreover, they have different masses that determine their wavelength. This principle is overlooked as mass and wavelength provide the same set of responses. Our problem is that the mass is so slight that we have been unable to measure it. The mass of photons is much smaller than that of a single xyzenthium crystal. Also, without mass, it could not respond to gravity.

The leading edge of an electron photon is the same as the gamma rays from nightsod i.e. jammerial influence leads. Only the strangeness attribute has these two ergitudes inverted. Jammeria causes the photons to dive into protons when they become trapped within an atomic subspace. After absorption, the juttorial force sends more "heat energy" to the electron. The electron gets excited and releases photons at closer intervals. The result of this activity gives us the added warmth to our senses.

Electrophotons have another attribute to consider. Despite not being formed at the beginning of Creation, they have the characteristic of a xyzenthium crystal in that all six diflohexets exist within the diflohexius. While the neutral faces of the photon align itself with the K-axis, the other four diflohexets also have their alignment. Since these four diflohexets promote magnetic energy, they align like the xyzenthium crystal. There are two positive and two negative charged diflohexets. Obviously, they continue the pattern: equal magnetic charges exist in opposite diflohexets placing the two opposite charges at right angles. The negative charged diflohexets aligns its axis with the E-plane within the framework of the electromagnetic wave. It follows that the positive charged

Chapter 13: Bondverse

diflohexets are aligns its axis within the B-plane. Because of these alignments, we get our model of the electromagnetic wave. However, we are not finished with the examination of an electrophoton. We have an issue concerning the nature of the formation of a photon within the electron.

After the electron is expelled from the nucleon and after each photon attains freedom from the electron, the electron's interior begins the photon formulation process. This entails the internal xyzenscape surface to form xergopath "feeler geysers" that move toward the center of the electron. The xergopath feelers divert from this activity as magnetic interaction, or the exchange between hypertips and hypotips, begins to control the directions in which the xergopaths connect. This energy flows back and forth between the internal crystals until magnetic energy engulfs the electron furnace.

As the electron "heats up," juttoria from the rayring intensifies, repelling the xergopaths to the K-axis. The K-axis is the kinetic direction of the forming photon, which is congruent with a line from the center of the nucleus to the electron's center. Ultimately, the rayring "pinches off" the xergopaths at the center of the electron initiating the miortex. In doing this, the hypotips from the ceiling are reflected back to the ceiling, and the hypertips from the floor are reflected back to the floor via right-angled duocollisions at the miortex. The result is that we have the two opposite sets of four neutral K-axial diflohexets formulating with magnetic energy maintaining its formulation laterally.

During this time, the miortex begins stacking its collection of xergopaths within its zeeds. The interior formation of the electron transforms. Instead of a radial formation of xergopaths to the center of the electron, we observe six streams of xergopaths forming along the dimensional axis of the diflohexius defining the photon. This alteration occurs because of the nature of the miortex. The right-angle duocollision, at the miortex by xergopath right-angle movement into and out of the miortex, formulates the pointed bases of the "raindrop" loop formation.

When the first magnetic xergopath stacks within the miortex, the magnetic pattern is established for the entire photon sub-crystal. The image is like a sine wave inscribed vertically around a circular path, such that, two wave exists. We will observe that there are two crests and two troughs. Each crest exists directly opposite from each other as do the troughs, and the crests and troughs exist at right angles to each other. Within this, each crest and trough contains the input of two magnetic xergopaths as depicted by the single magnetic diflohexet stream. Another pattern could exist of having twice as many crests and troughs, but it would create more stress upon the space/time continuum. Secondly, these waves will stack upon each other generating the least space. In other words, the crest of one circuit will fit into the crest of another circuit. If there were a trough sitting on top of the crest underneath, it would slide down into the trough. The result is the continuation of the magnetic alignment per magnetic zeed. Neutral diflohexets provide the K-axis.

Part 4: Creation of Atoms

The magnetic energy within a xergopath has a peculiar attribute about its formation. Imagine, for a moment, the image generated by the relationship between juttoria of a hypertip and jammeria of a hypotip within a magnetic xergopath. Recall that juttoria bends the space/time continuum upward and jammeria downward. The image is that of two chevrons facing in opposite directions representing the direction in which they bend the continuum. Recall that these two forces are meeting "head on" within the xergopath. In this, the bend of the continuum is facing only one direction as one chevron becomes inverted in respect to the norm via by its dimensional approach within the xergopath. In other words, the chevrons end up facing the same direction in relationship to the miortex. Positive charged xergopaths have chevrons pointing away from the miortex, and negative charged chevrons point inward to the miortex. From this observation, we see the correlation to our sine wave, positive being the crest.

Recall that the xergopath loops can rotate at the miortex without destroying the miortex or the six stream-like formations defining the "stems" of the diflohexius. This also occurs within the electron. In this, the spherical shape of the electron is not abandon by the xergopath alteration.

Unlike the original xyzenthium crystals having the external loops formed at the offset of their creation, the photon starts with the loops being tied into the xyzenscape walls. When the streams form from the center of the electron, they generate a loop-like formation near the interior surface wall of the electron. The xergopaths still are connected to their respective xyzenthium crystal. However, there is a conflict of vectors of the tavtips departing from the crystals and that of the tavtips returning to the crystals form their near loop-like pattern. That is, these vectors are at near right-angles to each other. This gives the looping pattern of xergopath the path of least resistance. By this, the internal most xergopath breaks away from returning to the electron wall to join its opposite counterpart xergopath to form a single loop via pure geometric alignment established at the miortex. Afterward, the xergopaths next and above also follow the pattern until the xergopaths meeting at the miortex become xergopath loops. The xergopaths emitting from the xyzenscapes of the electron interior rejoins to other xergopaths nearby reforming the electron's interior waiting for the next photon to be formed in a non-magnetic formation.

After the crystal-like miortex collects all the xergopaths determined by the number of xergopaths emitting from the electrosod ceiling, the photon furnace continues to input pure hypertip xergopath strands via the rayring. These collect outside the already established collection of xergopaths at the miortex formulating the basic photon. These xergopaths form loops internal to the existing photon structure. In this, the quantity of these xergopaths determines the wavelength of the photon. As the number of xergopaths increase, the length of the wavelength increases. The limit of the xergopaths allowed is equal to the total heat available to the photon. In other words, there is not enough heat to go beyond the natural limit. X-ray photons have the least of these pure

Chapter 13: Bondverse

hypertip xergopath loops, and naturally, infrared (heat photons) has the highest hypertip to hypotip ratio.

Redshift by Photons

As scientists look into deep space, they see a shift by light toward the red end of the spectrum. The redshift, attributed to be the result of the Doppler Effect, has another answer. There are two primary factors involved with the process. The first principle originates with the nature of the photon itself. The photon is not traveling at the upper limit of motion depicted by the Lorenz Transformation. If it were moving at this limit, the photon would be frozen in time and not be able to generate the redshift. The second principle of the wavelength expansion is that juttorial pressure exerts itself in finite space from the extra hypertip xergopath loops within the photon. As the repelling force is allowed to repel via time and the void behind the photon, it causes the wave to expand backward from the direction of the acceleration as there can be no gain of energy. Naturally, the longer that the photon exists in space, the redshift effect will become more pronounce. This is not to say the Doppler Effect doesn't play any role in the movement of stars. However, the general increase of the redshift via distance only means that the light has traveled through the relative void longer.

Another peculiar observation about the previous theory is that the earth would have to be at the center of the phenomenon. With the earth viewed as existing toward the edge of this galaxy, this can hardly be the case. As we shall see in the following chapters: even if the earth existed within the center of this galaxy, it still would not exist at the center of the universe. Even so, the idea that the earth exists at the center of the physical universe did not originate in the knowledge of G-d, but in science and pagan religions. While some of the earlier Christianity may have picked up this view from the prevailing pagan view, it had nothing to do with the Original Faith. Elohim created the earth for a specific purpose, and where HE forms the earth within the universe is of little consequence to the scheme of forming galaxies.

Color

We do not see the neutrons, protons or electrons of an atom singularly or in a collective manner. After examining the relationship between photons and subatomic particles, we realize that color can only come from photons shaped from electrons, with some extreme exceptions. This means color does not come directly from protons or neutrons. No stable isotope will give off photons (of visible light) from the nucleus. Furthermore, nucleons do not deflect photons; they absorb them. Despite this, protons and neutrons affect the electron's behavior concerning color. However, an electron does both absorb and virtually (appearance only) deflect photons. Actually, a photon can deflect an electron. This process can convert light into electricity. Even with the

Part 4: Creation of Atoms

photons formed from electrons, we only are able to view, or even need to view, a small fraction of their emissions as color or light. It truly is incredible, that after creating the neutrons and breaking about half of them into protons, electrons and a disintegrating neutrino, only photons enables us to see and feel warmth. Actually, we can only see electrophotons, but many of these are outside of color range. All the visible electrophotons are created shortly before being sent into the universe.

The color observed when a light source outside the object illuminates the object's surface comes from the accelerated energy levels occurring in the electrons surround the atoms comprising the compound forming the surface. Most of these electrons that emit this light are shared electrons. The reason the shared electrons are more apt to emit a photon is that they bounce back and forth between atoms. The added energy of the light are photons being absorbed by the nucleus and transferred to the electrons as they gather energy from the nucleus. The color returned for our vision from one of these electrons is determined by the journey the electron travels between atoms. When iron rusts, oxygen becomes bonded with the iron. The newly shared electrons give the reddish-orange color observed. Neither of these elements in themselves have any color of themselves. Color emerges by electron activity between atoms and from excitement from absorbing proton radiation.

Additive coloring and Subtractive coloring are inverse patterns. Notice the primary colors in additive coloring are the secondary colors in subtractive coloring and vice versa. Secondly the colors combine to make white in additive coloring and black in subtractive coloring. Subtractive coloring is the color seen in pigments and additive coloring in lights. Example red and green makes yellow in additive coloring. In subtractive coloring, yellow is a primary color. Many use to believe that red "fire engine red" was a primary color in pigments. In modern technology, we have discovered that magenta is the primary color in pigments. The color that we assumed to be a primary color is actually made up of magenta and yellow. This does not destroy the fact that yellow and "fire engine red" make orange. However, we cannot create magenta from adding any color to "fire engine red." Another coloring process occurs, which has nothing to do with the original coloring of the substances involved. This coloring change occurs with chemical interactions such as bonding occurs between the two substances and in some solution balances.

Sidestepping from biological encounters, we will look only at the subatomic responses sent to the brain. The retina of the eye has processing stations arranged in layers, the layer that most are familiar with consist of cones and rods. For our purposes, we will view the processing layers as a single module producing three sets of modulated electrical pulses to the brain from each photon it intercepts. As we know, light comes in various wavelengths. For our examination, we will use yellow. The wavelength of yellow is larger than green but smaller than red. Within the process measuring blue, no reaction is chemically produced as the yellow wavelength is beyond the range in which it

Chapter 13: Bondverse

can process. Within the process for green and red (orange-red), the responses are equal, manufacturing the same pulse magnitudes. If our color shifted towards orange the wavelength would lengthen giving a larger response to the red pulse. As we shift the color toward green, the red pulse becomes smaller and the green pulse becomes larger. After passing green and moving toward cyan, the red process finds nothing to work with as the wavelengths are out of range. But the blue process picks up as the wavelengths are short enough for interaction. Then we move into the purplish colors, which are shorter yet, the blue process continues to function as normal. However, the red-orange process picks up on the very short waves as well by another process. This gives us the illusion of a color wheel. Intensity of each pulse, which is the number of electrons activated within the pulse, determines the magnitude of the color.

Additive coloring concerns itself with the effect of combining colored lights or color radiation. This is not adding wavelengths but joining two photon inputs. If it were merely adding wavelengths, there would be only two primary colors instead of three as all other colors could be interpreted as approaching one color or another. Instead, we have three primary colors forming a color wheel. However, all the secondary colors are brighter than the primary colors as two pulses are generated instead of one. When all three pulses are fully expressed, without destroying the retina, we get white light. Black is the lack of pulse stimulation by the modules. Shining color lights on colored pigments annihilate color. Example, if we were to shine a red-orange light on blue or green surface, we do not get magenta or yellow. We get black, as the pigment absorbs all the red-orange light. This provides us with the behavior of pigmentation, it absorbs certain light, and "gives back" (reflects) light that it does not absorb.

Subtractive coloring is the method seen in adding pigments. This can be described as horizontal coloring. Horizontal coloring involves two different compounds mixing together to form a color. Each compound by itself emits a certain color. In adding the two together equally, we are actually dividing each color in half forming a new color. The number of photons emitting a certain color of one substance per second is now occupying twice the volume with no increase in photon count by the substance. This dims the ability to stimulate the eye by one half. The same is true with the second compound. Since the stimulus decreases, another illusion occurs to our sight. We think we are seeing a color of a different wavelength by the means of the dim input of both. Imagine a substance that has both yellow and magenta as its components. Yellow produces both a green and red-orange pulse, and magenta produces a blue and red-orange pulse. The result is that there are three pulses generated by our retina, blue and green and red-orange. The primary emitting color that emits is red-orange. However, all are divided by the area of the visibility of the substance. By this, we see subtractive coloring as not an illusion of the eye, but by color absorption between substances.

Opaqueness is another attribute of pigments. There are some conditions in which a substance may become translucent, which are partially transparent substances, or even

Part 4: Creation of Atoms

become transparent. Transparency occurs when the electrons do not deflect the photons at all. In fact, there is little or no absorption of the photons by the electrons. Most seemingly transparent substances actually do absorb a small amount of photon energy. Example: looking at a non-tinted glass pane though its edge, we observe a greenish color. However, true transparency indicates a substance in which the electrons absorb zero photon energy of visible light. Naturally, translucent substances partially absorb some photons but not all.

Forming Magnets

Iron has the unique electron configuration that gives it the ability to form magnets. If we have a rod of iron and a hammer and a compass to find Magnetic North, we can make a magnet. By pointing the iron rod north and hitting it with a hammer on the southern end, a magnet will form. Examining this process within the iron atoms themselves, we observe the basic principle in forming magnets. This principle can also be applied to superconductors.

We will start with the fact that the electron configuration of iron contains 26 electrons. In examining the levels and sublevels of electrons, we find that iron has two 1s electrons in the first level and two 2s with six 2p electrons in the second level. The third level has a slight variation. While there are two 3s electrons and six 3p electrons, it also has two 4s electrons with six 3d electrons. The two 4s electrons are collected at a lower level than the six 3d electrons. This leaves us with six 3d electrons, the weakest held electrons of the configuration. Moreover, the arrangement of the six 3d electrons is not arranged three on one side with three on the other. The arrangement is that five are on one side and one is by itself centrally located on the other side of the s "plane." The reason for the apparent lopsidedness of the atom is that the magnetic energy favors one side above the other because of the isobaric spin of the nucleus. The lone 3d electron position is centrally located on the opposite side of the s plane from the rest.

In order for there to be a magnetic change in the substance, there has to be a change in arrangement of electrons per atom or in some cases within the molecular structure. The external-most electrons occur in the 3d-electron sublevel. These are the most loosely held electrons within the iron atom. During the interaction, friction between atoms release electrons from their initial positions producing the overall magnetic charge of a given element of zero. There is some magnetic phenomenon that controls the location of each electron. For lack of better terminology, we will call it the magnetic peak and valley phenomenon. The magnetic peak, in this case, represents the greatest resistance to collecting a dislodged electron.

Within this phenomenon, each electron has a magnetic field of control outside the electron's physical shape; it is as in most energy fields spherical in nature. However, this shape can alter by external pressure exerted upon the field primarily by another electron.

Chapter 13: Bondverse

As stated earlier, there is another generally much larger field stemming from the nucleus of an opposite magnetic charge. The reason for the variation in size of fields is that an electron can only produce a field of -1. In the case of iron, the positive charged field equals +26. Naturally, there is spherical distortion upon the electron's magnetic field as the electron interacts with the central positive charge of the nucleus. This is observed as primarily underneath or toward the nucleus in relationship to the physical electron. Above the electron is still the general dome shaped field.

There is yet another twist to this phenomenon. That is that there is a twisted shape of magnetic line or xergopaths containing the positive charge. As stated earlier, this creates a favored and non-favored side of the nucleus. The image is that of walls built upon a sphere in a spiral clockwise from one rotation pole, the other counterclockwise without change in direction of rotation. Next, imagine a ceiling above the walls representing the magnetic hold. Within this assembly we will place a small spherical bead. On one side of the sphere the bead is pushed away from the rotational pole by the walls, while upon the other side the walls push the sphere toward the pole.

Magnetic fields generated by each electron will repel any other electron entering within their domain. Like gravity this field dissipates with distance. If there were only two electrons in the universe, they would push each other away forever, but more weakly as their distance increased. However, we will look at spherical shape from the standpoint of the distance of 1s electrons from the nucleus. Actually, their lateral influence is greater as the spherical shape pulls toward the nucleus by the magnetic attraction.

These fields generated by the electrons are the greatest in height directly above the physical electron. Naturally, the valleys form between electrons. As indicated earlier, electrons of different sublevels will sit, "gravitate," to the created valleys from those electrons underneath. By the time we reach the 3d-sublvel, the valleys begin to appear at different levels. Although symmetrical to the s-plane, one side dominates the "pull."

When the iron atom becomes magnetized, it exists at a state of imbalance. This means that the electrons are no longer arranged in "natural order." Some of them move to other valleys and become trapped there. Naturally, this phenomenon occurs with the outermost electrons which are the most loosely held by the nucleus. In doing this, the electrons become held in weaker positions occupied by the positive charge opening up stronger positive charged regions. The result is that on one side of the atom, there is an excess of negative charge; while on the other side, we have increased the positive charge manifestation via electron removal. Notice the balance, their added charges equal zero.

When we look at the altered electron configuration, we find the vacant location "seeking" for an electron. However, it is not able to gather one as the atom already has the 26 electrons that it needs and still has the same number in the outer shell. This imbalance is permanent until heat re-excites the electrons causing them to return to their original "natural order." Another reason that the electrons do not slip back is the

Part 4: Creation of Atoms

opposite "basket" formed by the s-plane is that the s-plane holds them back. It forms a ridge-like feature around the atomic "equator" prohibiting any natural movement back. The "holes," which the electrons left behind on the opposite side, allows positive magnetic energy to express itself in a greater manifestation.

When we examine superconductors, we have to look at the process in which they are formed. The primary factor, as much as the atomic elements themselves, is the extreme cold. When temperatures approach absolute zero Kelvin (-273.16°C or -459.69°F), the electrons become less excited and "wobble" in smaller circles and the electrons themselves become smaller as the repelling force inside has less energy to work with. The result is that the domain of the electron shrinks. However, we find that the limit in which the electron's domain will shrink is at absolute zero. On the opposite side, there is a plasmatic upper limit. This is the maximum heat that the electron can hold before the juttorial energy forces itself outside the electron while the electron is unassociated with an atom. We will discuss temperatures below absolute zero shortly. However, these extreme low temperatures make the magnetizing process work.

While the cold temperatures cause the electron domains to shrink, this alone will not make the atom magnetic. There has to be an external source forcing the electrons to collect on one side of the atom. At this point, any magnetic influence will cause this to happen. The elements used for the superconductors have many more electrons associated with the atoms. They yield results much greater than the difference of +1 and -1 making iron magnets seem weak. However, without altering their electron configuration, the atoms will return to equilibrium when the heat returns destroying its magnetism. Even after altering their electron configuration: if we were to add enough heat to the assembly of atoms, the electrons will become excited enough to dislodge and return to their natural nonmagnetic configuration. We then say that "heat" destroys magnetic energy as an observable phenomenon. However, we are not considering that heat generates plasmatic conditions that have magnetic properties. As within **plasma**, all the electrons separate from their respective nuclei.

Thermodynamics

Another branch of science is thermodynamics, which is the study of heat. There are three laws stated within thermodynamics. The first law concerns itself with the relationship of heat to work (movement of an object or mechanical energy). It states that some mechanical motion converts into heat and that a definite amount of heat converts into a certain amount of mechanical motion. The equation stating this relationship is that one calorie of heat equals 4.186 Joules of mechanical energy. The second law of thermodynamics states that the force of heat cannot transfer from a cooler object to a warmer one. The third law states that as an object approaches absolute zero (-273.16°C or -459.69°F) in temperature the amount of energy available also approaches absolute zero.

Chapter 13: Bondverse

First, we must establish some definitions. There are two kinds of calories. They are the small and large calories. The small calorie used by physics is the amount of heat needed to raise the temperature of one gram of water one centigrade. One gram weighs 0.0352 ounces at sea level, and one centigrade is 9/5 of a degree Fahrenheit. The larger calorie, usually symbolized by a capital c, is used in the food industry. This calorie equals the amount of heat needed to raise one kilogram of water one degree centigrade. A kilogram weighs about 2.2 pounds at sea level. The value of the large calorie is 1,000 times that of the small calorie. The calorie that we are using is the small calorie. One Joule is the energy needed to move one Newton a distance of one meter. A Newton by definition is the equivalent of about 3.5968 ounces. Therefore, a Joule is the energy required to move 3.5968 ounces a distance of 39.37 inches. This is the equivalent of moving one pound to a distance of 0.738 feet or about eight and 7/8 inches.

Returning to our equation, one calorie equals 4.186 Joules. We can calculate the amount of work one calorie can do into somewhat familiar terms. Within the metric system, this translates to the energy required to move 4.186 Newtons a distance of one meter. We can convert this to represent the work required to move 15.0562 ounces 39.37 inches. Putting in the standard pound format, it is the work of moving one pound 37.04 inches. However, if we were only moving an ounce, we would have to move it about 592.64 inches or 49 feet and 4.64 inches to expend exactly one calorie of energy.

The second law applies to the interaction between two objects of different temperatures. Primarily, this law states that the cooler object cannot cause the warmer object to become warmer. We may jump to the conclusion that no heat travels from the colder object to the warmer object. Essentially, this is the apparent behavior. The only item that gnaws at this logic is that the cooler object, unless it has the temperature of absolute zero, also emits heat. Any object that has any warmth (temperatures above absolute zero) is emits heat producing photons. Although, the cooler object will not overtake the warmer object in emitting heat, it still adds its heat to the equation, slowing down the heat transference process. The image is that of two objects shooting photons toward each other. Note: we are ignoring the temperature of the surrounding atmosphere and the temperature of the ground in which our imaginary samples are sitting. After the warmer side has lost some warmth, the number of photons shot across decreases. As the other side gains warmth, it is able to shoot more back. Eventually, as the two objects become closer in temperature, the heat transference slows down as they are exporting nearly the same amount of photons. Finally, when the two objects are of the same temperature, they exporting photons at the same rate canceling any appearance of transference.

As observed by scientists, heat is the active force. There is no cold energy radiation. Absolute zero (-273.16°C or -459.69°F) means no heat or light photons emit from the electron. This does not mean, however, that there is no juttorial energy. It only means

Part 4: Creation of Atoms

that the nucleus has no excess heat to transmit to the electron. Some claim that absolute zero cannot be naturally achieved because of man's contamination with nuclear energy. They overlook that all heat above absolute zero is excess to the need of the atomic structure. All our heat comes from the disintegration of the outer three layers of the neutron when it transforms into a proton. The heat transference of electrons are singular photons and partial in nature in comparison to the evaporation of about 8.4 billion xyzenthium crystals at two photons each per forming a single electron.

Reaching below absolute zero can be simulated within pressure chambers. The basic concept exists in boiling water. Water will boil at lower temperatures as we climb in altitude without being in a pressurized containment. The reason this occurs is that the atmospheric pressure decreases. The inverse is also true. Water will freeze at higher temperatures with increased pressure. By this, we are able to simulate temperatures below absolute zero.

At absolute zero, the juttorial energy existing outside the atomic nucleus generates an energy peak only equal to the plane of the "norm" within the space/ time continuum. However, outside this is a deep ravine surrounding the nucleus as the jammerial energy experiences the least resistance possible. The electron lays at the bottom of this ravine virtually motionless. Even so, juttoria is still present and the electron is spinning in by its own isobaric momentum without the wobble. The next observation is that the electron is still held away from the nucleus via microstrong energy of juttoria at the nucleus. In other words, the electron is not absorbed back into the nucleus.

Friction

This brings us to the concept of touching another object. We actually do not touch another object as the electrons in the outer shell of both objects repel each other. We only experience the pressure of resistance by the other object, and the amount of heat-photons emitting from the object as it relates to the friction that we are generating. All of this information transmitted to our brain cells though our nervous system gives us the sense that we are touching something in some manner.

Friction seems to be a contradiction as the surfaces react with a response of both heat and resistance. We notice that rougher surfaces offer more friction than smooth surfaces. In this, the structure of the material is then observed as being "bumpy" or uneven. The atoms with their electrons are more elevated in certain places on the surface. Heat manifest as the two surfaces apply pressure upon each other and move along the surface. This causes the repelling force of the electrons to remove an electron from its comfortable setting within the surface shell level of the atom. When this occurs, the electron gets cold upon its surface (by the increased distance from the nucleus) and releases a heat photon. This makes it more attractive to the nucleus so that it can find its way back via magnetic energy.

Chapter 13: Bondverse

A spark occurs as a particle becomes removed beyond the natural disengagement process by lack of cohesion, not necessarily bonded. The electrons of this particle become dissociated with particle far beyond the close range heat forming friction. The electrons release all of their energy to jump a shell level back to their original locations. This brings us to fire.

Fire

Illustrating the classic example of a flame upon a candle is more complex than we might think. We have the external fire source, both the chemical compositions of the wick and of the candle wax, and the chemical composition of the surrounding atmosphere. We will make it easier on ourselves and concern ourselves with the combustion process of the wax. We know that it requires oxygen in the atmosphere to burn the wax. The composition of wax is a complex arrangement of carbon, hydrogen and oxygen. The root of the molecular structure is carbon; most of the carbon is connected to hydrogen forming a hydrocarbon structure that connects to other external structures of carbon, hydrogen and oxygen. The spark that causes the fire, are photons created by the displaced electrons from the fire source, usually a match that is dipped in paraffin, which is another hydrocarbon compound.

We will begin our examination after the flame has drawn the wax up into the wick into the flame. The empowering force, that maintains the flame, is from displaced electrons. The flame, itself, is photons released as the electrons are forced to break bonds with the original wax compound realign themselves into new and smaller bonding components. The initial reaction to the added energy by the photons is that heat is transmitted to the electrons of the wax complex. This increases the molecular activity of the particular surface molecule. The molecule becomes somewhat ionic as the surface electrons are excited by an external source and leave their original position. The molecule tries to pull them back into itself but is unable to accomplish this task. The atmosphere has primarily triple bonded nitrogen molecules, double bonded oxygen molecules, and the inert gas of argon. The "weakest" held electrons are found in oxygen. These are grabbed by the wax molecule thinking this will resolve the issue. But it does not. The oxygen molecule draws the carbon atom from the complex to form carbon dioxide. This leaves a hole in the wax complex with more bonding issues, in which, the adjacent atmosphere is willing to supply.

That was looking only at carbon; recall that there was oxygen and hydrogen present in the outer segments of the molecule. The oxygen makes it easier for the other oxygen to interact, and the hydrogen collects on the oxygen forming water vapor adding to the depletion of the original wax molecule. All of this occurs in the first phase of the flame. The photon activity is minimal as only a small portion of the wax molecule is involved in the flame. Usually, we observe a bluish color in the flame. This shows that the electrons have to travel a short distance to bond again into a smaller molecule. At this

Part 4: Creation of Atoms

point, the excited wax molecule is beyond the grasp of the other molecules. Their heat causes them to expand into a vapor less dense than the surrounding air and they are pushed up. The expansion into vapor is seen as increased movement of the molecules, which hides the expansion due to its juttorial boundaries.

The second phase of the flame is to "eat" all the molecular structure surrounding the basic hydrocarbon. Being drawn up into the atmosphere, more sides of the molecule are exposed to the air. Oxygen is then able to be drawn to all parts of the molecule at a near simultaneous rate. This causes a bonding frenzy causing more electrons to leave their old bonds and forming new ones. Because of this activity of the electrons, more photons are released forming a brighter light seen in the middle band of light radiating from the wick as a yellow-white light is actually a collage of different photon wavelengths. More carbon-dioxide and water vapor formulates.

The third phase forms the final photons. These form the red part of the flame. At this point, all that is left of the wax molecule is the carbon core of the hydrocarbon. Six oxygen atoms attack a single carbon complex forming carbon dioxide. This process releases the most heat-photons (red) as the electrons travel the greatest distances to reach its final bonding. At this point, the wax molecule is completely dissolved into electronically stable carbon dioxide and water vapor. The vapors loose heat to the surroundings and become part of the atmosphere.

Reforming Plasma

In the formation of plasma, a ground state atom is exposed to extreme heat that excites the atom so much that electrons begin to jump off the atom's holding pattern. During this process the element becomes a vapor, which in turn becomes an ionic vapor. The term, "ionic", basically means that there is an imbalance of the magnetic forces within a given assembly, whether it is an atom or a molecule (two atoms or more). Ionic vapor, in this case, is composed of atoms that have an excess of positive charge. This is because the electrons are missing to balance the charge. After the last electron leaves, the element is in a pure plasmatic form.

The definition of plasma is a formation within any element containing nuclei without electrons attached to them. Since the nuclei have no electrons attached, they emit pure positive electromagnetic energy called a positive ion or a **cation**. The unattached electrons are called free moving electrons or (negative) ions. The resulting definition of plasma is a substance composed of cations mingled with free moving electrons.

Lastly, we look at the electron's condition within this plasmatic state. Since the electron is no longer physically able to cool enough to return to the atom, it is no longer produces the elemental visible light. These electrons are unable to get close enough to the nucleus for normal magnetic interaction. Microstrong energy has bent the space/time continuum upward far beyond the "valley" of attraction formed at ground

Chapter 13: Bondverse

state ("normal") via magnetic pull and gravity. These electrons still have cooling cycles as their distance from the various nuclei varies because of the electron's erratic movement. The electron releases low energy heat-photons as the electron's cooling and heating is slight and relatively gradual. Another observation is, occasionally red light is emitted. Within plasma, the "cool" electron quickly regains the heat expelled via the extreme juttorial radiation. The result is that the electron is pushed back away from the nuclei and remains disassociated.

Another observation presented in this article is that the plasma of itself radiates less heat than the heat needed to produce such a phenomenon. While this may be hard to prove within a lab, we can look to nature for proof. For this, we must look to the sun. More specifically, we're examining the black spots. From observations gathered of its nature, we know they are not black, but only appear so, as they are not as bright. We will illustrate that they "boil" up from a surface under the ocean of flames. This entire process will be explained in detail in following chapters, it suffices now to know that the surface under the burning ocean is hotter than the ocean surface. However, when scientists measure the heat, it appears cooler which coincides with it being darker. The electrons are unable to "open" to release the photons via lack of coolness. Since heat is measured by photon activity, its lack measures as coolness. Hence, we get black or deep red for the color of the sunspot. Scientists also know the sunspots result in magnetic storms. This is because the electrons can be magnetically moved toward an external positive charge and the nuclei toward an external negative magnetic stimulus. This gives the sunspots opposite polarity from the solar magnetic poles. As the electrons cool from their plasmatic state they release massive photonic energy both of heat photons and light photons giving the higher temperatures observed in the "grayish regions" around the sunspots. However, within the sunspots, primarily heat photons are released as they "cool" within the plasmatic state. This phenomenon is indicative of "dissolving" (cooling) plasma into vapor via other existing vapors and ionic "liquid" gases. Again, we are reaching beyond the scope of this topic.

Seven States of Matter

Matter has four basic behaviors defined by scientists. They are solid, liquid, vapor and plasma. However, we should be aware that there are actually seven states of matter in the strictest sense of the definitions that are defined by the four behaviors. Just as plasma is somewhat similar in nature to vapor, there are others that have similarity to a liquid or to a solid. First, we will examine the subatomic response to the common states of matter.

Plasma is the first state of matter experienced by formulated atoms. As expounded, the nuclei and electrons are completely disassociated from each other. Juttorial energy is so great that the electrons are unable to join to the atom. Intense heat-photons are emitted sparsely as the electron's photon gate is held shut by the intense heat (juttorial energy of

Part 4: Creation of Atoms

hypertips) radiating from the nuclei. Since the electrons are separated from the nuclei, plasma can and does respond to magnetic energy. As observed within the sun, the magnetic plasmatic phenomenon will deteriorate back into neutral plasma. **Neutral Plasma** is the occurrence of electrons and atomic nuclei within the same region. Actually, it is a composite charge of zero by positive and negative charged particles and not the existence of uncharged particles.

Vapors form in the cooling process from plasmatic ions. The atomic nuclei are able to collect their respective electrons and bond with other atoms. However, the magnetic energy existing between the electrons of any two adjacent atoms or molecules (atomic compound) do not allow them to collect together. The reason the electrons are able to do this is that their repelling force holds the particles farther apart than the gravitational hold of the particles. This is often attributed to the kinetic movement of the molecular structures, as heat and mechanical energy are the result of the same expression of energy. Gases seen today are vapors that retain this state at cooler temperatures. The reason for their existence is that the coolness needed for achieving condensation is much lower, as the electronic fields reach out with a greater magnitude than the gravitational force of the nucleus or molecules can exert upon each other.

Liquids form as the gravitational force of the nucleus is able to attract other atoms. In a liquid, this attraction is weak and the moving electric fields hitting other electric fields are able to push the atoms apart. This occurs because, as the electrons move in their orbitals, so is there movement of their magnetic fields. However unlike the vapors, the distance that they are pushed apart is not beyond the grasp of the gravitational field. The nucleus gradually, relatively speaking, becomes pulled again to the adjacent atom.

Solid forms are the coldest state of matter per isotope. Gravity has control, generally speaking, of the adjacent atom's location. They are held together with gravitational forces greater than seen by the attraction of an object being pulled to earth. This is because they are much closer to the source of the gravitation strength. However, the repelling forces of the electrons keep the two atoms from merging together or bonding.

There are three states of matter that exist that are not found upon the earth because of the temperature range in which we live. The behavior of matter actually exists in three states of vapor, liquid and solid. Plasma appears as a specialized vapor. However, plasma is unique from vapor, not only in temperature but in the condition of the atom, namely the electrons are totally disassociated from their atomic nucleus. The three additional states exist between plasma and vapor. These states are difficult to find using hydrogen as a model. The reason is that when the hydrogen atom collects one electron; it automatically becomes a complete atom of vapor that collects into molecules. This is not the case with larger nuclei.

Larger nuclei can collect one electron and still need to cool more before it can collect another electron. While this is occurring, the electron that was collected by a given

Chapter 13: Bondverse

nuclei still has attraction for bonding. Before cooling, the nuclei only repelled each other, but now, the electron pulls at other nuclei for bonding. Or more precisely, the two nuclei will bond as they fight over electrons because they continue to cool. This grouping of atoms, usually of the same element via different cooling indexes, form a denser complex than seen in plasma. These particles collect or condensate into a liquid-like matter called **lasma**, pronounced los-ma. Actually, it is a combination of two words, **la**va and pla**sma**.

Eventually the lasma collects into an ocean-like body. While this is occurring, electrons are still being collected. As the electrons collect, the attraction between molecule-like particles becomes greater. These particles collect and bond together, which are limited by the number of protons. This limit becomes important in smaller elements (as we shall examine later). The result of the collection is a solid, which is a single giant molecule. This is called a **molecuum**. This is pronounced mole-e-cue-um, from the words of **molec**ule and contin**uum**. If it were possible the result would be a molecule that fills the entire universe with the matrix of the element or elements.

Our next state of matter occurs when the atom thinks it has all the electrons it needs. In essence, this solid melts as it cools. The atoms separate back into smaller bonded molecules determined by their nucleus. These depending upon their composition will cool into a vapor or mist as the electrons repel each other between molecules. This state of matter is called lavor. **Lavor** (lay-vore) comes from the words **la**va and vap**or**. Below is a table listing the seven states of matter sequentially from the hottest to the coolest as the bondverse started with plasma.

Seven States of Matter

1. Plasma
2. Lasma
3. Molecuum
4. Lavor
5. Vapor
6. Liquid
7. Solid

Notice that all the liquid forms start with "L." Along the same line, they resemble the state of matter next to a known state of matter. The alternate solid state is easy to remember as a giant molecule that would be a continuum by itself, if there were enough atoms. Of the seven states of matter, the only state that proves hard to find is the

Part 4: Creation of Atoms

second, as we can find the other six in nature today. The second form of matter exists buried inside the molecuum formation like an egg-white inside the shell of an egg, the yoke being pure plasma.

There are examples of these plasmatic vapor states existing within today's universe. Like plasma, except for the electric arcs between wires, we are unable to get close to examine their composition. Only lavor can be seen today as the fiery ocean surface of the sun. Molecuum, can only be "seen" as the exposed "ocean floor" of a sun spot. These states of matter play an important role in the formulation of celestial bodies.

Chapter 13 Quiz

1. What is the first electron sublevel?
   A. "s"
   B. "p"
   C. "c"
   D. "d"

2. Where does color come from?
   A. Neutrons
   B. Protons
   C. Neutrino
   D. Electrons

3. An example of a nucleophoton is:
   A. X-rays
   B. Visible Light
   C. Gama Radiation
   D. None of the Above

4. What is a rayring?
   A. Ring of Light
   B. Strangeness
   C. Ring inside an Electron
   D. Ring made by Neutron Evaporation

5. What Plasmatic state of matter resembles a solid?
   A. Molecuum
   B. Lasma
   C. Lavor
   D. No such thing

6. (T/F) Nuclei have five levels and seven sublevels.

7. (T/F) Electron orbitals are spherical in shape.

8. (T/F) Juttoria / Jammeria determine the wavelength of light.

9. (T/F) The electron become cold and expels a photon.

10. (T/F) Electrons determine the number of neutrons in an atom.

Part 5
Creation of the Macrocosm

Chapter 14

Creating Galaxies

We can now return to the progression in which Elohim created the macro-finite physical universe. We need to step back to the point in which the neutron cubes were beginning to separate. Recall that the resulting form is spherical at its optimum. We are going back a little before the optimum is reached. If Elohim does not intervene: after the neutrons to reach the optimum under the laws previously set for the physical universe, they would still be isolated in equilibrium. After reaching the optimum, they have about fifteen minutes before they all turn into protons forming Hydrogen-1. Thirteen minutes after that, the neutrino completely "evaporates." If this were to occur at this point in time, there would be no neutrons left in the universe able to generate the different elements.

Obviously, Elohim intervenes again in behalf of HIS Creation to attain the desired results. At this point Elohim divides the waters and does so by putting a firmament in the midst of it. Again, we are looking at a fractal of this action. While the scriptures relates to the final crescendo of the phenomenon, we have several occurrences of the fractal. The first was in the creation of a neutron from a cube by erasing "strings," via absorption, of xyzenthium crystals for its spherical form. This particular phenomenon is the second manifestation. There are yet two more occurrences yet to manifest within the creation of this physical universe.

The word firmament is a very interesting word, both in Hebrew and English. The word in English comes to represent the sky as Elohim calls this firmament, heaven. In it, HE puts the stars, sun and moon. By this, we observe something beyond than just our atmospheric shell. When we examine the word in English, it is hard to visualize an ethereal sky; instead, we visualize something hard. It is only logical since the word derives from a Hebrew word, which has a root meaning of an object being hammered thin. The word is raw-key-ah in Hebrew. The spelling of this word is also important. The spelling in Hebrew is Reysh-Quuf-Yod-Ayin. As noted earlier, each letter represents a picture: Reysh is a man's head; Quuf (or Qoph) is the back of the head; Yod is a hand, and Ayin is an eye.

In analyzing the Hebrew, we form two two-lettered words generally the second letter is like an adjective to the first. This reflects in Hebrew that an adjective follows a noun. The first two letters form of this four-lettered image provides the initial concept, flatness. Imagine being aware of your head as a whole; now, imagine the back of your

Part 5: Creation of the Macrocosm

head. Compare the relationship of the back of your head with your whole head. Viewed as a relationship to each other, we visualize the second or the modification image as being flat, hence flatness.

The second two-lettered image of Yod and Ayin give the pictures of a hand and an eye. The first picture is a hand modified by the image of an eye. Actually, in Hebrew, the Yod, in this instance, acts like a vowel marking. However, the vowel markings also have a pattern of meaning. The hand usually represents work in this case by Elohim. The eye represents knowledge. This shows skilled work that created an outcome with a purpose beyond the result, as all work has a desired result. The concept is that of creating a hammer for the purpose of hammering to gain another result. The second two-letter image modifies the first. We attain the concept of attaining thinness by skilled work such as hammering.

The actual Hebrew root has three-lettered image. The last letter added is Ayin. This can be discerned analytically by knowing the nature of the Hebrew language itself. The Ayin (eye) is added to the end of the first two letters. The eye illustrates "seeing to it" or "making it" flat. Perhaps, it is better put "knowledge (how to make) flat"; hence, "to pound flat" is the meaning of the word. By this, we can understand the need for the Yod. The Ayin or eye also connotes centering. Imagine something being pounded from both sides. The central mass is flattening. When someone strikes a pliable substance against a hard surface, the same phenomenon occurs. The difference is that people do not often associate the hard surface as counter pressure acting upon the pliable substance. Hence, it takes pressure from both sides to make the central pliable substance flat.

Firmament Subset Phenomena

The second fractal of the firmament phenomenon has three subsets apart from the mathematical process determining its limit. First, we have the coagulation of neutrons, which will become atoms of various isotopes of elements. The second subset is the forming of different plasmatic storms within the shell; these plasmatic storms formulate galaxies, stars, solar systems of planets and their moons or satellites. The last subset concerns itself with the inverse phenomenon, which exists because of the many shells of galaxies formed within the universe.

Firmament- the Implosion Phenomenon

To grasp easily the process invoked, we will divide the illustration into two parts. The initial illustration is that of a group of isometrically placed spheres throughout the entire universe, ignoring the magnetic breakup of neutrons. Our second illustration will modify our initial image with a couple of factors involved within the process. However,

Chapter 14: Creating Galaxies

the basic principle of the first illustration is exact. The gravitational pull reacts uniformly between spheres keeping them stationary. Actually, all the forces are isometric in nature promoting the same stationary results. Since we are viewing the neutrons after their spherical formation, we are placing ourselves at the beginning of the plasmaverse, instead of the neutronverse. However, we will need to return to the neutronverse before the analysis is complete.

Imagine moving a single sphere in any direction even slightly. This sphere would naturally become closer to other spheres and would fall into them. Our second observation is the effect upon other stationary spheres in which the sphere leaves behind. Since the removal of the sphere from its original location, there is another alteration to its gravitational pull. The gravitational pull is weaker to those spheres behind the moved sphere, and it stronger to those in which the sphere approaches. If we were to draw circles (representing the holes in space) that connect adjacent spheres, we will find the same scenario. The circle directly behind the moved sphere grew larger, and the one in front shrinks. The result is that the standard distances between the spheres are smaller than the larger hole generated by the single sphere movement; they begin to fall toward each other away from the position of the removed sphere. As these move away a central hole formed behind the group of displaced spheres grows even larger.

The following illustration will further explain the phenomenon. Take a thin piece of rubber (example: a section cut out of a deflated balloon). Stretch it tightly over the mouth of a jar and tie it with a string to hold the position of the rubber. Next, take a pin and poke it in the center. The rubber immediately recedes to the edge of the jar's mouth. The same happens in the transforming of the neutronverse into the plasmaverse. Now, imagine this process happening three-dimensionally a spherical image forms. For now, we will examine only one occurrence of the phenomenon. However, as we shall observe later, there are several different locations within the universe that this phenomenon occurs.

Isotope Coagulation Shells

We will start at the center of the phenomenon. As the neutrons are pulled away from external gravity of the universe, they coagulate. The means by which they coagulate is also gravity, at a different level, that forms the shell. Our examination of this process is very slow in comparison to the actual time lapsed within the phenomenon of this subset. As the neutrons from the center move out, they exert a gravitational pull upon the neutrons that they are approaching. Pulling neutrons away from the external mass of the formulated neutronverse, it marks the end of this phaseverse and the beginning of the plasmaverse. As these neutrons collect, they begin to interact. While they are interacting, they are moving away from the center. As the external mass continues to pull, they form isotope shells, which collect to form a **multigalactic shell**.

Part 5: Creation of the Macrocosm

There is another interesting observation of this internal process. Notice that four neutrons are involved in the initial contact between neutrons from the very core. This will eventually form larger multi-nucleon isotopes. By this, we see that hydrogen has no major formation process within the multigalactic shell core. This will become important as we examine the nature in which the earth forms.

In essence, there is a ripple like phenomenon occurring during the exodus outward from the center. Later there will be a counter ripple effect occurs with the external neutrons moving toward the multigalactic shell. The ripple develops in density of neutrons. Similar to a ripple in water, these ripples progress in size the further they form from the core up to a limit. Unlike the ripples in water, the active force is not in the center but by the collective external gravitational force of neutrons. The collection of neutrons forms a mathematical sequence from smaller values to larger values. In water, the waves move through the medium. In the plasmaverse, the waves collect material as they move leaving a growing void. Within the plasmaverse, neutron "clumps" collect as the gravitational force fluctuates within their proximity before the innermost "ripple" reaches their location. Again, these "ripples" are three-dimensional shapes; hence, sub-shells or isotope shells.

The initial speed of the neutrons is naturally zero. However, they quickly accelerated to near the initial speed of light. The "speed of light" signifies a natural limit of speed per given set of parameters rather than an actual velocity. The parameters in essence become the ratio between juttorial and jammerial energies. Within this period, there has not yet been much loss of energy/mass by the protonization process. We should note, however, the limit that they reach is not their final velocity.

Our first factor of required alteration is that a neutron will only exist as a single neutron for about 15 minutes. At the present speed of light: If the central neutrons started at the sun, they would have gone only a little beyond the orbit of Mars before hydrogen starts to form. Actually, there is not a need for a different process, but a difference of timing. Imagine this collection process begins before the neutrons achieve any major distance from each other. The result being that before the neutrons could shrink away to any major distance, they start collecting. Elohim divided their isometric balance by pushing a hole between eight cubical spheroids. The arrangement of these neutron cubical spheroids forms a three dimensional cubical crystalline matrix. The movement can be visualized by taking a cube and cutting it in half along each of its dimensions to form eight cubes. The fissure of separation starts at the location where these eight cubes join, placing the separation at the center of our chopped cube. This gives the initial eight neutrons that are out of balance, and consequently, generating the imbalance within the entire universe.

The added difference makes it possible for neutrons to join before completely transforming into a sphere. This gives added time to the galaxy forming process. However, as we shall soon discover, the process is still extremely rapid in nature. The

Chapter 14: Creating Galaxies

reason that this gives added time is that it takes time for the shrinkage process to form completely a spherical neutron. Then there is time for the spherical neutron to shrink to its final size. The latter process is actually longer than the initial shrinkage of the neutron. The reason this adds so much time is that the time of shrinkage from the relatively large neutron cube into a smaller sphere is so much longer than the external movement by the much closer neutrons. Secondly, the shrinkage process internal to the neutron slows down as it approaches the final state, while the external process accelerates through time. Additionally, the third result is that the neutron collects into larger isotopes before the "inner wave" reaches them slowing their breakup. Fourthly, this widens the distance in which the isotopes will form in front of the multigalactic "mass-wave." The magnitude of this distance is also extreme in ratio to the 15 minute **"Mars wave limit."**

Another item to observe is the size of these neutrons. A conservative exponent estimate of the size of the cubical neutron in relationship to the resulting neutron is about 1,000,000 times per dimensional direction or 10E18 in volume or one million trillions times larger. This extrapolated estimate comes from the size of a "cold neutron." The actual exponent may even be larger, making this estimate infinitesimal to the actual relationship in size. Imagine all the neutrons existing in uniform distances. If the neutrons shrunk into spheres before coming together, the larger isotope nuclei could not form as easily. The reason involves the time taken by the neutron to reach its destination.

Particle Elimination by Collection

The second subset phenomenon derives from the elimination by radial addition. On a piece of paper, place dots in uniform rows and columns to fill an entire page. From a point in the center and centered in between a four dot square, place the point of a compass and draw a circle maximizing the number of dots within the circle without going beyond the dimensions of the page.

The objective now is to eliminate dots from the center of the circle to its circumference. First, we are going to eliminate the central columns and the central rows. From the chosen four-dot square, take a ruler align its edge from one corner of this square to the circle's edge encompassing an entire column of dots. Draw a line from the chosen corner of the four-dot square out to the edge closest to that dot. Do the same for the other side giving two parallel lines. Repeat the process using the remaining two corner dots forming two more parallel lines projected in the opposite direction. Note: all of the central dots are used, and the form maintains a sense of symmetry. We are now ready to eliminate another set of dots equal spaced between these lines. Since the central dots are used and a neutron cannot move in two separate directions simultaneously, we must go to the next closest dot. Draw these lines to incorporate their receptive row segments out to the edge of the circle. The result is that we have

Part 5: Creation of the Macrocosm

the circle divided into fourths. To give a better feel of the process shade the areas between the lines.

Next, connect the dots that are a 45° angle from the center of our selected square. Repeat the process on each side of this line making three lines total. Do the same for 135°, 225° and 315° (all divide the original right angles in half). Counting the dots on each line, we find that the second group (45° angles) has fewer dots per line. Finally, we shade in between these sets of lines.

Notice that the angled lines at 45° are always found by going vertical direction one dot and horizontally one dot. We might think that a 30° or 60° to be an obvious choice, or 22.5° (half of 45°). The next selection selects by the next closest distance between dots aligned radically. In our next selection, connect the dots that form line segments by moving two dots toward the edge and over one starting with the next closest dots to the center. Again, form three parallel lines and shade.

Lastly, we will group the spaces between the shaded areas. Draw lines parallel to all the shaded areas using the dots. Again, we shade between all nonparallel lines. Actually, these wedges further divide into smaller areas. From examining these shaded areas, we find the process.

The first observation that becomes evident is that the last lines are closer to the 45° angles than the original right angles drawn; this generates uneven spaces. The second observation is the distances between the connected dots are often further than dots lateral to the direction. Eventually, as the process continues, the lateral dots (neutrons) will be drawn into the system of neutrons moving straight out from the center. These two observations form the first two factors involved in the phenomenon.

With our previous example, we cannot really see the isotope sequences that become element sequences. Imagine each dot that we have connected being a segment of an isotope ring (a cross-section of a shell). First, we observe some storms have "thick" element sequences and wide domains. The there are others with thin domains with "thick" element sequences and wide storms with "thin" sequences of elements. Thick element strings are the isotopes with many neutrons; implicitly defined, the thin element strings have few neutrons per isotope. The nature of these strings then can be described symbolically using numbers. The first string of numbers represents a thick sequence and the second a thin sequence. The actual sequences are much larger, far beyond trillions in number, and much more complex in numerical values. Actually, we are a little ahead of ourselves with the numerical illustration below, as we have not described the total process of the isotope strings.

$$4\ 4\ 7\ 9\ 7\ 4\ 2\ 2\ 1\ 1\ 1$$
$$3\ 5\ 3\ 2\ 1\ 1\ 1$$

Chapter 14: Creating Galaxies

The largest number represents the central layer of the shell that will exist at the largest isotope level within the shell for a given general angle. Using the left most number to represent the side of the shell facing towards the center, we observe the largest number closer to the internal numbers or left most number. The reason for this imbalance is illustrated by the fact that there are more neutrons external to the developing "shell" than there is composing the shell in its initial development.

In our drawing, we used only two to three lines per direction. Actually, there are multitudes of lines entering in a particular direction starting from different locations. Imagine a ray that expands as it moves away from the center, incorporating even more lateral neutrons. This makes the variation between storms even greater. Within these rays, they divide again into smaller rays. Each "ray" will eventually develop a plasmatic storm. Now, we need to back up again and examine the growth of the shell.

The innermost "wave" or "shell" generating the multigalactic shell, or shell of multiple galaxies, starts out as a thick layer moving outward. As it moves outward, it thins via greater surface area. However, there exists a counter process. The next isotope shell above it pulls the shell via gravity. The shell gains more mass and more acceleration because it is being pulled by the shells in front of it. After merging with the shell above it, again, the shell thins and gains mass by the shell above it. This process continues until the shells begins to merge in groups simultaneously as all shells approach the limit established by the speed of light. Each shell increment contains an increase in isotope size.

There will come a point in time of its growth that the gravitational pull of the shell will overcome the gravitational pull by the external plasmaverse and the plasmatic shell will pull neutrons into it from the external space. Speaking purely from the standpoint of earth, this occurs at the isotope of element radon. In other locations, this element will be different. Initially, the effect is that of neutron snow, hinted at earlier viewing the plasmaverse in a microcosm perspective. These neutrons collect to the isotopes upon the surface of the multigalactic shell instead of being collected by the external larger isotope wave or shell. This increases the size of the isotope upon the surface and ends the isotope growth per shell above the multigalactic shell. Later, isotopes formulate external to the sphere and are pulled into the shell adding more mass. Not only is there an implosion outward toward the shell there is an implosion inward to the shell. Truly, matter hammers into a thin sheet by empty space as the Hebrew word indicates within the Torah.

Neutron Snow

Despite the dimensional sizes of the nuclei of these elements, they can fit within a cube that is six nucleons in measure. Six cubed is 216 plus 12 equals 228, and the leap from the isotope of Astatine-210 to Radon-222 does not maximize the capacity. By this, we

Part 5: Creation of the Macrocosm

observe the initial neutron clump of astatine, before the involvement of centrifugal force and magnetic reactions, fits into a 6x6x6 cube with six neutrons missing. The image is a 6x6x6 cube with all the corners missing except two. They exist on opposite corners on the bottom side. Even so, the plain of neutrons above the nuclei has to be 6x6 neutrons for a total of 36. This is 24 more than the 12 needed. The next item to understand is that this clump is shrinking fast due to the shrinkage of neutrons into a spherical format. It is only able to draw twelve neutrons from the plane.

On the opposite side of the plane there is another 6x6x6 cube offset by three. In other words, the facial centers of one set sit upon the intersection of corners in the opposite set. With a plain of neutrons, one neutron thick, separating them. They each take twelve neutrons leaving twelve neutrons per 6x6-neutron square within the plane. Within this particular scenario, these twelve separate into Helium-4 and Hydrogen-1. It will even formulate slightly larger isotopes, as it grabs "unused" neutrons from the nuclei bands surrounding it.

Another observation of the scenario is that one of the twelve forming neutron snow becomes a proton giving the formulation of an isotope within the next element. By this, it becomes evident that the transition into the next element was not by the progression in neutron clump size formulating the next multigalactic shell. Within this shell of isotopes, 210 neutron clumps could formulate other elements.

Our next observation is somewhat not as obvious. This is that the entire phenomenon of the transition from the multigalactic shell being pulled into the external universe to the multigalactic shell pulling the external universe to itself also has a wave-like behavior. Recall that the central plane that feeds twelve neutrons to two different sets of nuclei bands. This implies that there is an empty space between the next set of two bands. In other words, the plane of neutrons existing between the two bands of nuclei appears between every other band of nuclei. This alternating pattern exists within a large range numerically speaking; in comparison to the thickness of the multigalactic shell, it is very thin. Actually, the layer forming the neutron snow appears to be a layer one neutron deep between those pulled away from the center and those attracted to the shell.

We now need to examine the formulation of the natural occurring isotopes from Radon-222 to **Uranium**-238. Recall that we saw Astatine-210 nuclei become the foundation for Radon-222 by collecting twelve of the 6x6 neutron square from a central plane. Now, we are going to maximize the phenomenon. Within this scenario, all the neutrons within the 6x6x6 neutron cube remain giving 216 neutrons. Since the 6x6 square shares its neutrons to two sets of nuclei, not all of the 36 neutrons are available to either band of nucleons. Since 238 – 216 = 22, we know that Uranium-238 needs 22 of the 36 available neutrons. When we subtract 22 from 36, we are left with 14 neutrons. We can divide the remaining neutrons to two opposite corners at seven neutrons per corner.

Chapter 14: Creating Galaxies

Imagine for a moment that we have four sets of 6x6 squares arranged into blocks of 2x2. Let us arrange the seven neutron corners, such that, the corners that form the central intersection of the four 6x6 squares contain the seven neutron corner. We now have four sets of seven neutrons connected together giving a total of 28 unused neutrons. However, the counter-sync Uranium-238 above also needs 22. This leaves six unused neutrons. We could say it forms an isotope of Lithium-6. But, we would also be saying that there is a nucleus of 216 that is not radioactive, which does not exist within the framework of the elements formulating our solar system. The reason that the 216 isotope would form is that on the opposite two corners of our 6x6 neutron square has zero unused neutrons to be gathered by the opposite nuclei band pattern.

Now we are in for some more neutron manipulation. Instead of seven neutrons for two opposite corners, we are going to distribute them among the four corners. The pattern of neutrons becomes a complex formulation of corners containing one, two, five and six neutrons. It is an alternating pattern between 6x6 squares with ones, and sixes on opposite corners and squares with two and five neutron corners. Moreover, the corners containing five and six "unused" neutrons join together, and the corners containing only one and two neutrons are joined together. This generates alternate intersections containing six and 22 "unused" neutrons. Notice that 216 + 6 = 222 and 216 + 22 = 238. The result of this configuration is that one layer contains an alternate pattern of Radon-222 and Uranium-238. Both are radioactive elements. Composition, between the two bands, is 75 percent Uranium-238 and 25 percent Radon-222. These are the central most bands of crust material. There are even denser bands, but they are, up to this point, improvable as they are not found in the earth's crust. These are more centralized within the plasmatic storm that forms earth. After these bands the isotopes shrink in size. The next sets of element bands form Uranium-235 and other radioactive elements. They still have the 216 core neutrons but acquire less neutron snow. Eventually, the "snow" plane disappears as the nuclei sizes shrink to 210 and below per band.

The numerical counts of the nuclei only give a portion of the picture. If we were to look at the counts alone, we could easily assume that a gradual transition occurs between isotopes giving such isotopes as Uranium-236 and Uranium-237 before the occurrence of Unamium-238. Generally speaking, this assumption is fairly safe. However, we are not talking of a neutron rain, but a snow. There are other factors involved, the primary factor that deviates from this assumption is the geometry involved between neutrons. Within this phenomenon there are two major considerations, the macro-geometry and the micro-geometry.

We will first examine the macro-geometry. Recall that the neutrons form from a cubical format, and that the multigalactic shell is spherical in nature. A conflict occurs between the cubical alignment of neutrons and the spherical gathering of neutrons to the multigalactic shell. This can be seen by drawing a circle on a piece of graph paper and

Part 5: Creation of the Macrocosm

shading in each square that the circular line crosses. As a result, we get a circular-like object with a very rough edge. The same is true within this phenomenon, except it is three-dimensional instead of a two-dimensional object. The result is that there are some neutron centers that are outside the actual definition of a circle and others that are inside with a few sitting exactly on the defining line. This gives a wave-like appearance upon the surface of the sphere.

Thus, we observe Uranium-238 and Radon-222 sharing the same neutron snow plane. When we look at the nuclei bands surrounding a neutron snow plane generating Uranium-235, we observe that Thorium-231 forms half the nuclei within these bands plus the formation of Hydrogen-1. Then there are bands in which only Thorium-231 forms releasing even more Hydrogen-1. By the time we reach the last neutron snow plane, we find that not only Hydrogen-1 forms, but Helium-4 forms as well. It is even possible for elements up to Carbon-12 to form as our 6x6x6 nuclei core dwindles down to a 210 neutron clump. By this, there is much Hydrogen-1 forming within the formation of these super large nuclei both within the neutron snow and without. This will become important, when we examine the components of our sun.

The micro-geometry is the change in shape in the relationships between neutrons. Again, we need to return to the cubical representation of the sphere. This time, remove all the engulfed cubes that do not make-up the surface of the sphere. Next, we will replace the solid cubes with a stick-figure of the cube, i.e. very thin wires joined together defining the cubical edges and corners. Underneath this framework, we will place a thick spherical balloon-like object the size of the sphere. We then increase the air within the balloon, such that, it applies pressure to the external assembly. As the wire cubes give room to the growing sphere, their structure alters into a smoother spherical appearance. For a simple analysis, join four equal sized pencils with tape to form a square and pull two opposite corners. This draws the two opposite corners closer together. In our case of a cube, it would draw the other six cubical corners together. If we push the two opposite corners of a cube together, the other six corners move further away. This gives us a jump from a 216 "core" neutrons to a 210-neutron core.

Moreover, when we move to the end of our multigalactic forming scenario, we find that the neutrons appear to reach a purely spherical shell arrangement, such that, it appears as a flat plane to the surface of the sphere. There are different patterns of collection that follow the basic principles of the neutron flake patterns. Some patterns break-up into smaller patterns like the five and eight neutron pattern. After diminishing to purely the formulation of Hydrogen-1, the single neutron pattern, the arrangement of neutrons becomes insignificant at the micro-geometric level. It becomes a process of just collecting hydrogen nuclei.

Another micro-geometric level phenomenon concerns itself with the closeness of the neutron snow plane to the two bands of nuclei. Their closest encounter occurs naturally at the centermost layers of isotope waves or bands. This brings us back to the

Chapter 14: Creating Galaxies

formation of uranium and its leaping in nucleon counts. Uranium-238 is the central formation of the crust forming elements. As the nuclei bands become further from neutron snow plane, the neutron collection from the neutron snow plane per nuclei core becomes smaller in count and more neutrons lost to form Hydrogen-1 and other elements.

Isotope Rings and Islands

There are two variations to the layer pattern of isotopes. The first of these sets of isotope rings. Imagine a vase pouring out multicolored sand unto a flat surface. Within this image, let the initial pouring be that of magenta upon a red plane. Afterward, another color pours out in color sequence violet. This is followed by indigo. After the sand has been poured, it settles into a flat plane. If we could take a side view of the plane, we would observe a central magenta segment surrounded by two segments of violet, which is surrounded by segments of indigo. Actually, the magenta is a central disc surrounded by violet and indigo rings. Each color represents a different isotope, more importantly a different element. Our example used three colors, but the number of elements is not limited to three. The actual number may be more or less depending upon the length of the isotope strings.

Now imagine more sand pours of blue and cyan and of small amounts at different locations. This forms islands of isotopes and elements upon the surface of the multigalactic shell, as the pouring is not continuous upon the shell by the last few elements. The exception is the last and massive layer of hydrogen. There are islands that form within the interior of the multigalactic shell after the breakup, but most element islands form externally and before the breakup. We will discuss these when we examine the formation of earth.

Multigalactic Shell Breakup

Now let us look at this process a little more detail. At the beginning of the movement outward by the neutrons, inertia modifies their velocities. In other words, they had to accelerate from motionlessness to a velocity that is compliant to the force acting upon them. Unlike an object falling through the atmosphere, there is no friction by other particles. This allowed them to approach the initial speed of light, which was much greater than today. After reaching nearly the initial speed of light for a period of time, their accumulating mass started to draw external neutrons to themselves. After the neutron snow, the shell eventually pulls down isotope waves into the growing shell. As the exterior shells pull into the multigalactic shell, two different phenomena occur. The first phenomenon is that the crashing mass gradually slows the growth of the multigalactic shell down. The second phenomenon is that shells collected by the gravity of the multigalactic shell add gravity. The added gravity diminishes the neutron

Part 5: Creation of the Macrocosm

collection pattern forming isotopes as more matter pulls away from the external nuclei collection process. The external isotope pattern reverses the sequence with smaller increments between isotopes compared to its internal collection of isotope shells as described earlier.

If this external collection process were to continue, the velocity of the multigalactic shell would be halted, or even reversed. However, as seen historically the shell was neither halted nor reversed. The process exhausts itself of neutrons and consequently isotopes. We will describe the reasoning later. Another observation that demands our attention is that many of the final external neutrons arrive too late to combine with other neutrons. This creates a large quantity of Hydrogen-1 isotope in the outermost layer. This does not mean that the entire process took only fifteen minutes, as there is time needed for the shrinkage of the neutron.

The result of the deceleration is the growing multigalactic shell has slightly merging isotope layers via pressure. The heaviest compression expressed within layers form the "central most shells" within the complex. These are not truly centrally located. Their location is somewhat lower toward the internal void of the multigalactic shell. However, this effect does not destroy the purity of the majority of the isotope layers, as they are much thicker than this influence of this phenomenon.

The multigalactic shell is slightly uneven in mass per region via different sized strings of isotopes. The initial image is a sphere with spots. These spots are a collection of star "spots" forming galaxies. After all the external mass has been collected by the multigalactic shell, the shell keeps on expanding. This thins the shell. Gravity is not enough to hold stars and consequently galaxies together within the shell. The reason being the thickness of the original multigalactic shell would have pulled the gravitational centroids apart as gravity is continuous throughout the shell. If it were to continue, the shell would become paper-thin and there would be very few galaxies, stars, planets or asteroids, primarily just dust. By this, we know another dimension enters the picture.

After all the hydrogen collects upon the multigalactic shell, the velocity of the shell expansion becomes a constant and continues expanding. The cooling process begins. Recall that there are seven states of matter: plasma, lasma, moleculum, lavor, vapor, liquid, and solid. Plasma is the initial state of the expanding shell. Before cooling into vapor, the atoms must collect all of its electrons. The three somewhat unfamiliar states of matter occur during this process. The primary principle of these three states is that as the nuclei collect their electrons, some of the "grabbing" occurs from other atom's collected electrons. The result is plasmatic bonding that occurs between atoms.

Plasmatic bonding does not follow the general rules for bonding. This gives the unique opportunity to form molecules that would be nearly impossible to achieve within the normal bonding processes. This will become evident as we examine the formation of the earth and planets within our solar system. We call it plasmatic bonding because it

Chapter 14: Creating Galaxies

occurs only in the cooling process of plasma into a vapor. When we heat an atom from a vapor into plasma, the "grabbing" process of electrons does not exist. The reason is the atom "sheds" electrons instead of collecting them. Another interesting phenomenon concerning the shedding of electrons is that the resulting positions between nuclei are farther apart than in our initial plasmaverse. The cause of this difference is that the escaping electrons from the atom push the atoms apart, while the newly formed plasmatic multigalactic shell has nuclei pounded together by the "firmament process."

Within plasmatic bonding, lasma is the initial bonding that occurs with the cooling nuclei within the multigalactic shell. Here we observe two nuclei grabbing the same electron. Lasma forms when the nuclei gains electrons connected with other atoms and that collection is much less than half the electrons needed per number of protons existing in the nuclei. This state of matter generates a greater repelling force between atoms forming the magnetic liquid-like appearance than found in liquid. The liquid-like formation occurs as the bonded atoms have electrons creating a polarity within the molecule, and the substance is still cooling desiring more electrons. While the nuclei have attraction to the electrons existing in an adjacent molecule, their extreme positive charge holds them further apart. However, this ability to repel diminishes as more electrons enter into the assembly via cooling of the molecule. Another factor that contributes to the liquid-like formation of lasma is that the molecules are denser in makeup than a single plasmatic nucleus. The bonded molecule holds the adjacent atom/atoms closer together than their plasmatic occurrence.

The exception to the lasma formulation is hydrogen. In this, hydrogen needs only one more electron for the two nuclei to complete the hydrogen molecule. This factor promotes the quick formation of hydrogen gas. Another factor is that hydrogen is the external most layer of the shell touching the external void near absolute zero degrees in temperature. Internal to the multigalactic shell, the hydrogen phenomenon is virtually nonexistent as there is almost no or incidental hydrogen present. This absence is also true for the lighter metals as lithium and beryllium. As far as the matters concerning the earth, nitrogen is the first abundant element found facing the internal void. In other regions of space, we find other elements facing the internal void. However, within our solar system carbon is the lowest isotopes found internally to the shell, as we shell observe. Again, the temperature of the void is near absolute zero.

This brings us back to the cooling process of the shell and the seven states of matter. Initially within the lasmatic state, the nucleus cools enough to collect a single electron. As noted earlier, there is no hydrogen lasma. In other elements, we find that the lasma thickens as it collects more electrons. Lasma occurs initially within the interior of the multigalactic shell, as hydrogen exists upon the surface as a kind of atmosphere protecting the layers underneath its covering. Since lasma is a liquid-like substance, it has a tendency to collect into a mist. The mist is actually a molecule-like particle. The

Part 5: Creation of the Macrocosm

reason they collect is that the electron creates a polarity with the nucleus. Using oxygen as an example, one side of the polarity is +4 and the other -1 having a composite effect of +3. Even though the composite is +7, the -1 charge of the electron attracts other oxygen +7 atoms. The vibration of the movement of the electron and the excessive positive charge of the nucleus keeps the substance as a liquid.

Returning to our plasmatic shell, the thick hydrogen layer gradually turns into a gas protecting temporarily the external layers of the shell. Meanwhile, the bottom layer turns into the liquid-like lasma being of isotopes greater than eight nucleons. As the liquid-like lasma forms, gravity becomes a factor. The liquid tends to gravitate towards the various centroids existing within the shell. Besides the central gravitational pull of the centroids of the heavier core material, a lateral gravitational and magnetic fields between galaxies and the stars and their satellites storms also collects. These lateral forces form "arms" connecting adjacent storms together. The resulting image is that similar to a web-like image with random-like weavings.

As the shell grows, the layers of the multigalactic shell become thinner promoting more cooling. The connecting arms also stretch into thinner strands. The cooling lasmatic atoms collect more electrons, and the lasma thickens. Eventually, the lasma freezes into a molecuum and the shell shatters. The primary shattering occurs between galaxies. Individual shattering between stars occurs afterwards, as the interstellar centroids move away from each other by moving within their expansion vectors established by the multigalactic shell. After that breakup, planetary platelets break around the star. Again, the catalyst is the different vectors and mass distributions.

Molecuum occurs when the bonding between atoms reach their maximum count per atom. Again, we are not observing any normal bonding pattern. Most of the lasma and molecuum form molecules of the same element except at the edges between layers. The maximum number of bonding-like connections between atoms is twelve, involving thirteen atoms, within the framework that the atoms are all the same element with equal number of electrons. Within this pattern, we observe a collection of six "equatorial" atoms to a central atom and three polar atoms above and below the seven. The three polar atoms are a part of another equatorial collection of atoms. The result is there are interlacing hexagonal patterns throughout the element. Each hexagonal patterned layer is offset by an half a hexagon. The element exists as one single colossal molecule.

Variations exist between different molecuum of elements. Consider the elements that have less than twelve protons. The molecuum experienced by these elements do not fully develop. These elements are, in sequence: hydrogen, helium, lithium, beryllium, boron, carbon, nitrogen, oxygen, fluorine, neon and sodium. Magnesium is the first element to complete the molecuum pattern. Some of these elements are able to imitate the complete structure; these elements have four or more protons (beryllium upward to sodium). These molecuum are less dense containing less atomic connections, therefore weaker. We will examine some of these as we analyze the formation of the earth. On

Chapter 14: Creating Galaxies

the other side of magnesium, we have elements with more than twelve protons. The result is that the time spent as a molecuum becomes longer depending on the number of protons. Lastly, this process has no regard for noble gases, gases, metalloids or metals. For example, krypton and xenon are noble gases, their molecuums last longer than iron molecuum because of the number of protons describing their elements.

Returning to our multigalactic shell, we observe the breakup occurring upon the surface underneath the multigalactic shell instead of the top because of the hydrogen protection phenomenon. The formation of nitrogen lasma is our next issue. It forms with the rest of the interior surface of the multigalactic shell. Initially, a plasmatic mist forms. The lasma gradually forms a liquid surface layer, internally speaking, within the multigalactic shell. As it joins with lasma forming at other locations, it appears as a vast ocean. Unlike the ocean of the earth, it has no floor. Another difference is that the gravity under the ocean is uneven. The initial image of the ocean is that there are locations in which it swells via the gravitational centroids. The fluidity of the lasma gives gravity a chance to interact with the substance. The "water" moves toward the centroids along its general "plain." Before total evacuation to the centroids, the lasma freezes into a molecuum. The composite surface molecuum is then seen as a solid with a variable thickness. When the molecuum shatters, the thin regions control the breakage locations. In comparison to the interior surface of the multigalactic shell, the molecuum plain is as a brittle ice sheet thinner than paper. The image is that of flatten conical platelets with irregular edges containing a swirling central mass with a large hydrogen covering composed of both gas and plasma. While it is true that the hydrogen gas would eventually move around to the opposite side and mingle with other elements. This is found to be minimal as the hydrogen cools first and falls into the elements directly under it.

The edges of the broken molecuum plain expose more plasma to cool, which is closer to the central layer of the shell. The edges between galactic storms drift apart as the multigalactic shell continues to expand. More freezing and breakage occurs. The resulting image that of walls forming at the edges angling away from the plasmatic storm center holding the plasmatic material within the storm.

Next, we observe that eventually the molecuum melts into the next state of matter. Lavor is a liquid-like state in which the element believes that it has collected all of its electrons. The atom still exists in an extreme heat. These liquids would still need to cool to form a vapor. It is within this state of matter that a final molecular form of the element solidifies. This molecular structure is responsible for the **ozone** (O_3) and some mineral formations. When the atom cools into the final states of matter, the coolness modifies the structure until all the bonding electrons are collected at ground state.

After the primary surface elements reach vapor, gravity begins to play an important role in the formation of the solar system, planets and inter-stellar responses. The heat is loss to the void via photons. However, we will need to return again to illustrate the

Part 5: Creation of the Macrocosm

formation of our solar system and the earth. This is only an overview of the phenomenon forming earth. We will soon examine in detail the process.

Gravitational Divergence

Gravity, as we know it, behaves as centroids. The actual gravity force does not originate from the center of the object, but it is a collection of the gravitational force exerted by every atom within the form. Knowing this, we then realize that a sphere weighing ten pounds and a sphere weighing five pounds falling through a vacuum do not fall at the same rate. While the heavier one does not fall twice as fast as the smaller weight, it does fall minutely faster. The reason is that gravity is not a "one-way street." Both objects possess gravitational forces. The total computation of the gravitational pull equals the addition of the gravitational force of the two objects. The problem is that the ten-pound sphere and the five-pound spheres gravitational force is almost zero when compared to the gravitational force of the earth. The phenomenon is greater than adding a 1/16 of an inch or a 1/8 of an inch to a length that is exactly 100 miles long. For all practical purposes, we are adding only zeroes; the gravitational force will pull objects upon the surface near sea level at virtually the same rate.

Along with this, gravity decreases with distance by the inverse square of the distance ($1/d^2$). As an object moves farther away from the source of gravity, the force becomes weaker. The reason that objects near sea level does not express this noticeably is that any distance above the surface is minute in compared to the average distance of the gravitational force gathered by the earth. If the earth was 8,000 miles in diameter, the effect of distance has upon the equation then appears as $4,000(1/(4,000+d^2))g$. Note that when $d=0$ that $4,000/4,000=1$ and $1*g=g$. This means that our gravity constant is unaltered at sea-level as it is already 4,000 miles from the earth's gravitational centroid. We multiply this result by "g," the constant representing the acceleration of gravity, which equals 9.8 meters per second squared or 32 feet per second squared. Relate this to a car accelerating to 60mph in one second, which equals 88 feet per second. The acceleration by gravity is only a little over a third the force of this acceleration.

However, for our purposes, we can annihilate some of the complexity. For instance, we will not concern ourselves to the gravitational force in respect to the surface, but to the centroids. In doing this, the equation becomes a gravitational constant multiplied by $1/d^2$. For our gravitational constant, we will use 100 units per second squared. This will get us out of the dilemma of using the feet or meters and provide a nice even number. Now, we can deal with the issue of finding the effect of gravity as density changes by the change in volume. The basic concept is that of taking a large sponge and compressing it into the smallest possible form. This translates to taking an object that is less dense and compressing it into a smaller denser object. Gravitational measurements between the uncompressed sponge and the compressed sponge prove futile. The gravitational effect is immeasurable upon the surface of the sponge as the

Chapter 14: Creating Galaxies

magnitude of the earth's gravitation overtakes the entire process. We can isolate the process using mathematics.

We will use three particles that have the same attributes such as mass and density. We will call our particle of reference: particle-x. The other two particles will serve as samples of the adjacent material; particle-p represents the closest particle of the material to particle-x. Particle-b represents the backside of the forming object from the initial material. We are going to move particle-p and particle-b closer together toward the midpoint between the two particles. Our general equation of the effect upon the gravitational pull is the following: $G=K(1/p^2 + 1/b^2)$. Note, this is not the total attraction as particle-x also contributes to the total, and the equation would become more complicated, ($G=K(1/p^2 + 1/b^2 + 1)$). Finally, the K is our predefined 100 units per second squared, and p is the distance of particle-p is from particle-x, and b is the distance particle-b is from particle-x.

Within our illustration, we will initially set particle-p one unit from particle-x and particle-b ten units away from particle-x. Applying our formula, we get 100 multiplied by the addition of divided by one squared and one divided by ten squared giving a net gravitational pull of 101 units per second squared. When we moved the two particles one unit apart at the midpoint, we get the following equation: $100(1/5^2 + 1/6^2)$. This gives us the answer of about 6.8 units per second squared. Notice that the average distance of both particles remains the same, namely 5.5 units. This will become important as we continue our examination of the process. Another observation is that particle-p has the greatest gravitational impact upon particle-x via proximity as described by the equation relating gravitational pull to distance.

Our second equation reflects the same phenomenon, except that we moved particle-b two units away from particle-x. This gives the equation of $100(1/1^2 + 1/2^2)$ or $100(1 + 0.25)$. Naturally, the result is that of a greater gravitational pull by the two particles. Their net influence becomes 125 units per second squared instead of 6.8. Next, we need to compare this with our original setting of a 101-unit pull. Again, particle-p provides most of the gravitational pull upon particle-x.

Using our information, we can graph the process of the division of mass from a single unit into separate entities. As the central density of a star, for example, draws mass into itself from its surroundings, the star's density increases. This increases the gravitational force emitting from the surface. However, it decreases the pull upon the external mass not pulled into the star due to the distance increase from the mass. If we were to move farther away from the star, the difference in gravitational pull diminishes by squares of the ratio of extended distance to the original distance. This brings us to the actual dividing process.

The centroids continued movement in a ray-like fashion causes the initial weakness within the structure from the time of the shell's origin. As the multigalactic shell

Part 5: Creation of the Macrocosm

expands, the heavier mass drift farther apart by the means of the increased surface area of the spherical shell. When this occurs, the lateral gravitational force hold between regions of the shell weakens between masses and the masses become more centralized by the phenomenon previously described.

Antishell Subset Phenomenon

We are now on our last subset phenomenon to present before painting the picture of galaxies. We have only hinted at this particular phenomenon within our previous analysis. Recall that these multigalactic shells formulate within different areas of the universe. To illustrate this phenomenon, we will draw non-intersecting circles upon a piece of paper. Each circle represents a multigalactic shell. These circles are not always equal-distant intervals as Elohim chooses different distances for different results.

Imagine each circle growing to touch each other. Notice, there is space remaining outside the circles. Recall that there were neutrons forming various isotopes, and finally Hydrogen-1 gravitates onto the shell after a certain point. The circles represent the full extent of this phenomenon. Meanwhile, outside this phenomenon is another process. Gravity pulls these neutrons inward away from the shell. Note: There is no motion of the assembly in general other than the implosion away from the multigalactic shell process. Despite the movement by the neutrons, the net movement is near zero. Therefore, the **antishell** galaxies are primarily still. Their movement originates by gravitational interactions with other galaxies and stars or via uneven matter collection. These masses generally form large non-spinning gas clouds and neutron stars, which may have a rotation depending upon the mass these systems. In most cases, the galaxies formed near these regions via the multigalactic shell will pass through these regions.

There are two extreme formations within the universe; they are the black holes or neutron stars and quasars. Each forms from opposite processes. **Neutron stars** form in the antishell phenomenon, and the quasars form within the multigalactic shell. They differentiate by the manner in which the neutrons collect, as we shall soon observe.

Quasars and Neutron Stars

Quasars formulate within multigalactic shells. They exist because of the favorable neutron alignment during the formulation period. Collecting along a linear alignment, allows gravity to accelerate an extremely large mass near the speed of light. The space between neutrons is at a minimum within a direct collection alignment generating the mass. One of the distinguishing features other than its velocity being near the speed of light is that they have a separated tail. This tail is conic in shape. The reason for this is that the quasar is like a neutron star. In other words, a quasar is of an extreme mass. The protonization process is also massive. Light radiates from the massive atmosphere

Chapter 14: Creating Galaxies

of electrons. Since the quasar is moving near the speed of light, light deflected in front of the quasar's path bends as it would around a neutron star. The unique phenomenon involved is that light slings behind the quasar before gravity can draw the photons back into the quasar. The image is similar to air moving around a wing of a plane. With the exception, the vacuum exists behind the object instead of above it.

To illustrate this further, draw a circle and enclose it within a symmetrical teardrop. Place the teardrop's center of curvature in the identical location of the circle. Let the pointed end of the teardrop meet at an angle near 30°. Extend the lines beyond the intersection. Shade the interior of the circle and the region between the extended lines. This is the image similar to that seen of a quasar. The shaded area between the extended regions exists for the following reason. The light follows the path of the teardrop to its point as nearly congruent light (laser-like), invisible unless in direct alignment. The laser-like light crashes into itself and shatters. Its nearly congruent photons disperse. The shattering gives the visible separated tail. Next, add several congruent teardrops of near sizes with extended lines. Furthermore, imagine these lines exiting as a rainbow. Red exist the furthest out (of visible light) and violet the closest. In essence, there are bands of laser-like planes bending around the gravitational field of different colors colliding at the rainbow's "tail."

Additionally, we know that these planes are not purely laser-like in nature as the quasar itself is visible. In order to be a true laser plane, every photon would need to move parallel to the quasar's surface. This would leave no photons to reach our senses except for those at the tail. The reason for this visibility is that some of the newly formed photons collide with the congruent moving photons.

Hyperons (baryons of the second and third generation: charm, strange, top, and bottom quarks) do not occur in quasars nor neutron stars. The reason is that the process forming them requires distance for the transformation of the split. In the quasar, the protonization process formulates one giant **hyperatom**. However, it is not always one atom, but many isotopes ironically bonded. In the neutron star, the protonization process is severely impeded. Recall the inverse galactic process that forms them. The neutrons collected nearly simultaneously retarding the protonization process via being previously connected. Unlike the quasar being recollected isotope nuclei, the neutron star is an extremely large set of neutrons attempting the protonization. In this, very few protons and consequently relatively few electrons generate as the large mass slows the timing of the transformation process immensely. The photons generated are trapped into a decaying holding pattern just above the surface. Their photons normally become absorbed into the star's main mass. Only the nucleophotons escape from the surface, these "escapees" are projected nearly straight out from the mass.

Another interesting feature that distinguishes a quasar from the neutron star exists in further analysis of their surface composition. Both are extremely large nuclei objects but they have different configurations. While the neutron star has a surface primarily of

Part 5: Creation of the Macrocosm

neutrons, the quasar surface has a proton surface. This polarity exists because the last isotope that quasars engulf is Hydrogen-1 plasma. The gravity is so strong that the Hydrogen-1 nuclei (protons) collect upon the surface covering its neutrons. By this, there is an intense presence of microstrong energy upon the surface pushing the electrons away from the gravitational pull of the nucleus. This gives the electron the movement needed to produce photons that are in the range of visible light. Meanwhile, the neutron star's electrons have little microstrong force to push them away from their extremely large nucleus. The result is that they rest upon the neutron surface being in constant contact with intense heat. The electron's outer shell rarely contracts to emit any photons. The intensely compressed neutrons that experience the protonization process generate the extremely short wavelengths found in scientific observations.

Novas

Once we realize that there were no photons to give sight at the beginning of creation, we will know that we will never be able to witness the uniform state at the "birth" of creation. Secondly, all of the created galaxies formulate within the same timeframe because of the tendency of neutrons to form hydrogen isotopes. The concept that stars are of different ages and placing our own sun as only medium old star has some problems. While it is true that we are looking at light that has traveled as much as trillions of our light-years away the youngest stars being the one farthest away, we should not conclude that stars are of various ages and die, implode and become novas. There is no star within the universe eons older than our own.

The phenomenon that creates novas is that of two stars crashing into each other. This phenomenon would be less likely to exist if there were not more than one multigalactic shells created within the universe. If there were only one multigalactic shell, the stars would only continue to generally separate from each other, but this is not the case.

When two stars come into close proximity to each other, three possibilities occur. One, they have enough kinetic energy to continue their path away from the other star. Two, they become trapped by each other and form an orbital path around each other as a binary star. Three they crash into each other and form a nova. The later, naturally, is our focus.

Imagine two objects traveling in space moving in vectors slightly under parallel to each, eventually the paths bringing them into close proximity to crash. Now, imagine that they are the only two objects in space. The advantage of the second condition is that there are no reference points to mark their velocities. The result is that the two stars velocities will be relative to each other. In this, they can only record that they were drifting toward each other by gravity. As they come closer together, they accelerate to a greater velocity as falling into each other. Before they can meet, they pull gases and material from each other's surfaces. By the time they meet, both have reached their greatest velocities. The initial contact creates a flat circular plane that projects outward as the two surface layers meet at equal force. Later the kinetic energy of the impact passes through and removes matter from the opposite surface giving a double splash effect. The fanning effect forms as the stars tear apart to form a flat plane altering their kinetic input. There are more factors involved, but we would be getting ahead of ourselves. Lastly, their original kinetic velocities move the assembly in the direction that is a composite of their two vectors in relationship to their impact.

Chapter 14 Quiz

1. What caused the Multigalactic Shell to form?

   A. Big Bang
   B. Neutrons Collecting Process
   C. Expansion of the Universe
   D. Contraction of the Universe

2. What causes the Quasar "tail"?

   A. Collision of Electrons
   B. Bending of Space
   C. Movement of the Quasar
   D. All of the Above

3. Neutron Stars form because?

   A. Antishell Phenomenon
   B. Neutron shelling process completed
   C. Both a and b
   D. Neither a or b

4. Nova s form by:

   A. A dying star
   B. Two stars crashing
   C. Three stars merging
   D. None of the Above

5. How many Multigalactic Shells formed?

   A. 1
   B. 7
   C. 12
   D. Countless

6. (T/F) Our galaxy is part of a double implosion phenomenon.

7. (T/F) Firmament means to hammer thin.

8. (T/f) Neutron snow formed when the plasma cooled.

9. (T/F) Quasars are neutrons stars that formed by implosion.

10. (T/F) Novas are stars that died.

Chapter 15

Creating Earth

The bondverse occurs in the second day of creation. This phaseverse ends the dividing of waters fractals involving the initial creation. Within this, the actual end of the dividing of physical "waters" ends with the actual division of hydrogen-hydroxide (water). Elements become cool enough to bond for the first time since their formation. The cooling process starts as some photons hurl out into the void never to return via finite speed and a finite time that the creation is to exist. After the breakup of the multigalactic shell, the cooling process accelerates. Rather than recreating the wheel, we will give a very brief overview of the bonding process. For a more complete examination of the process, return to chapter 13 in the section on bonding atoms.

Bonding occurs as different atoms join themselves to other atoms by collecting their electron as part of their own makeup. Covalent and Ionic boding exist at two ends of a spectrum of bonding determined by electronegativity. **Electronegativity** is the affinity for an atom to collect an electron. Zero being totally covalent in nature and 4.0 being totally ionic, we give 2.0 the breaking point of being ionic. Covalent bonding in general follows the octet rule, as most of the elements that we are concerned with exist within the first 18 elements with the exception of iron. As the octet rule operates, we find noble gases that do not bond as they already have eight electrons in their outer shell. Plasmatic bonding follows a different set of rules using the same principle. Even plasmatic noble elements will bond because they grab for electrons to fill their shell "void" regardless of the electron's location (attached or unattached).

There is an opposite phenomenon to electronegativity. This phenomenon, lack for a better word is **electrodonativity**. This is the tendency for some atoms to donate their electrons to other atoms. If we were to look at the electronegativity table and reversed the values per period, we will have a good representation of this phenomenon. For an example, lithium would have an electrodonativity of 4.0. This is the most extreme case of the phenomenon. If we look at lithium, it is the most active element within the normal chemical replacement process. When we look at the radius of lithium, we find that it is about four times as large as hydrogen. On the outermost layer is a single electron. This electron's magnetic field has more exposure than any electron found in other atoms and is held the weakest of any atom. It is easily sucked into the electronegativity of another atom, which is the exposure of positive magnetic energy. With the exception of noble gases, the table's middle offers the most resistance for chemical replacement interaction, at least within at the ends excluding the noble gases.

Part 5: Creation of the Macrocosm

Formation of Our Solar System

Our galaxy, the Milky Way, formulated within one of the multigalactic shells. As noted by scientists, our solar system and subsequently our planet exist closer to the edge of the galaxy rather than toward the center. The outer stars move close to the same rate as the inner stars, unlike our planets. The primary reason for the low variation in movement is that their established speeds were acquired while existing as a smaller shell before the breakup. Lateral gravitational considerations at that time were weaker than the radial pull. There is a secondary vector, the vector of separation stems from its expansion established by the growing shell before it broke away from the shell. Galaxies separate as they develop gravitational borders between shell regions, aided by the increased shell surface area generated by the shell growth. Finally, gravity between stars and the center of the galaxy is too weak to require more speed. Before our solar system became a separate entity within our galaxy, the cooling process begins.

First, we must examine the breaking away by the solar system from the galaxy and the planets from each other. All of this occurs before the hydrogenation process of the planets. The initial image is that of magnetic/gravitational arms forming between star storms. The basic concept is similar to taking a piece of taffy and pulling it apart. Unlike the taffy, it is not solid and has many locations as centers of the pull. The resulting image is that of a complex collection of webs existing at three levels—galactic, stellar and planetary webs. Central to these webs are spherical forms that expand or pull further apart. Each web intersection is a plasmatic storm. The Central region is actually twisting like a huge hurricane. The arm strands are like feeders to the central storm. These feeders break at a location that is the midpoint between the storms determined by the mass between the two storms. Example: a larger storm would have a larger arm than the adjacent smaller storm. After breaking away, the arms become part of the storm and begin to twist with their plasmatic storms.

Eventually, the external tails of these spirals break away from the storm's center. This leaves the central storm as a disc. These plasmatic tails will formulate a multitude of smaller plasmatic discs, as their local gravity develops, from various denser parcels within the heavier central mass. These newly developed entities later formulate the **Levanastorm**s (le-vah'na-storm(s)) Phenomenon. Levana is a Hebrew word meaning moon (full). The reason for the separation originates in the conflict between gravity and the continued expansion generated by the growth of the multigalactic shell. Eventually, localized gravity supersedes this expansion as will be seen.

Before the contraction of plasma into the spherical centers, all space within the plasmatic storm of our solar system was occupied. The image appears as a multitude of hurricanes of various sizes existing around a central hurricane. The central **solar disc** also was a hurricane that has three major **substorm**s on its solar disc edge. Beyond these were the "no-man's land **microstorm**s" of the asteroids. The levanastorms also

Chapter 15: Creation of Earth

formed disc-like shapes external to the asteroid belt. Note: Jupiter and Venus are not counted as part of the solar system; they entered later as we shall observe. Jupiter is a mini-solar system unto itself. The solar disc substorms were Mercury, Earth and Mars.

As the external hurricanes' arms twist, they bend around forming a ring-like form engulfing an entire band of plasmas, vapors and liquids. These rings become concentric to the central disc. These arms engulfed minor storms. In the case of the larger outer planets, the arms engulfed levanastorms. Another observation is that the gaseous and lighter elements dominated these storms as the central heavy-core elements, primarily radioactive, centralize to the initial formulation of the solar disc.

The initial solar system had several major planetary storms three internal and five known storms were external counting the tenth planet (see Beyond Pluto in the Appendix). Using the internal planetary storms as a guide, we observe three major levanastorm arms formulating our solar system with possibly four as we move away from the central location. The reason for their existence is that at one time they were part of the "web" connecting adjacent stars. By this, we observe several more planets at regular intervals formulate with atmospheres primarily of hydro-boron compounds.

After these planets begin to develop, the near uniform gravity no longer surrounds the solar disc, and the solar disc begins to contract. As this occurs, three major storms overtake the heavier plasma at the hurricanes' edge. Again, regular intervals appear. Within this occurrence, the distances between intervals are 53 million miles. Note that the base is 36 million (Mercury). The earth's central location exists near 89 million miles instead of 93. We will expound upon these locations later in chapter 20.

An asteroid belt formulated between the planetary storms rings of Mars and Saturn. The width of the external planetary rings is one billion miles giving 500 million miles per side from the storm's center. Saturn's center is an 886 million making the Martian edge of the storm at 386 million. The interior planetary storms being 53 million-mile widths are roughly 26.5 million from center to the radial edge. However, the length was much larger. Since the Martian center is at 142 million, its outer edge is about 169 million miles from the solar system's center. This leaves a width of 217 million miles of **ministorm**s of levanastorm material. This is enough distance for four more interior planets to formulate. However, the thickness of the material was too thin to coagulate into any major plasmatic planetary storm. Ceres was the largest of these attempts.

Formation of the asteroid belt can be mathematically represented by some geometric manipulations. We need to look at the xyz-coordinate system. Using the y-axis in its traditional vertical position, we look at the central xy-plane as the cross-section of the plasmatic storm fiber of our solar system. From the location where the dimensional axes meet, using the xz-plane, we can measure out a circle that engulfs the outermost orbit of Pluto. Using the equation $y=1/x$ on the xy-plane limiting $x>0$, we define a line pattern that runs close and nearly parallel to the x-axis at Pluto and bends away from the

275

Part 5: Creation of the Macrocosm

x-axis upward as it approaches the y-axis. As the line gets very close to the positive y-axis, it begins to run a near parallel course to the y-axis. For convenience, let us limit the movement up the y-axis to be the same length as Pluto is from the y-axis. Now, let us rotate this line around the y-axis forming a funnel under the xy-circular-plate. Let all the space between the plate and the funnel become plasmatic material. For initial simplicity, let us view all the plasma as being of the same atomic structure; let us say iron. Now, let gravity become the sole factor action upon this structure. The strongest initial response is that the material forming the funnel's tip would collapse toward the plate. Gravity increases at the center of the plate pulling material toward the center. Gravity is peculiar, in that, it is a collective force of each atom and that its force decreases by the reciprocal of the distance squared. This gives the external mass of the plate a gravitational force of its own by sheer mass. From this, a global center forms with an external ring. The external ring gives added denser mass to the planets like Uranus, Neptune and Pluto. The set of points at which the separation occurs also forms a circle between these two masses. The circle is not just a two-dimensional image; it is as a wall cylinder. Again, there is gravity associated with this "wall". Being of tiny mass, comparatively, it is able to hold only tiny plasmatic storms forming the asteroid belt. Each plasmatic storm forms from a "funnel like" input from the multi-galactic sphere. Instead of one funnel in our analysis, there are many and of different sizes incorporated with the one main one forming the solar system. In our equation, this occurs at $y=1$. After the central breakup, the remaining central solar disc starts with a diameter of 19 to 22 million miles. Even so, much of Mercury's matter was drawn into the solar disc before Mercury could collect it. If this were not to occur, Mercury would be several times larger. Even though Saturn is the largest "natural" planet within the system, it is also the lightest, as observed by scientists. Similarly, Mercury, being the first planet beyond the solar definition, is lighter than that of the projected density increase. Even so, its mass is denser than any other planet.

There is one final factor concerning the formulation of the planets within our solar system. The expanding process initiated by the multigalactic shell comes to a halt via central gravity. When the external planets were forming, the millions and billions of miles seen today was a much smaller distance. The radial movement of the masses gradually nullify via the central gravitational pull of the solar plasmatic center. The planets establish orbital equilibrium. At one time, most of the planets outside the asteroid belt had rings including Pluto until Charon arrived. Mars, Earth, and Mercury never did, as the solar disc absorbed all the heavier unincorporated materials.

Natural Hydrogenation on Other Planets

As in all solar systems, we find that there are layers of plasma. The heaviest layer found in the center plane of the spiraling disc, putting the lightest as the external most layers. Each element exists primarily within a pure isotope collection, except for the minor

Chapter 15: Creation of Earth

mixing by the spiraling spin. This is truly minor when compared to the mixing during the cooling process. After the central layers of our solar system developed and before the hydrogen invasion, there was a pouring of lighter elements upon the solar disc. The image is like pouring out different color liquid-like substances at distinct contacting intervals. Naturally, the heaviest ones pour out first and the lighter ones afterward sequentially. These occur as the "falling strings," similar to the occurrence internal to the shell forming process. The falling string process also has centers, the isotopes involved are fluorine, oxygen, nitrogen and carbon, although there does not seem to be much fluorine detected within our solar system. The isotope rings show the central solar disc as a distribution center as it develops its own entity from the rest of the shell mass. Afterward the light elements poured thin layers nearly uniformly before the massive hydrogen rush to the mass.

The forming of elemental rings occurs around each star within the multigalactic shell. The heavy core elements are primarily found in the sun. These do not always form heavier matter densities. However, the number of nucleons is much greater than found in the lighter core material arbitrarily marked by the limit of iridium, which is very dense as a solid. The lighter core material starts with iron and ranges down to osmium, which also forms a very dense solid. The crust materials are also inclusive of iron and before neon. The atmospheric gases start at neon to the surface at helium. Hydrogen is an atmospheric gas gathered after protonization has occurred. There is a progression of atmospheric elements from a neon and fluorine solar base to oxygen moving out over both the earth and Mars, nitrogen stretches over Saturn, and carbon engulfing planetary centers out to Pluto. Another observation is that there are some light core material past Saturn. Actually, even Saturn may contain some crust material. Note: the terms of crust and core are used in reference to the role these elements play on earth.

If Mercury could maintain its original atmosphere, we might find fluorine in it. However, this information is lost, as the atmospheric gases ripped away before Mercury developed into a planet. The solar disc pulled off the gaseous elements; during the time that the solar disc shrunk pass the orbit of Mercury. As scientists observe, the atmosphere of today upon Mercury is the result of gasses trapped from the solar wind from the sun. Some even go as far to say that Mercury has no atmosphere.

Over earth and Mars, we get **Hydrogen hydroxide** (H_2O, water). This is because oxygen is the primary element below hydrogen and above the surface "soil" plasma of their respective crusts. Over Saturn this hydrogenation process forms **ammonia** as the hydrogen mixes with nitrogen (NH_3), an element one proton lighter than oxygen. Further out to Uranus, Neptune and unto Pluto the hydrogenation process formulates **methane** (CH_4) with the carbon, which is one element lighter within the same layer. Outside these elements rings, hydrogen continues to interact with rings of lighter elements of boron, beryllium and lithium. Externally, there are other asteroid belts and an oort containing sulfur instead of the iron found in the first belt, as will be observed.

Part 5: Creation of the Macrocosm

Another interesting note, Saturn is the lightest and largest planet originally assigned to this solar system. Recall the wide collection of thin strings of isotopes that can occur near thick isotope strings. Since it would float on water, it is safe to estimate the heaviest element in the core to be no heavier than aluminum. By examining the rings of Saturn, we can ascertain the general chemical composition of the core. As the core develops, it draws in the mass faster than the local gravitational force can release the material of the thin layer. Imagine a core that has aluminum as its heaviest element. Saturn is the first planet beyond the solar disc. It is the collection of a multitude of centroids that formed late in the multigalactic shell process. The reason for the lateness is not by slowness, but by ability to establish its own collection after the major solar storms collected neutrons, consequently smaller isotopes, for their central bases.

Upon the surface of Saturn, we will find granite-like substances containing light elements. Lithium is a very light soft metal weighing half as much as water, could very well compose much of the rock under a frozen ocean of ice. The metallic compositions would be with other light metals with light gases such as lithium nitrite Li_3N and frozen beryllium nitrite acid $HBeN$ among other compounds rare or not found on earth. This is true for all planets; even Mars has some compounds that are slightly different from that on earth like iron with large quantities of oxygen attached. On Uranus, Neptune and Pluto, we will find rocks that are primarily carbon and aluminum and even sodium and magnesium, with cores containing large quantities of silicon, sulfur, phosphorous, potassium and calcium.

Earth's Plasmatic Storm

Rotation of the earth's plasmatic storm before breaking away from the solar mass, as with all planets, was opposite of the central solar rotation like two touching gears. As stated earlier, the natural hydrogenation process started before complete solar-mass separation from the earth. Trapped hydrogen interacts with the exterior shell fragment surface of our solar system. Below that, there are light metals and more gaseous isotopes, under that are the lighter surface metal and metalloid isotopes, with the heavier core isotopes centralized under them. The term under then takes a twist as we move away from the center, we find more of the same surface metal and metalloid isotopes. Afterwards, we enter another layer of atmospheric gases, and then we exit into the interior void of the multigalactic shell.

Imagine being present during the formation of earth's plasmatic storm, we observe many small storms swallowed by a larger central storm, others roll via friction. For this reason, the orbital paths of the earth and the planets move in the same direction as our present day sun. Looking down upon the earth's Northern Hemisphere, the movement is counterclockwise. Orientating ourselves to the Southern Hemisphere, this movement would be viewed as clockwise, by the means of the inverted perspective as the formation that is to be land and Magnetic North is in our Southern Hemisphere.

Chapter 15: Creation of Earth

Recall the fractal example. The development of the planetary storm is the next event within the fractal expressed in Torah that Elohim divided the waters. The formation of separate plasmatic storms is another division of the waters to the earth. There is yet one more occurrence to this action. Here, the expression of waters is refers to plasmatic substances. Even water (H_2O) will formulate from this interaction.

Before looking into the hydrogenation that took place as the earth developed, we need to examine the element bonds that did not occur. As dubious as this may seem, it is important. As seen within the atmosphere of Saturn, there is ammonia (NH_3), yet we have water vapor (H_2O) and nitrogen (N_2) in the atmosphere. Neither are their large quantities of other nitrogen, oxygen or hydrogen mixed compounds.

The nitrogen side of the plasmatic eye had very little else involved. Recall that the central isotope masses had to thin out as the multigalactic shell expanded. This means that the original internal material had to thin, as there was no more material to add to maintain its thickness. This is unlike the external makeup of the multigalactic shell, which was continuously gaining material as it was forming. The reason the nitrogen was able to maintain its mass was its proximity to the central metallic plasma forming the "eye" of the earth via gravity. This gives the lack of large quantities of nitric compounds in the atmosphere, such as: nitric oxide (NO), nitrous oxide (N_2O, laughing gas), nitric dioxide (NO_2), and even indirectly nitric acid (HNO_3). Coming from a plasmatic state, nitrous oxide and nitrous dioxide would easily form.

The carbon layer is also not prevalent on the bottom side of the storm. We observe this as the lack by abundance of cyanide (CN). The only compound prevalent in our atmosphere formulated by nitrogen is nitrogen gas (N_2). The formation of nitrogen gas occurred simultaneously with the hydrogen gas upon the exterior. Once this gas forms, it becomes very resistant to reforming into other compounds. The reason for this is the triple bonding with itself. The only metal known able to cause nitrogen to react is lithium. By this, we know that nitrogen within our earthly plasmatic storm was primarily absent from the other side of the storm, as the compounds in which lithium exists lacks nitrogen. If lithium were to pass through any quantity of nitrogen even as a gas, it would have grabbed them on the way to the earth surface.

There is a reason for this lack in occurrence. Recall that the solar disc once engulfed the plasmatic storm of earth as far as the asteroid belt. During that time, the central layer of our plasmatic storm was untied with the solar storm. Eventually the heavy core elements separated because of the density developed by the storm centers. After they separated, the elements above them also began to separate. Recall the nature of gravity and density. While the gravitational force increased within the storm, gravity decreases external to the storm because matter moves away from the edge to the center. Despite all of the gathering by the core, there are elements still continuous throughout the interior storms. Argon was the last remaining layer of isotopes to maintain their unity. Eventually, argon becomes the noble gas observed today. However, in the plasmatic

Part 5: Creation of the Macrocosm

state, it is an isotope layer heavier than silicon and aluminum. By this, we see two layers of silicon existing even after the separation from the solar plasmatic storm: one above the central core material, and one under the central core material. However, aluminum existed primarily on only one side of the plasmatic storm of the earth, which we will be expound the reason later. Argon may have even been a thick layer within the earth's plasmatic layers before forced to stretch out thin known as the **argon plain**. The argon plain keeps nitrogen and oxygen on two separate sides of the storm as a molecuum. By the same reason, carbon and hydrogen are also kept away from nitrogen.

Now, we are ready to back up in time and present the entire process to the plasmatic storm forming our planet. The solidification into molecuum would have stated with hydrogen, but hydrogen only needs one electron to form a complete atom and can only form a solitary bond with another atom. At best, it could form strings temporarily when the electrons are held in the seventh shell and shared with an adjacent atom. In this, we have electron, proton, electron, proton, electron…proton, electron. Each atom thinks it has two electrons. The next phase is the dual element molecule. The string breaks up as the cooling moves the electron inward to the innermost shell level. In essence, the radius of the hydrogen shrinks. This changes the dynamics of hydrogen string. There is a shift in pull magnitudes and the electron is freer to move about the atom. This motion along with additional electrons destroys the "hold" of hydrogen into a string. This covers the entire transition from being plasma into a vapor for hydrogen.

On the opposite side, we have nitrogen. This element first forms into a lasma with the rest of the interior of the multigalactic shell. As the lasma cools and condenses, electrons are needed to fill the atom's demand. During this period, electrons attached to the adjacent atoms are attracted and the atoms bond temporarily into a molecuum. The molecuum does not completely form as nitrogen only holds seven electrons. As stated earlier, only elements with 12 or more protons will form a complete molecuum structure. However, the nitrogen **elementuum** (a single element molecuum) does form a solid that joins with the bottom layer of solar system. Before this surface melts into lavor, more molecuum forms underneath. Eventually, we reach the central layer of argon. This layer is dimensionally very flat and thin in comparison to the other layers. The argon elementuum lasts longer because it needs to cool enough to collect 18 electrons as opposed to nitrogen needing only seven.

Upon the upper (outer) surface, molecuum begins forming with the light metals of lithium and beryllium and later into boron, and carbon. Unlike the nitrogen layer that completely covers the earth's plasmatic storm underneath, the light metals form discontinuous clouds over the central storm. This leaves other elements directly under it exposed. Beyond these layers exists a thicker and larger centralized island of carbon. Directly under the carbon island is a vast sea of oxygen covering the central plasmatic storm. Below this sea of oxygen is the argon elementuum floor. Above the oxygen sea exists a large cloud of hydrogen invading into the sea mingling with the oxygen. Near

Chapter 15: Creation of Earth

the argon floor, a thin oxygen elementuum formulates. Notice that nitrogen forms about 78 percent of the atmosphere while oxygen forms only 21 percent. We may tend to think that there was more nitrogen plasma than oxygen. This is until we look at all the oxygen absorbed by hydrogen to form water. The hydrogen invasion occurs throughout the entire upper surface of the plasmatic storm. As peculiar as it might seem to us, the hydrogenation of oxygen even begins before the earth collects all the atmospheric gases within its domain.

Another interesting item about the argon elementuum floor is that it is highly fractured. Moreover, the argon plain does not connect to the twisting plasma near the core of the plasmatic storm. A primary reason is the conflict of force vectors between a stationary outer mass and a rotating mass. The plain hangs around the earth as the rings of Saturn. Unlike Saturn's rings, the argon plain is solid and does not have a smooth circular edge. The edge of the argon plain surrounding earth is anything but circular but jagged. Outside this edge is broken pieces of the argon elementuum of various small shapes. This plain once connected with Mars and Mercury as liquid plasma (lasma). It broke while crystallizing between planets. As the planets moved around the solar center, they collected the some of the pieces, while the rest eventually falls toward the solar center.

As noted earlier, the hydrogenation of oxygen occurred away from the central plasmatic storm of the earth. The cool hydrogen melted the elementuum of oxygen to form on the surface of the oxygen sea. Only at the argon elementuum floor and under the umbrella of light elements did any oxygen elementuum form. The oxygen elementuum under carbon decomposes or "melts" into oxygen molecules as it cools. The hydrogen march to the argon floor halts as the water vapor forms into heavier molecules than the hydrogen gas. The result is that the water vapor forms a barrier preventing any more hydrogen from entering into the oxygen sea. The oxygen that hydrogen melts from the moleucuum cools rapidly into a gas via the "cool" water vapor. The word "cool" is a relative term in relationship to the plasmatic conditions and not to present atmospheric conditions. Another location that oxygen becomes absorbed is at the crust. The primary crust material is initially silicon forming SiO_2.

The ozone forms in a phenomenon unique to the formation of the planet. It forms as a chemical reaction between oxygen plasma and water vapor. The oxygen replaces the hydrogen of the water molecule. The reason for this occurrence is that the oxygen electronegativity of the cooling plasma is greater than hydrogen. This is easily seen as oxygen as an atom has the electronegativity value of about 3.4 while hydrogen is only 2.1. Actually, while cooling from a plasmatic state, all elements acquire higher electronegativity values otherwise no electrons would be attracted to the element. This also places the ozone higher into the atmosphere than the resulting oxygen formation.

Let us return to the phenomenon occurring at the storm's center. Now, it becomes important to know exactly the elements involved. Starting from the external most layer of hydrogen, we will work down to the core material. There are some traces of helium,

Part 5: Creation of the Macrocosm

but not of any significant mention. Under these gases is a thin layer of lithium, beryllium and boron. Under this is a thicker layer of carbon. There is no real measurable nitrogen on this side of the plasmatic storm, possibly even totally absent from this side of the storm. The reason for its near absence is that oxygen mass pushed nitrogen away beyond Mars, which pushed carbon out to Uranus and beyond within this isotope layer. If we could examine beyond Pluto, we would find external bands of boron to lithium. For this reason, we have a very large layer of oxygen. Moving toward the sun from earth, oxygen is then preceded by traces of fluorine and neon. The presence of fluorine and neon is not realized as Mercury's atmosphere is destroyed by the absorption of the gravitation of the sun and the blast of the solar wind.

This brings us to the surface forming elements of the earth. The first layer is that of sodium, and magnesium. Each of these layers by themselves is larger than the lithium, beryllium, and boron layers combined. Under this is a good-sized aluminum layer followed by a very large layer of silicon. Under silicon, there are traces of sulfur and phosphorous. Then the gases of chlorine and argon layers exist. Of the two gases, argon is the most plentiful. However, argon does not play any major role within this region of the storm. Our next notable elements are potassium and calcium. Then we have layers of trace elements leading to a large layer of iron. These make up almost the entire crust of the earth. There are other heavier elements under these, but most of them form the original plasmatic core. The plasmatic core changes as the earth forms as we shall soon observe.

Examining the elements working from the multigalactic formation underneath, we find that they are less complicated. The external most layer underneath is nitrogen. This large nitrogen layer precedes a large layer of silicon, with immeasurable traces of other elements in between. We need to note that the sodium and magnesium are also trace elements on this side. After reaching silicon, the elemental arrangements become more consistent with the topside elements. The reason for this is the forming of the central spheroid of the earth by the heavier core elements occurring before the storm completely separates from the solar disc.

Returning to the upper side of the storm, we see a sequence of metal and gas elements layering between each other. Obviously, hydrogen is the top gaseous layer, followed by a thin metallic plasmatic layer of lithium, beryllium and boron; the latter is a metalloid. Under that, we have a large gaseous layer containing the nonmetals of carbon, trace nitrogen, oxygen with a trace of fluorine and neon. Our second metallic layer contains primarily of aluminum, silicon, which are surrounded their neighboring elements. Iron does not immediately join to them; there is another gaseous layer of chlorine and argon. Under that, we find iron with all of its neighboring elements. Beyond this there are other alterations involving krypton and xenon. However, the latter two remain primarily within the plasmatic core. Our primary concentration will be on the outer three sets of alternate element dispositions. Within these the layers, our surface

Chapter 15: Creation of Earth

definition was determined. The primary response is the "rain" or "hail" effect depending upon the particles being fluid or solid in nature.

Before getting started in our raining phenomenon, we need to examine the boiling and melting points of the primary surface elements involved. The elements are listed below by their initial plasmatic arrangement indicated by "P." There are two other sequences involved. Represented by "B," is the boiling point sequence and "M" the melting point. "L" is the liquefied groupings and "S" is the solidifying groupings. "G" represents a general grouping of elements changing states based upon their melting and boiling points. The letters represent a sequence of cooling. Some of the listed temperature data are exact, others are approximate, and scientists have extrapolated the boiling point of carbon. We are working backwards in comparison to the chart. The scientists composing the chart were heating up elements to reach the plasmatic state. Within the scenario of creating the planet earth, we are cooling the plasma to a solid. Therefore: if a substance has a melting point of 100° F and another of 50° F, the 100° F substance would solidify first. The reason is that as we move from 200° F downward to zero. We will reach 100° F before we reach 50° F. The elements that vaporize last are the first to condensate into a liquid. Inversely, the last to liquefy will be the first to solidify.

The table below does not list the natural gases found within the temperature range found upon the surface of our planet. Within the range of this table, they are: hydrogen, helium, nitrogen, oxygen, fluorine, neon, chlorine and argon. Two other elements are missing from this table, they are sulfur and phosphorous. These two elements are trace elements comparatively to the other elements. Sulfur does seem to remain lower in the crust. Example, volcanic fumes contain sulfur vapors. There is also a leap to iron. The reason is that, in our planetary storm, iron is the primary element under the large silicon layer. Iron has a much larger input into the crust formation than all the transitional elements preceding it. It is also the last element that exists in any great quantity within the schema of elements formulating the crust of the earth.

M	B	P	Element	Melting Point		Boiling Point		L	S	G
9	8	1	Lithium	357°F	181°C	2,447°F	1,342°C	D	E	B
5	3	2	Beryllium	2,348°F	1,287°C	5,378°F	2,970°C	B	C	A
2	2	3	Boron	3,798°F	2,092°C	7,236°F	4,002°C	C	B	A
1	1	4	Carbon	6,426°F	3,552°C	8,721°F	4,827°C	A	A	A
10	10	5	Sodium	208°F	98°C	1,621°F	883°C	E	F	C
8	9	6	Magnesium	1,200°F	649°C	2,025°F	1,107°C	D	D	B
7	5	7	Aluminum	1,221°F	661°C	4,472°F	2,467°C	D	D	B
4	6	8	Silicon	2,570°F	1,410°C	4,275°F	2,357°C	B	C	A
11	11	9	Potassium	146°F	64°C	1,400°F	760°C	E	F	C
6	7	10	Calcium	1,542°F	839°C	2,703°F	1,484°C	D	D	B
3	4	11	Iron	2,795°F	1,535°C	4,982°F	2,750°C	B	C	A

Part 5: Creation of the Macrocosm

Carbon within this table has the highest boiling point. There are a few other elements that have similar high-end boiling points. However, they are heavier core elements like osmium putting them deeper within the plasmatic storm further from the cooling effect of the external void. The massive hydrogen layer above the lithium, beryllium, boron and carbon layers retards the cooling effect of the external void. But underneath the storm we have only nitrogen with some trace elements protecting the crust material of silicon and iron. Notice that aluminum, silicon and iron have similar boiling points, hence, similar solidifying points. This will become important after the spherical molecuum forming process passes beyond the iron layer. The carbon molecuum forms a shield above the spherical form of the core from the cooling hydrogen. Formation of the lasma within the spherical storm of the earth is the beginning of the crystallization of the outer crust of the earth. Lasma first forms within the layers of aluminum, silicon and iron layers as they have similar solidifying temperatures and in the outer regions of the crust. As the spherical lasma forms into a molecuum the nitrogen molecuum and the carbon molecuum applied pressure upon the surface. This pushed oxygen, sodium, and magnesium lasma down into the aluminum, silicon and iron plasma. In doing this, magnesium and sodium were also pushed down below the aluminum layer. Sodium and chlorine later collects to form salt. Because of this phenomenon, we find the lighter element of magnesium making heavier bonding formations under the generally lighter formation of aluminum of the **Felsic** rock layer (density 2.6 to 2,7). Nitrogen molecuum pushes primarily nitrogen lasma down into the opposite polar location.

Forming of the Crust

After the initial galactic breakup and as the solar disc evacuates, the earth breaks from the solar disc core material; gravity takes the semi-flat central plasmatic elements and pulls it into a spherical form. The image is that of the solar core contracting from the asteroid belt pulling plasmatic core material from a flat disc into a spherical storm. As it draws the central core material to itself, pockets of heavier core material separate and begins its own spherical formation. Being of heavier elements, the spherical plasmatic center formed giving the external lighter core and surface material a place to amass.

Aluminum, silicon and iron are the next "surface crust" materials to alter; they change into lasma as it collects electrons via cooling. The lasma collects and forms into a magnetic mist. The mist is heavier than the pure plasma under it, and it falls as droplets. The mist heat up again into plasma and rises. In doing this, the plasma underneath loses heat and eventually begins to form more lasma mist. At this point, there is a near ellipsoid mist like cloud forming the surface of the earth. The clouds of lasma were actually formulating into two bowl-like shapes. A flat crystalline formation by the elementuum of argon temporarily prevents the two sides from joining.

As both the clouds of aluminum, silicon, and iron thicken on their respective sides, they gain more weight. This draws them closer to the storm's center via gravity. In turn,

Chapter 15: Creation of Earth

they become denser. Repeating the scenario of forming droplets, falling, vaporizing and cooling more material, they eventually form liquid sheets. The gravitational attraction between the edges of the two "bowls" pulled them together pinching off the external argon plain from argon inside the storm. Afterwards, the ellipsoid became more spherical. The lasma mist surrounds the central plasma as a shell. As the lasma mist thickens, a liquid-like formulates a bubble. Heat prevents the bubble from collapsing. The reason is that the lasma liquid could only fall so far until it "vaporizes." The atoms are pushed back to the surface being lighter in weight. As they move upward they meet other droplets coming down. Iron cools by this invasion of silicon and aluminum lasma formulating heavier lasma raining deeper toward the core. Actually, as they move back out, they collect elements and move upward as well. By this, a certain amount of mixing occurs between oxygen to magnesium and aluminum to iron and silicon.

The bubble gains more mass via collecting mist from within. The extra weight causes it to fall toward the core generating more thickness and creating internal pressure. Eventually, the aluminum, silicon and iron lasma mingled with oxygen to magnesium lasma forms a large ocean with no floor. While this is occurring, the surface "freezes" into a molecuum. At this point, the plasmatic surface of the earth stops shrinking. We should now recall that aluminum was primarily found on one side of the storm. This means that aluminum existed primarily on one side of the "equator" of the earth, the side facing outward toward the hydrogen, as will be observed.

The silicon molecuum mixture sinks, but not in the traditional sense. The surface melts into lavor while the lasma below the molecuum freezes. The lavor above the silicon molecuum surface mingles with the elements above it, while the lasma underneath mingles with the plasma under it. Underneath the oxidized silicon molecuum shell, silicon now is the catalyst for cooling iron. Iron begins misting and collecting upon the internal ocean of oxidized silicon within the earth. Iron is a large layer. Now, we have three processes working. The upper process begins to formulate the light crust (felsic, density around 2.7), the molecuum maintains the **mafic** (3.0) composition, while underneath the **ultramafic** (3.3) crust formulates. Iron elementuum eventually forms.

Pressure of the core underneath this process decreases slightly as matter is being absorbed into the process cooling and requiring less space. However, the pressure is still quite great. Despite the strength of the established molecuum, the earth shrinks a little via added weight altering the formation of new molecuums.

The plasma underneath is now under extreme pressure. This causes the plasma to behave more like vapors in a pressurized container. Just as each vapor occupies the container uniformly, the plasma begins to function in the same manner. This is important as it brings heavier elements to the interior lasma ocean. These elements then collect upon the surface and sink into the ocean-like lasma forming a very heavy layer under to the iron elementuum. This will become important later as we examine gravity. The molecuum continues to sink or formulate even below the iron element.

Part 5: Creation of the Macrocosm

Meanwhile back on the surface of the spherical form the cooling process continues. The surface moleecuum turns into lavor; lavor turns into vapor and is becoming a liquid again. The mist turns into rain, and the rain evaporates before it hits the liquid-like lavor surface above the moleecuum. Eventually, the oxidized silicon rain and iron lavor mixes and remains a liquid above the liquid-like lavor. It is similar to the vapors of the atmosphere resting upon solid matter without necessitating the need for a liquid surface to be in between. In other words the compounds are, such that, they can rest against each other without conflict of state of matter. In our case, the collecting process of electrons for iron is much more extensive than that of silicon. The iron atom holds 26 electrons while the silicon atom needs only 14 to become a complete atom.

Aiding the process of the liquid state of matter existing next to the lavor is that compounds, which needs to be cooler to remain as a liquid, will evaporate leaving other compounds that remain as a liquid with the heat of the moleecuum. The result is that we have added another layer sealing the internal heat from the substance above it. This provides the cooling process of the surface an added enhancement. Eventually all the surface (crust) compounds liquefy.

Another factor involved in the formation of the earth, and for all planets, is that the liquid-like surfaces are not still, there are waves. While the dimensions of these waves are unnoticeable looking at the earth as a sphere, at ground level they appear. The primary effect is that elements that float to the surface are made discontinuous.

The resulting surface of the earth is a molten planet with some metallic vapors near the surface mingled with oxygen and nitrogen. The nitrogen will in most cases remains unchanged being triple bonded to itself. The lighter elements are still raining to the earth. Now, we are ready to examine the formation of the felsic layer.

Most of the silicon bonding with oxygen results in SiO_2. This compound composes about 60 percent of the earth's crust. Along with this detail, about 75 percent of the crust is composed of oxygen and silicon elements. Within this, oxygen composes near 50 percent of the crust. Oxygen and nitrogen both are seen to develop the atmosphere against the surface during the time that the surface was still misting into a liquid ocean. Both elements manage to mingle with the surface elements without losing their initial bonding in their gaseous forms; this is especially true with nitrogen being triple-bonded.

Formulating the Felsic Layer

The factor that permeates the phenomenon is the rotation of the plasma, vapors, liquids, and semisolids. As noted earlier, the storm rotates around an eye forming the planet. Also noted earlier, the whole assembly is shrinking as it cools. By this, there is a twisting involved within the assembly that alters by the cooling. This further mixes the elements within layers. It also leaves its mark upon the surface, which we will explain in detail shortly.

Chapter 15: Creation of Earth

Imagine two particles sitting on a solid disc at different distances from the disc's center. Let the disc rotate at a speed so the particles remain upon the disc. The particle closer to the disc's center is moving slower than at the edge, when we consider its linear velocity. This means the outer particle is traveling over a larger distance per second than the one closer to the center. With the same information, we observe the two particles having the same angular velocity. This means both particles both particles make one complete revolution at the same time.

Next, imagine a large cup of water being stirred with a spoon. After taking the spoon out of the water, place two floaters at different distances from the vortex of the moving water. Observe that the floater closer to the vortex has a greater angular velocity than the one toward the edge of the cup. This is a similar scenario to the case with the plasmatic eyes that form galaxies, stars and planets. Unlike the floaters, the masses of these objects exert great amounts of gravitational and magnetic energy generating interactions between objects.

From this, we can set up an imaginary whirlpool with different properties. Imagine one in which both floaters were moving the same linear speed. The angular speed of the outer floater is slower than that of the internal floater. As the floaters become drawn closer to the vortex they gain angular speed but not linear speed. This is magnified, if the external floater linear speed were to increase as it approaches the center, as in the case of acceleration due to gravity. Even without this increase, the angular speed will become greater than that of the internal float because of its linear speed.

Another way to look at it: Imagine a person on a bicycle traveling in a circle of 100 feet diameter at 5 miles per hour and another at 200 feet diameter at 7 miles per hour (both paths are concentric in respect to each other). The person on the inner circle will make one revolution in a little less than 45 seconds. Meanwhile, the external bicyclist will complete one revolution in a little over 1 minute. This gives the outer bicyclist an angular velocity slower than the angular velocity of the inner biker even though the outer one is traveling at a faster linear rate. Note: In order for the outer bicyclist to travel the same angular speed, its linear velocity has to be 10 mph. But, consider this bicyclist moves inward to the circle that the inner biker is traveling maintaining the same speed, the angular velocity increases beyond the inner bicyclist. This is the nature of the shrinking plasmatic eye, as the plasma moves toward the storm's central eye, it gains angular velocity.

In essence, the central region of the earth moved as the sun's surface before solidifying. Scientific observers have found that the central region rotates one in every 25 days while the polar region rotates once every 36 days. The difference is even more dramatic when examining their linear velocities. Even at a point 30 degrees from the polar region, the calculations would give an extreme result. At 30 degrees, the surface has only half the distance to travel viewing it a perfect sphere. Therefore, the angular velocity at the equator using this particular polar linear velocity gives a rotation once in every 50 days.

Part 5: Creation of the Macrocosm

The same is true with the earth at the equator; it has a faster angular velocity during the time of being a plasmatic storm. The linear velocity of matter coming into the planetary storm from greater distances equatorially increased the angular velocity of the equatorial region of the earth or any planet during its plasmatic existence. This factor also plays an important role during the solidifying of the surface.

Even before the surface solidified, gravitation played a role in separating the molten liquid minerals. The heavier liquids sink pushing the lighter minerals to the surface as expected. However, centrifugal force also has a role. If it were only gravity, the felsic cap would never develop. Instead of a cap, the whole layer would cover the earth. This layer, naturally, would be much thinner. Centrifugal force forces the heavier material toward the equator. This forms a liquid ridge. This ridge is shallow by our standards of a mountainous ridge; never the less, it was a ridge.

The ridge was composed of mafic material holding the felsic material in one of the polar locations. As we shall see later, this polar region was in union with the magnetic north. However, the magnetic north was in direct alignment with the South Pole. As we examine plate tectonics, we will see why this is so. Returning to the scenario, the felsic material was literally pushed away from the equator by the mafic material. In fact, the felsic layer formed a mountain of its own. The central polar region having the least centrifugal force involved offers the least resistance to the felsic push back. Strangely enough, gravity aided in the formation as well. Gravity seeks to form a perfect sphere. Just as the plasmatic eye bulged in the middle as gravity manifested, so does the felsic liquid bulge. This makes the felsic cap thicker in the middle, later it will become a single central mountain within the landmass with extremely gentle slopes. The mountain is more as a circular plain that bulges in the middle as the lower heavier layers pull more.

Adding yet another dimension to this process, we observe that the equator is being accelerated more than the polar locations. By this, the liquid shell at the equator is moving at a faster pace than the polar locations. As this shell cools into a singular object, solidified bands of materials formulate into layers.

The object in this case, earth, has to reconcile the differences in motion between the masses of crust. Since the larger mass is at the equator, the Polar Regions become accelerated. By this, a twist occurs between the different compositions of crust. All of the felsic material becomes twisted to a certain degree. The mildest twist occurs at the polar site, and the most extreme existed at the lateral edge of the felsic caps mass. By this, concentric fracture-like formations occur at the edges between the felsic and mafic material.

The felsic material did not sit on top of the mafic material as pieces of paper. It was more like a boat displacing some of the mafic material. We might say, the felsic layer was somewhat embedded upon the mafic surface. The felsic mountain protruded approximately 100 additional feet above an imaginary projected level felsic plain, which

Chapter 15: Creation of Earth

is virtually undetectable by sight. This is minor in comparison to the rise of the felsic cap out of the **mafic plain**, ocean floor, as we shall soon observe. The measurement is assuming that the felsic material forms a perfectly flat pain, which it is not.

It should also be noted that the crust did not solidify evenly. In fact, the first layer to solidify was the felsic layer. The reason is that the lighter material was raised out of the central heating system provided by the internal plasma. Lighter material, in general, cools faster than denser material. By this, we observe the felsic cap cooling first. The felsic cap floated upon the mafic molten ocean. This further establishes a weaker union between the felsic cap and the mafic layer. The buoyancy of the cradled felsic cap caused it to rise slightly above the mafic ocean.

The general density of the felsic layer is from 2.6 to 2.7. This means: by putting in 26 pounds of felsic material into water, it will displace 10 pounds of water. The density of the mafic layer is about 3.0. Now, imagine the molten felsic cap floating in molten mafic material. Since the felsic cap is approximately 14,000 feet thick (close to 3 miles), at a density of 2.7 times that of water its buoyancy in the molten mafic is 0.9 (2.7/3.0). This means: if a flat slab of felsic material of 10 feet thick were sitting in the mafic fluid, one foot would exist above the surface. Applying this to 14,000 feet, 1,400 feet is lifted out of the mafic plain. At a density of 2.6 times that of water, over 1,800 feet is above the plain. This measurement is still much below the general 1.5 miles (over 7,000 feet) of water that will eventually cover the surface.

The felsic material does not completely cover a hemisphere of the earth. From the appearance of the ocean floor, there is not much residue of felsic material indicating a broader band of felsic mass beyond the disc. The image is that of lighter plasmatic materials falling to the earth came from one direction. As such, they fall upon the sphere/ ellipsoid mass via gravity. For the light material to reach the equator, the cloud-like layer raining to the surface would have to be infinite. The reasoning is found in the nature of angles tangent to a point on a parallel plane. If we were to establish a line perpendicular to the two parallel planes and project lines from the intersection of one of these planes to the other at intervals of five degrees each, we will find that each angle, even though equal in size, intersects the opposite plane at increasingly larger intervals as we approach $90°$. Consider the distance generated between $85°$ to $87°$, about 17 units, and the distance generated from $87°$ to $89°$, which rounds to 29 units. The intervals increase even more radically between $89°$ and $89.9°$. Actually, increasing the angle from $89.9°$ to $90°$, we find that we have to leave "our" finite space.

This is the basic description of the felsic cap. As noted before, the felsic cap is not flat; it bows outward in the middle and tapers. The circular disc is about 9,300 miles in diameter and nearly 3 miles thick. Before there was an ocean, its central most point existed approximately 1,600 feet above the mafic plain. This elevation tapers similar to a bead of water upon a polished surface. Note the edge of the greatest elevation change is at the edges, with only slight elevation changes at the central region at the bead's top.

Part 5: Creation of the Macrocosm

This is similar to the felsic cap. Since the cap felsic material cooled first, it spent time floating upon the molten liquids of the mafic plain. The cap edge is not an abrupt 1,400 foot cliff. It slopes down to the mafic plain to about the 23rd parallel from the equator of the formation. The felsic elevation raises the water level by about 700 feet.

Finishing Felsic Formations

Meanwhile, the surface of the storm is not idle. We need to backup in time again and examine events occurring within the external layers. Hydrogen remains a cooler gas touching the outer light elements. When an atom cools enough from being in purely a plasmatic state, it becomes a lasmatic liquid. Initially the atom captures one electron from the plasma, and sometimes shares the electron with an adjacent atom. As the lasmatic liquid continues to cool, more electrons are gathered. As noted earlier, before the substance is able to cool enough to be a vapor, plasmatic bonding has already occurred. The elements form plasmatic unions without regard to bonding rules of complete atomic structures. The initial results of this bonding formulate a molecuum of the element. However, this state ends in lavor liquidation as the substance cools toward a stable vapor formulating a semi-ground state atomic structure.

Carbon Rain. Carbon within the earth's plasmatic storm responds differently than those seen external to the asteroid belt. External to the asteroid belt the carbon atoms come in direct contact with the cooling hydrogen. The carbon atoms of earth are temporarily protected by a shield of lithium, beryllium and boron. This delay gives the carbon atoms time to collect among themselves forming their own elementuum. The reason that carbon formulates first is the response to collection of electrons was greater than the upper elements. During this process, it forms into lasma droplets. As they continued as collect to form into a molecuum, gravity pulls some of the larger droplets toward the earth. These large lavor droplets become spherical; they cool and drop toward the earth. These droplets formulate not by a few carbon atoms, but by literal multitudes of atoms. Some of the larger earlier ones fall and go deep into the silicon layer forming today's diamonds. Having a higher boiling point as a collective, the carbon droplets do not dismantle in the silicon liquid lavor.

Meanwhile, hydrogen is working its way through the eroding "shield." The cooler hydrogen does not collect upon the lithium layer, as lithium has a much cooler boiling point and stays as plasma much longer. However, this does not stop hydrogen's "march" toward the earth surface; it only provided a temporary delay. The warmer carbon atoms heat the molecule and take the hydrogen as they return to a plasmatic vapor. These atoms being lighter than carbon are pushed back upward via buoyancy. Eventually, the hydrogen collects to the carbon forming hydrocarbons. These are less numerous than the multi-carbon units are, as hydrogen "clogs" their bonding sequence. As they fall through the oxygen plasmatic vapor, they collect oxygen via ionic bonding by the cooling oxygen to form crude oil. Later, silicon cools (thickens) and more lavor

Chapter 15: Creation of Earth

forms, hydrocarbons rain down collecting oxygen and splash upon the lavor floor. These do not fall far into the sodium and magnesium "plate" above the silicon layer and become seed pools for Elohim.

Beryllium Rain. Between lithium, beryllium and boron, beryllium is the first to coagulate after hydrogen. Today, we find that beryllium pockets formed ($Be_3Al_2Si_6O_{18}$). The initial response of the cooling beryllium is to collect electrons, and bond. Beryllium does not form true molecuum. The reason that this does not occur is that this element can only collect four electrons by definition of its nucleus. The result is a form of a multitude of beryllium sheets stacked upon each other. This condition starts to breakup as the atom cools enough to collect another electron. At this point, the first shell level requires two electrons that not shared with other atoms. This leaves the outer shell, needing only one electron in which it is "willing" to share. The final form is that of a trio-beryllium complex (3Be). In the case of beryllium, each beryllium atom has only one electron in its outer shell shared with the adjacent beryllium atom. Within this structure, each atom forms two bonds so that each atom "thinks" it has two electrons in its outer shell. Note: there is not enough room for eight, as two electrons do complete the outermost sublevel of the atoms. This neutralizes the molecule as it falls through the plasmatic gases to the liquid surface. While the trio-beryllium complex may have collected some oxygen on its course to the surface, there is no evidence indicating that they were able to keep it. The surface liquid engulfed any collected oxygen.

As the beryllium atoms collect electrons they bond. The first bonding occurs with silicon connected to four oxygen atoms (SiO_4). The initial reaction is that the beryllium breaks up and replaces an oxygen atom of two silicon molecules giving $BeSi_2O_6$. However, in doing this, beryllium loses its outer electrons to the silicon atom. This causes beryllium to seek to more electrons. Since this process is occurring to three beryllium atoms simultaneously, the response is one answer. The answer is to collect two aluminum atoms and bond with them. In this each aluminum atom shares its three external electrons with each beryllium atom, one each. By this, the balance is restored.

Another interesting note is that the two aluminum atoms exist at right angles to the rest of the molecule. This gives both aluminum atoms the ability to share electrons with the three-beryllium atoms. After accomplishing this task, the beryl molecule complex is temporarily the heaviest molecule in the liquid, having an atomic mass of about 537.5 u. This is nearly equivalent of joining ten iron atoms. During this period, other beryllium complexes fall into liquid mixture. The same process occurs again. The resulting beryl molecules collect forming large masses, in terms of the count of molecules. These later form the beryl crystals. Other beryllium particles fall as individual atoms collecting six oxygen atoms on its way to the surface and react in other ways.

Boron Rain. Boron fell in a similar fashion. The composition in which we commonly find this element is **borax** ($Na_2B_4O_7 \cdot 10H_2O$). There exists another common variation that is **kernite** ($Na_2B_4O_7 \cdot 4H_2O$) without involvement with **hydroxide** (OH). Again, we

Part 5: Creation of the Macrocosm

observe that the base configuration has no nitrogen or hydrogen. The only hydrogen found exists because of the collection of hydroxide or water vapor. The elementuum lattice of boron is still a sub-formation of true moleculum. Instead of twelve bonds, only five are possible, by the dictate of having only five protons in the nucleus. The image is that of thin sheets of boron connected alternately between layers. The result of its termination is that instead of three atoms as in beryllium, we have four. The basic plasmatic effect affected the boron configuration. We have atoms "believing" that they have acquired all their electrons by bonding with adjacent atoms and sharing electrons. Unlike the beryllium scenario, the boron atom runs into a problem. Imagine a square that has a boron atom at each corner. Place an electron between of each of these atoms. These electrons are shared by the two corner atoms defining the square's edge. The result of this process, assuming the first shell has already attained its electrons of each atom, is that each boron element now "thinks" it has two electrons in its outer shell. However, the atom needs three electrons for the total of five as demanded by the nucleus. For this, we place an electron in the center of the square. Now, we have a problem. All four of the atoms cannot share one electron, only two can. One set of opposite corners share the electron, leaving the other two atoms needing one more electron. They are able to collect their private electrons, one each, as the plasmatic condition cools into a vapor. Secondly, the two corners that share the central electron are drawn closer together by the electron.

These fall to the surface liquid of the planet. Their development into a more stable molecule is less complex than that of beryllium. As they fell through the atmosphere, they collected oxygen. Unlike beryllium, the conditions are cooler making the demand for electrons greater. Let us return to our square configuration of boron. Place an oxygen atom for each location of the electrons holding the boron crystal together. The cooling plasmatic oxygen vapor grabs the electron only to become part of the boron scenario. The result is that it "fakes" the original condition by bonding with both boron atoms, which actually becomes more stable. Meanwhile, external to the crystal, the two boron atoms having their own private electron, ends up sharing it with an oxygen atom. However, these two oxygen atoms are only half satisfied. They still need another electron. At this point, the four boron atoms have collected seven oxygen atoms. Eventually, they make their splash upon the earth surface. The oxygen is now able to acquire the needed electron by bonding with two sodium atoms.

As far as the bonding process is concerned, the process is complete. However, they have some "hot spots." These are the locations in which positive magnetic energy continues to dominate. The two sodium atoms become positive charged as their bonded oxygen has primary control over the electron. The other four "hot spots" are still involving our original boron atoms. Again, oxygen is the culprit; oxygen has a tighter hold on the electrons. After water molecules develop, they collect to the "root" molecule. Unlike the collection of hydroxide, the water molecules do not bond with the developing molecule. Unlike hydroxide, the water molecule is not an ion. This means

Chapter 15: Creation of Earth

that the water molecule is complete in its formation both hydrogen and oxygen are satisfied. Hydroxide, on the other hand, is incomplete as the need by oxygen is not totally met, as it needs another electron. Because hydroxide bonds with the developing molecule, the physical and magnetic characteristics change. By this, fewer water molecules will collect to the assembly as fewer "hot spots" formulate to attract the water molecules. Borax formed first before the "root" molecule could settle into a packed format in which the "hot spots" were covered leaving the maximum of four exposed "hot spots."

Lithium Rain. Lithium is the last light element to liquefy over the earth's atmosphere. It is still over 2,000 degrees Fahrenheit. Like carbon, lithium gathers into huge strings. Unlike carbon, there is no cross bonding between strings. The reason is that there is only one electron in its outermost shell and it can only hold two electrons. The image is that of placing an electron between each atom of a huge row. After collecting into strings, the strings fall into themselves forming large spheres of lithium. The term large means in atom count, not as in sizes as boulders, perhaps to the size of a grain of sand. Even so, these are very light in comparison. These strings can have lateral ties as they cool; these breakup as other elements come into contact via electrodonativity.

As they fall to the earth, the surface lithium of the "ball of string" reacts to the contact of hydrogen and oxygen molecules before they have a chance to deflect. The result is that forms of lithium oxides and lithium acids appear. The lithium acids join with more oxygen and fall to the earth. The lithium rain pocketed the semi-liquid surface giving the future felsic layer the final touches. Another interesting item about the lithium stringed ball is that it breaks into string segments as it hits the surface. Silicon and iron begins to solidify during this time giving a soft mush-like surface rather than a thin liquid. These strings rapidly disunite as they interact with other compounds. Lithium exists within the earth in pegmatite a coarse form of granite. Granite is a three-dimensional patchwork between dark and light colored minerals. Lithium also exists within both in plants and animals.

Initial Atmosphere

Up to this point, nothing has solidified, except for the previously mentioned molecular formations. We have one more stage to go into within the liquid account of earth's history. During this time, hydrogen interacts with oxygen. The hydrogenation process of oxygen was the only phenomenon occurring outside the earth with any great vigor. Eventually, the hydrogen ran out of energy, leaving a small amount of hydrogen gas in the upper atmosphere, unable to reach the argon plate as there is resistance by the vapor by weight. Another interesting note is that there is still oxygen left in the atmosphere. If it were not for lithium through carbon, we would have very little oxygen left for our atmosphere. The only other major item to occur is back upon the surface of the earth. Calcium and fluorine are combining as calcium boils to the surface.

Part 5: Creation of the Macrocosm

Nitrogen on the opposite side of the storm has already formed into gaseous molecules. Just as oxygen gas becomes trapped inside the felsic side, nitrogen also becomes trapped in the opposite side. The actual supporting observation seen in the nitrogen mining off the coast of northern Chile and southern Peru; however, at this point, this fact is not self-evident. Note: South America was not always next to this region, as we shall observe later. The argon elementuum slowly melts into lavor. The lavor has freedom of movement and runs toward the surface destroying the solid barrier between the two storms. As the barrier disappears, the two storms merge into a single storm.

Ozone, water vapor and nitrogen flow with the argon vapor toward earth. There is not much variance in temperature upon the surface as the primary source of heat is from the earth itself. This temperature is well above the boiling point of water. Within this near uniformity, there are no winds, just heat. There is another trace gas that exists, that is carbon dioxide. This will sink under the argon layer, as it is heavier. This leaves us with the ozone and water vapor. The ozone is heavier than the oxygen gas molecule in mass, but is less dense via structural formation. The initial ozone (O_3) formed like the plasmatic vapor form of beryllium, except it was trying to attain eight electrons fulfilling the octet rule. In gathering the electrons, each oxygen atom shares one electron per adjacent atom. The single bonding between two atoms is weaker, allowing the other electrons to repel each other generating larger distances between atoms. For this reason, the ozone layer floats above the troposphere in today's atmosphere. This leaves us with the water vapor. Only hydrogen gas and helium are lighter than water vapor within our atmosphere. The water vapor forms an extremely large atmospheric layer above the ozone. The shape of this layer resembles two saucers turned toward each other. It is as a very flat ellipsoid surrounding a near perfect sphere of troposphere and ozone. Naturally, outside the water vapor is a thin layer of helium and hydrogen respectively. All the other lighter elements, lighter than water vapor, have generally fallen to the earth; their residue would be almost, if not, immeasurable.

The water vapor migrates toward earth forming a smaller less flat ellipse via gravity, this places more pressure upon the sphere as the centrifugal force of the storm diminishes within the vapor via closeness to the planet. The result is the earth's surface forms a much smaller sphere. This liquid spherical shell continues to shrink. As it shrinks, the atmosphere becomes denser. The liquid shell becomes thicker. A crust begins to form upon the surface of the liquid shell. A saturation point is reached, when the plasmatic pressure under the shell equals the external surface pressure. At this point, water vapor dominates the atmosphere touching part of the surface. The atmospheric pressure is nearly 1,730.4 pounds per square inch or a fraction under 118 times denser than present our day atmosphere. This makes the pressure on the surface of Venus today a little small being 93 times that of earth. Again, this shows that Venus formulated from a smaller system. If Venus suffered the trauma of Mars, its atmosphere would be extremely thin like Mars. Originally, Mars had an atmosphere comparable to that of pre-flood earth. We will expound its nature and destruction later. We should note

Chapter 15: Creation of Earth

however, on Mars, the "bottom layer of gas" was not nitrogen as seen on earth. It was carbon. Carbon has one less proton than nitrogen. The result is that carbon dioxide formulates giving the initial atmosphere of Mars moderate temperatures comparable to those on earth today via the greenhouse effect. This is also indicative of the lack of a barrier between oxygen and carbon.

Before continuing on, we need to examine the relationship between the atmosphere of Venus and that of the initial atmosphere of Earth. Earth's pre-ocean atmospheric pressure is about 117.7 times that it is today. Gravity of Venus is 91 percent that of Earth's. If we multiply 117.7 by 0.91 we get the pressure if the Earth's atmosphere existed on Venus. If we multiply by 91 percent again we would get the amount of atmosphere Venus would have in proportion to Earth's. This atmospheric pressure equals about 95 that of present-day Earth. The resulting double division is consistent with gravity being described as acceleration. Notice, that the calculated 95 atmospheres is a little over 105 percent of the actual atmospheric pressure of 93. The reason for this variation could be the evaporation experienced during the time spent as comet-like state. Secondly, Venus is closer to sun making it closer to the source of the solar wind's blasting effect.

Let us return to earth's atmospheric pressure of approximately 117.7 atmospheres. We derived this figure by the following information: If the earth were perfectly smooth, its entire surface would be under slightly less than one and a half miles of water. A gallon of water weighs 8.34 pounds and has a volume of 231 cubic inches. When water turns into vapor, it does not lose mass. To find the pressure per square inch, we need to examine only one column of water with a one-inch square base and 1.5 miles high. The math then becomes a matter of finding the number of cubic inches there are in a square inch pillar 1.5 miles high (1.5 x 5,280 ft x 12 in). Divide this number by the cubic inches there are in a gallon (231); multiply that by 8.34 pounds. Add to that answer 14.7 the atmospheric pressure of today. Then we need to divide by 14.7, the pressure per square inch defining one atmosphere. While this gives a very precise number, we can only make rough estimations of the actual value. The figure of 1.5 miles justifies a precision of only two digits; this gives 230 as the computed figure was slightly over 233.4 atmospheres. Input the acceleration of gravity in respect to distance, we divide by 2 giving 116.7. Add 1 for the present atmosphere, this gives us 117.7 or 120 atmospheres allowing for only two significant figures

Forming of the Ocean

Recall the purity of the atmospheric layers. The Troposphere gases did not encompass the whole planet after the first crust formed. Most of the felsic cap extent well above the severely compressed Troposphere. Utilizing the value of 40,000 feet for the Troposphere height derives the following calculations. We divide this value by 117.7 atmospheres. This gives a height of approximately 340 feet. From this figure, we

Part 5: Creation of the Macrocosm

determine the felsic land extended roughly 160 feet beyond the altitude of this base air mass of nitrogen, oxygen, argon and trace compounds. Above this layer was the ozone, the layer directly above this was the water vapor. As indicated, the water vapor was an extremely large layer before the ocean formed.

The ocean was not formed by rain. The Torah states that it did not rain on the earth for there was no man to till the ground. Recall that all possible particles have already been pulled into the surface when it was plasma and vapors. The way the ocean forms was by condensation. It could only condensate upon a surface of which that it had contact. By this, we observe the felsic cap needed to bow at least 360 feet high. In fact, it needs to be taller than that to initiate and hold the process in motion. Note that the ozone also does not completely cover the surface by reason of the height of the felsic cap. This partially explains the ozone hole at the "bottom of the earth" above Antarctica grows and shrinks by seasonal change in the atmosphere.

Since the atmospheric pressure was near 120 atmospheres, the condensation point of water was much higher. The condensation point of water became in the neighborhood of 400 degrees Fahrenheit. The water that formed collected as due on the felsic mountain and ran off onto the mafic plain. The image is that of a hot rock gathering condensation from the surrounding steam.

The mafic plain is much hotter than the felsic cap. As the water ran off onto mafic plain, the water vaporized. This accomplishes two tasks. The mafic plain is cooled faster; but more importantly, it stirred the atmosphere. The rising vapor mixed with the Tropospheric gases making it as one air mass temporarily. The water gradually ran further into the mafic plain before it turned back into vapor. Eventually, all the mafic plain was covered with water, thereby cooling the mafic plain enough that water does not evaporate.

Water running off the felsic cap, did so as we observe water collect on the surface today. It gathers into streams that run downward into the basin. The result is that grooves formed in the felsic cap. Unlike the land of today, there was only one mountain located in the center of the felsic cap. Recall that the felsic cap has a twisted surface (not to be confused with Havilah, soon to be defined). The grooves formed spiral out from the mountain. The nature of the spiral is gentle. Some of the grooves broke away from the gentle spiral to go straight down to the mafic plain. The main stream entering into the region of Havilah broke into several smaller streams forming a large delta-like formation as indicated in the name given to it in the Torah.

We should note there was one primary spiral groove that formed as the felsic cap twisted. Just as all twisted material, there is at least one valley with one corresponding ridge. As expected from reading the Torah, the polar region has the gentlest twist has only one valley. However, as we move away from the central region more and more valleys appear. No matter how shallow the incline may be, water will find the lower

Chapter 15: Creation of Earth

elevation for drainage a system. Note: there is only one river coming out of Eden, breaking into four heads.

As the water rises forming the ocean, the atmospheric pressure decreases. The whole assembly cools. Eventually, the water covers all the land. After the steam stop rising from the water, the atmosphere began to become still again. The water vapor continues to collect upon the surface of the ocean as long as it was touching the water. However, the tropospheric gases eventually dominated the surface. The reason for the separation can be seen as gravity's effect upon the molecular weights of the gasses. The oxygen molecule weighs around 32 atomic units, a nitrogen molecule weighs 28 units and argon nears 40 atomic units. Comparative to water vapor weighing in at 17 units, tropospheric particles are heavier forcing water vapor upward. After this occurred, the ocean ceased to collect water. All the other water vapor was held away from the surface by the tropospheric gases of oxygen, nitrogen, argon and other trace vapors. This is the end of the fractal described in scripture as being the dividing the waters below from the waters above.

At this point, the entire surface is under about a mile and a half of water. The atmospheric pressure is about 10 atmospheres or about 147 pounds per square inch at sea level (we will expound on this latter). The troposphere divides into two air masses, the northern air mass and the southern air mass. These two air masses separate by the lack of Coriolis Effect (twist via rotation) at the equator. This small belt of atmosphere known as the initial doldrums regulates heat exchange between the two. However, heat exchanges were minimal at this point. The entire surface is still very hot, near the boiling point of water (212° F). Even so, the water provided another barrier to the heat of the plasmatic core. This increases the atmospheric cooling rate via the interaction with the external void. This, in turn cools the water.

Chapter 15 Quiz

1. The Earth's Plasmatic Storm is a substorm of:
   A. The Asteroid Belt
   B. The Sun
   C. Mars
   D. Venus

2. The initial Atmosphere of Earth was:
   A. About 117 Atmospheres
   B. About 234 Atmospheres
   C. About 2.5 Atmospheres
   D. About 10 Atmospheres

3. He Felsic Cap exists In which of todays' Hemisphere?
   A. Northern
   B. Southern
   C. Eastern
   D. Western

4. Diamonds started as:
   A. Hail
   B. Seeds
   C. Snow
   D. None of the Above

5. After the Ocean formed, the Atmosphere was:
   A. As it is today
   B. 2 times that of today
   C. 10 times that of today
   D. 100 times that of today

6. (T/F) Our solar system formed by three passing stars.

7. (T/F) Argon fell to the earth after it formed.

8. (T/F) The felsic cap was initially divided into two sections.

9. (T/F) The initial atmosphere was very thin.

10. (T/F) Atmospheric pressure after forming the ocean was ½ of today's.

Chapter 16

𝕴nitial 𝕱ormations

And Elohim said, "Let the waters under the heaven are gathered unto one place, and let dry land appear: and it was so. And Elohim called the dry land Earth; and the waters called HE Seas: and Elohim saw that it was good.

(Gen 1:9-10)

Churning of plasma does not determine the magnetic polarity; otherwise the solar magnetic flip would flip the solar rotation. Differences in temperature between the plasmatic surfaces outside and inside the multigalactic shell determine the charge. Since the interior surface of the shell cooled first via exposure to the internal void, the nuclei collected electrons before the outer surface. The external side becomes magnetic North regardless of the rotation of the magnetic storms because of extra electrons. The plasmatic charge exerted by our solar disc was much greater than that of today reaching beyond Pluto. Other stars were also exerting the same charge. However, only a small handful affected our sun. Out of this handful, only one had any major impact as the planets formulate. Proxima Centauri is the closest star today, the star involved could be Alpha Centauri, Sirius, Altair, Procyon or even Arcturus. However, there were other stars closer but were too small to have much of an impact. One star caused our solar disc to flip about $173°$ without reversing its rotation. However, the sun does move to a tilt of $7°$ (180-173). The reason for this tilt is that the moleucum shell of the sun has cooled. The friction of the electrons moving to their new location by a twisting flip caused the moleucum shell to shift. Since the moleucum shell represents most of the mass of the sun, the entire assembly ends up rotating in the $7°$ angle.

This occurrence is the second flip of the earth's magnetic axis, the first being when the earth's magnetic storm became a separate entity from the sun. The first flip is similar to the effect of splitting a bar magnet along the magnetic axis. However, the first magnetic flip did not cause a flip in the rotational axis. The reason is that the moleucum shell was thinner and hotter giving electrons freedom of movement. Therefore, it is because the earth's moleucum had cooled and thickened that caused the earth to flip its rotational axis. The result of the rotational flip of the earth is that the earth now rotates in the same direction as the sun. This is true with all the planets involved. Another reason that Venus was not part of the original system, it rotates in the opposite direction. If the first flip had generated the necessary phenomenon to create dry ground to rise to

Part 5: Creation of the Macrocosm

the surface, Elohim would not have spoken to cause the second flip if dry land was a natural sequential event.

The magnetic emissions by these solar (or stars) discs are beyond any phenomenon seen today. The only clue to such an existence exists in our examination of the sunspots. We will describe them in more detail later. However, the required observation is that these relatively small (occupying a very small fraction of the surface) magnetic disturbances generated by the sunspot's storms has been observed to affect the needles on compasses. Imagine the effect when the whole surface is involved with no obstruction.

As the result of this colossal magnetic interaction, the physical spins of all the planets shifted. Note that the star was much closer during this time than it is today. In fact, one observable planet was deeply affected by the interaction. Uranus was caught in the middle of this magnetic interaction. Note: The alteration affected the spin primarily. Uranus and Neptune were affected further by the magnetic interaction. Even so, Neptune's interaction was primarily with our solar disc.

The crust structure of Uranus and Neptune were already more complete than the other planets by being further from the solar center. By this, a "magnetic bar" was completed under the surface. This bar bent at an angle nearly 60 degrees on both planets. As we shall see shortly, this bar is not a bar in the true sense of being a bar, but it behaves as one. This bend is in response to internal integrity rather than an external interference.

Neptune and Uranus both had softer foundations as illustrated by the angle of the magnetic axis in relationship to the earth's. More precisely, the ratio between the structural integrity of their crust to their magnetic strength is less than that seen with the earth. Note: it is highly unlikely that these planets' cores have any resemblance to the chemical make-up of the earth's central core. The reason is that they formulated external to the solar disc having a different isotope pattern.

Observe also that Neptune is denser than Uranus. Uranus has a density of about 1.2 (water is 1.0) and Neptune has a density nearly 1.7. Compare this with earth having a density of 5.5 and a mafic layer of 3.0. Density does not equate directly to hardness. For example, water is denser than ice. The denseness of Neptune illustrates, as it should, that Neptune is further from the phenomenon that generates the planetary lightness seen in Saturn. Therefore, instead of having large quantities of silicon-oxides in their crust, it could have beryllium-carbonate compounds and frozen hydrocarbon like compounds. Saturn's composition, however, would be even more different having nitrogen compounds instead of carbon (observed on Neptune or Uranus), something like trilithium nitride where one nitrogen atom are single bonded to three lithium atoms. There would also be countless ammonia mixtures. Note: trilithium nitride is virtually an unknown substance on earth; however, this compound is possible in large quantities in such an environment found on Saturn.

Chapter 16: Initial Formations

Saturn, Mars, Earth, and Mercury had the change only of rotational axis by the magnetically excited plasma under the surface. Neptune's axial shift extends as an extension of the same phenomenon with our solar disc. The shift in rotational axis by the earth was a near 157 degrees via magnetic rotational shift. The shift in rotational axis becomes almost 144 degrees at Neptune. Uranus flipped only 83 degrees because of interference by other stars, i.e. Alpha Centauri. The image is that the stars behave similar to cloud eddies around the major storm, namely the Milky Way. They never move far from their location in circular-like patterns that dissipate with expansion.

Notice that the Earth did not tilt near 23 degrees, but it was near 157 degrees. As stated earlier, the pattern is highly afflicted. Note the sequence of the pattern without the afflictions starting at those closest to the sun: earth tilts at 23 degrees, Mars 25, Saturn 27 and Neptune 29 degrees. If the magnetic effect was solar orientated at twenty something degree angle, the angles should get smaller as we move away from the sun; however, this is not the case, as the inverse is true. The magnetic source is outside the solar system. This external influence could either be close to Neptune exerting a minor influence or on the opposite side exerting a large magnetic response. If it were the scenario that promoted the least magnetic response, then Uranus would not have its rotational axis at near 90 degrees. This would be in response to being close to the pull.

Jupiter and Venus were in a different galactic sector. While Jupiter was a star in its own right, it is much smaller. Venus was a planet collected by Jupiter before entering our solar system. Its density and atmosphere suggests that it was part of a solar system similar to ours, but not ours. We can derive this from the heavy concentration of carbon dioxide (96%) and the **sulfuric acid** (H_2SO_4) clouds above it. The sulfuric acid occurred after the formulation of the planet and its atmosphere. Originally, the sulfuric acid was absent from the atmosphere, instead there was water vapor. When Jupiter collected Venus, it passed though the sulfur asteroids surrounding our solar system. As this occurred, the sulfur was pulled into its atmosphere. The sulfuric acid never reaches the surface as its droplets of "rain" evaporate by the intense heat radiating from the surface. This further reinstates itself as Venus rotates on a similar axis as Jupiter of approximately 180-3 degrees, which is 177 degrees. The image observed is that of a planet interacting with Jupiter like a comet and it thrown toward our solar system. During this time, the axis flipped to match Jupiter's massive magnetic field. Along with this, the density of Venus is slightly less than that of earth, instead of being slightly heavier being closer to the sun. This is more about Venus in Chapter 17.

These interstellar magnetic reactions provide a dynamic effect upon the earth. The shift of axial rotation generated different velocities upon the surface. That which once was stationary is now moving and that which was moving became stationary. Let us look at the latter scenario: the motion twists into a new rotational axis by an external force. Since this information is obscure by other planetary developments, let us turn to Mars for a reference.

Part 5: Creation of the Macrocosm

By examining the Martian Polar Caps, we observe that there is a spiral like twist at these locations. This corresponds directly to the same phenomenon on Earth. Similarly, in analyzing our moon's surface, the twist at a polar site also occurs. However, this region formed by the initial cooling process. On earth, within the **Felsic** region, density of about 2.6 to 2.7, it formed a region called Havilah in the Torah. This Hebrew word has a root meaning to be twisted circular (spiral). We are getting ahead; we have not finished examining the forming of dry land. One last note: there is an opposite spiral in the mafic plain, as we shall see.

The magnetic poles do not reformulate within the mafic crust of the earth. The magnetic forces of the earth exists as a bar magnet beneath the surface affecting the ultramafic layer. The felsic crust acts as dead weight to the magnetic movement. The felsic material moves with the mafic material as it is partially engulfed. It breaks free during this process. There is a double slippage process under the crust. One is known as the **Moho** by scientists and another is a place between solid matter and molecuum matter (described in detail later). While the molecuum matter flipped, the rotational axis of the earth the surface remained nearly unchanged in its rapid movement and the stop in response the solar magnetic interaction with the passing star via a well lubricated barrier. Afterward, the surface changes direction to match the core movement.

Equatorial shifting is our next focus. The elliptical shape of rotation shifted about $23.5°$ as well. This cracked the felsic bands reach the equator. The result is that a concentric crack formed $66.5°$ around the old rotational center. Tension was released which formed by the cooling mafic plain via cooling water. As the mafic plain contracted, the plain pushed the $133°$ (diameter of 9,190 miles) felsic cap out of the water.

Not only, does the felsic cap push out of the water, the Magnetic North moves back in alignment with the southern rotational axis. **Havilah** forms during the time of the upper crust changes directions by the means of friction. The earth begins to rotate at the new $23.5°$ angle. Friction eventually overcomes the momentum of the upper crust.

We need to note that not all the felsic cap rose out of the water at once. The rate at which the felsic cap ascended to and beyond the oceanic surface was not linear. The image of velocity upon a graph looks similar to a graph of a square root function. After the cap came close to the ocean's surface, the velocity reduced slowing in rate as the velocity approaches zero. It appears first as a large disc of near white and red material below the surface. Steadily, the cap moved toward and began to overtake the surface. Again allowing the water to drain in the grooves formed earlier, the water even formed new patterns via surface irregularities formed by movement. If we could see the disc rise out of the water, it would appear red. Any white coloring would appear red via solar plasma. The Hebrew word for ground is "a-da-mah," which is the root for red.

One observation to note is that the mafic layer "remembers" or records where the old magnetic axis was. The new axis generated its own magnetic system, as the solar

Chapter 16: Initial Formations

magnetic energy was still very strong. This information will become important when we study the crust of the earth today. The new axis forms by the alternation of the original magnetic axial existence.

Another interesting phenomenon is that wrinkles appear upon the surface as the felsic cap rose. These wrinkles formed by the contraction of the mafic layer under the felsic cap. They formed the mountains of Noah's (No-ach in Hebrew) time, which were various heights even up to about 32 feet high. Today, we would call them small hills. These gradually formed upon the surface as the mafic layer shrank underneath the felsic cap.

Initial Land Features

For this cause, we examine scripture in the second chapter of Genesis; even though, we are not done with the first chapter. In Genesis 2:10-14 of the Torah, it talks of these rivers. The Garden of Eden was east in **Eden**, the central region of the felsic cap. A river goes into the Garden from a location outside in Eden. This river waters the Garden (which is ahead of our point of examination) and flows on out.

At its departure, it splits to become a head of the Pison River into the land of Havilah. Pison is a Hebrew word that has a root meaning of being scattered, or spread out flat. The next river to branch out is Gishon. This river encompasses Cush (Ethiopia). Gishon means to gush in Hebrew. From archeological data acquired, this river flowed west of today's Ethiopia and flowed down the Amazon River when South America was connected to Africa. The next River is called Hiddekel. This river has been attributed to the Tigris River. Hiddekel stems from two different Hebrew root words. Hiddekel is composed of four Hebrew letters: Chet, Dalet, Quuf and Lamed. Chet and Dalet form the root word- sharp. Quuf (or Qoph) and Lamed form the root word for being light (weight) and swift. By this, we observe Hiddekel as a minor river that branches off the main river at a sharp angle. The last river is Perat (Euphrates). Both Hiddekel and Perat travel through Assyria. The Hebrew letters formulates Perat: Pe, Resh and Tav. Again the meaning is gushing. However, notice the Resh in the middle of the word. This means that this river is the main one; whereas, Hiddekel has a Quuf indicating in Hebrew that it is in some way least.

By this, we observe a river started in the central most location of the felsic cap and moves in a gentle spiral down to the ocean. Before it leaves the central location four branches occurs. The first branch goes eastward into the region that was disturbed by the new rotational axis. This scatters the river into more branches that pass around this region. The next branch breaks off in a southwest direction external to the spiral like the first one. The third river branches on the inward side of the spiral leaving the main stream. Meanwhile, the main stream continues its course to the ocean. The Euphrates and Tigris rivers were named after the flood for these latter two rivers. Even though

Part 5: Creation of the Macrocosm

the courses of the two rivers were somewhat different, there were locations away from the mouth that resembled the old ones.

First Signs of Life

Let us return to the original felsic cap. Thus far, no life formulated upon the earth. The inorganic realm has been our only concern. Life comes from the Ruach (spirit). This involves another dimension, which is beyond the definitions provided thus far. As noted earlier, the physical universe uses only a very small fraction of the space allocated to it. Within this framework, many other dimensional frames can interlace without being involved directly with the physical material. Actually, some scientists have said that there are 10 dimensions associated with this universe. Six of which are unable to manifest totally within this present universe. Imagine another set of xergopaths interlaced with six xevim facilitating life. We are not referring to the decayed life of today, but to the pre-fall life. Within the pre-fall phenomenon, Heaven and Earth could exist together simultaneously.

If it were possible to observe all the atoms of a living organism before and after it dies, we would find no physical change in the chemical composition between the two instances; there are no disappearing "life" isotopes. However, after death the chemicals follow their "natural death" course of decay. Natural death stems from the fallen state that we observe today. The only purpose of subatomic matter is to provide life with vehicles to function within this realm and sustenance for the vehicles' maintenance.

There is another process that occurs at this point to bring life. This is beyond our scope of the natural creation on earth; we will limit ourselves to the natural creation. Natural creation is defined as being that which is under the authority of the natural laws set forth by Elohim thus far inorganically.

When we look at the surface of the planet, we have observed that the felsic cap has only partially risen out of the water. If we were to examine the first layer of the felsic surface, we would see it to be relatively thin and soft. It is composed of light minerals almost as a damp collection of dust. Under this are the layers of organic granite. Both layers engulf the entire felsic cap both above and below the water. The water waves are almost nonexistent. There is no moon to generate tides. The solar disc has not yet formulated nor "ignited;" yet it is a gravitational centroid.

On the third day, Elohim created plant life. This occurred in the organic granite layer just under the surface. The pockets of organic material would never formulate anything by themselves. Elohim transformed these carbon compound pockets into **seeds** by using the surrounding inorganic material as shells. Within these shells, all the chemicals needed for forming life forms was present. Elohim formulated plant life both in the water and out of the water. The warmth came primarily from the earth, as the sun is not there to divide day and night.

Chapter 16: Initial Formations

Plants existed without daylight, which separates night from day for a thousand years. Consider chlorophyll, which contributes to the greenness of a leaf. Chlorophyll exists in two basic chemical compositions; chlorophyll a's composition forms as $C_{55}H_{72}MgN_4O_5$ (green), less common, chlorophyll b's composition is $C_{55}H_{70}MgN_4O_6$ (blue-green). In the second form, the extra oxygen displaces two locations for hydrogen. Note the heavy carbon and hydrogen input into the composition. In both cases, the general color is green. The color green means that the plants are absorbing red light as it coverts photonic energy into energy that sustains the plant.

Looking into the sky during this time, it would be pitch black except for the solar disc, which exists in various shades of red. Remember, the sun has not yet "ignited". The image of the sun is similar to Saturn, except not as well defined. The rings were as thick collapsing clouds of red that became thicker as they approached the central mass. External plasmatic vapors partially obscured the central spherical surface. The image was much larger than the sun of today. However, the sun was constantly shrinking as it was cooling. In essence, the earth existed in something similar to a photo lab's dark room.

Igniting the Sun

The solar disc cooled much like the earth. The external gases cooled the surface, only it took longer. There are a few differences; Actually, there are three primary differences. The central core composition is of nuclei that far exceed the natural occurring uranium on earth. The central plasma keeps the sun keeps the surface from cooling to stabilize electrons around atoms into the ground state. The core naturally is very radioactive releasing heat producing gamma rays. The radioactivity produces some other unusual effects as well. Perhaps the most retarding principle is the large amount of hydrogen external to the solar disc. This slows the approach of the cooling effect upon the elements designated to become the fiery ocean via state of matter, lavor.

The term "igniting" is somewhat misleading. Igniting usually involves heating a substance to a point of combustion. However, in the case of the sun, the cooling process simulates the combustion-like process from a hotter state. As noted earlier, plasmatic elements have their electrons stripped from them. These electrons are still attracted to the nuclei but are too "hot" to settle to any nuclei. They still have their cycle of warming and cooling. Because of this, the red light that they do give off is possible. Most of the time, they give off infrared light. However, the rate of these occurrences is less than found in regular light, as the intense heat holds the electron's "photon door" shut. Since we measure heat via photonic activity, we get an illusion that the plasmatic material is cooler when it is not. The prime example of this phenomenon is the sunspots observed within the framework of today's phenomena. We will discuss this in more detail later; we still have not formed the purely spherical or ellipsoid sun.

Part 5: Creation of the Macrocosm

As the solar disc cools, it pulls all the external arms into itself. The spherical shape will become the only remaining feature. However, we have not yet achieved the spherical shape; the sun is a flatten ellipsoid and is still red. The elements begin to settle into their layers. These layers are not well-defined elemental layers, but a composition of like materials. The lower mingling elements are composed of a slightly wider variety than that found on earth. The primary account for the larger variety is the central location of the mass. The cooling of the elements allows the electrons to gather around the nuclei. Just as going from the ground state into plasma, these electrons give off visible light as excited electrons. Hence, we have an ignited sun giving off light that will separate day from night. The purpose of the larger array of elements is to provide a brighter and "whiter" light via heat release.

Unlike the earth, the solar surface stays as an ignited liquid ocean. This ocean exerts pressure and the solar disc contracts into a spherical form. Like the earth the plasma underneath behaved similar to a multitude of different gasses in the same pressurized container: they disperse throughout the internal space of the container. Unlike the earth, the sun has a larger set of elements. Man has created elements that have 103 protons, but the sun has large quantities of elements with somewhat larger nuclei.

Recall that the standard model of the electron shell structure can hold 280 electrons. An atom with 280 electrons would have 280 protons and over 280 neutrons for a nucleus of 560+ nucleons when compared to the 260 nucleons of Lawrencium. These neutrons tend to form a shell-like surface to these nuclei. Considering this tendency for the existence of extraneous neutrons in larger nuclei, the number of nucleons within the nucleus could reach approximately 1,000 (estimated from a single layer of neutrons surrounding the balanced set of 560 nucleons). Just like the elements beyond and including some below uranium, they are extremely radioactive. This gives a large number of extraneous neutrons within the nucleus. This is working within the framework of the model limit.

Actually there is no limit to the size of the nucleus, as the number of protons increase beyond 280, new electron shells would develop to handle the electrons. In the latest statistical model, the cloud of electrons hovering over the nuclei would increase in size. However, in dealing with the standard formulation as indicated by the electron limit, this would be the approximate size. The required observation is that these nuclei dwindle in size as the protonization process subtracts Helium-4 nuclei from the nucleus until the transforming neutron is not in contact with a proton or until the element become stable (nonradioactive). However, within this solar system; at least upon the surface, the shell level count is seven.

Initially, these ultra-heavy solar nuclei collected under the cooler fiery lasma ocean of the lighter external elements. Like the earth, a thick liquid-like bubble forms exerting pressure upon the core. The molecuum shell forms first as an external shell that melts upon the surface into a lavor ocean as the shell of molecuum sinks deeper into the

Chapter 16: Initial Formations

mass. As this occurs, lasma continues to form underneath the shell via condensation forming another ocean on the inverted side of the molecuum shell. Because of the added density surrounding the molecuum shell, the shrinkage decreases as its heat becomes somewhat stabilized. Eventually, the ocean floor forms as a heavy-element molecuum solid. The unique phenomenon concerning this shell is that the molecuum plasmatic solid is that the electrons are given little or no chance to release photons via the intense heat (as noted earlier) giving the floor a reddish color even black via the surface ocean. The molecuum solid has a density nearly liquid because of the loosely held electrons as opposed to embedding into adjacent atomic shells. Any movement of the electrons equates to converting back into a liquid lasma. The general thickness of the floor is very thin in comparison to its diameter. Even so, it is hundreds of miles thick.

The fiery ocean of lavor behaves more like a liquid as it moves back and forth through the temperatures of the molecuum and the temperatures of the external vapor. The key factor is that the lavor heats up as it moves toward the molecuum shell. In this, it loses (repels) electrons freeing molecular "bonding." Instead of gathering loose electrons as the lavor cools, it gathers electrons attached to other nuclei generating denser molecules. In this, the lavor behaves as boiling water.

Entrance of the Moon

There are four basic theories about the moon's origin. One: it broke away from the earth's crust. Two: The moon's creation was as a dual system to earth. Three: the earth captured a **planetoid** (a small planet). Four: It formed by a metallic ring that condensed into a sphere. Out of all of these, most scientists favor the fourth one.

However, the fourth one is not without problems. Despite that, it has the density that we would expect from such a ring; it lacks some basic elements. If the ring formed into a planet-like moon, it had to occur during the plasmaverse. The reason is after the ring becomes a collection of solid materials. There no longer exists any means of uniting the ring into a single spherical solid. This is because the rocks have virtually lost their magnetic polarity from their plasmatic state and local gravity locked into a "kinetic prison." This "prison" construction occurs during the cooling of the plasmaverse. Otherwise, within the plasmaverse, the matter would have collected. We have examples of this ring phenomenon around the sun and of all the outer planets except Pluto.

If it did become as another plasmatic system, it would bring us back to the second theory mentioned. The problem with the second system is that the density and elemental composition do not match. The density of the earth is 5.5 while the moon's density is 3.3 (Again, water is 1.0). It is first reported by samples of the moon's surface, there is oxygen but a lack of water. However, water ice is found by later missions to the moon. The first theory is destroyed by the fact of searching for water.

Part 5: Creation of the Macrocosm

Beyond this, the moon has virtually no atmosphere. If the moon's formation occurred along with the earth, it would have atmosphere. Yet, there is oxygen within the rocks. The gravitational pull of the moon is approximately one sixth that of the earth's. Recall the atmosphere of the earth before the ocean forms being around 220 atmospheres. Even if the moon had 1/100 the atmosphere of the original earth, it would be twice that of today's earth. Along with this, recall the virtually total lack of hydrogen connected with oxygen. If the moon was to form within the framework of our planet, hydrogen has to be present forming large volumes of water. The low amounts of hydrogen indicate that the atmosphere of the moon was stripped from its surface, because hydrogen would have collected during the multigalactic shell process.

Lastly, if a collecting process formed the moon, the spiral-like geometric structure of the "South Pole" of the moon would not exist. This requires a rotational axis and a unified layered surface (not a conglomerate of various sized rock pieces), which at one time the moon had. Then there is the reason for the pieces to be separate in the first place -the lack of gravity to pull together as observed in the rings of Saturn and the asteroid belt.

This brings us back to the third theory in which the earth captures the moon. Scientists have ruled this as highly unlikely as the odds of such an occurrence are highly unlikely. This ruling is quite accurate, if it occurred purely by chance. The Torah states that Elohim placed the sun, moon and stars (B'reshyit, Gen. 1:17). In other words, HE selected their distances from the earth. In the case of the moon, HE brought it in.

The phenomenon Elohim used that brought the moon to the earth is part of the encounter with another solar system that altered the magnetic axis of our Solar System. As mentioned earlier, another star passed close to ours; with the star exists a system of asteroids and planets. These asteroids were much bigger than ours were and much further out. Along with this, there were many smaller particles as well. The image is a multitude of planet like asteroids mixed with even a greater multitude of partially formed planets. Like many of our asteroids, they cooled before gravity could pull them into spherical shapes.

We need to note that not all of these invading moon-sized (although some were bigger than the earth) attached themselves to planets. Actually, most of the planet-like moons seen in the outer planets were already in place. The pre-existing moons formulated before the planet's collection of them by the means of levanastorms. Note: often these moons are made of different materials than the planet they orbit. Sometimes, they are even denser than the planet than they orbit. This is due to their tie with the central solar plasmatic storm. However, the invasion added new moons to our solar system. Some crashed into planets. As noted earlier, our moon was not initially between the asteroid belt and the sun.

During this time, Charon joined with Pluto altering its course forever. Another interesting observation is that the two actually rotate around each other. Another

Chapter 16: Initial Formations

observation made by scientists is that one mass is darker than the other one. Pluto is the lighter colored mass containing like-materials of Uranus and Neptune. It is an extension of the atmospheric process, which formed our traditional outer planets.

Further inward, Neptune catches Triton (a moon). Triton is the only moon that moves retrograde according to the standard rotations of other satellites. Because of its retrograde motion, it will eventually be pulled into Neptune. There was another one of these asteroids attracted by Neptune's gravitational field, which crashed upon the surface. After crashing into the surface, it plunged through the crust. This created a hole to the planets plasmatic core. Unlike Mars, Neptune's atmosphere had already cooled to near absolute zero. The plasmatic heat pouring out of the hole turns the surrounding surface "ice" instantly into a liquid that attempts to cover the hole. It forms a vapor, which generates the observed hurricane. The same phenomenon occurs on Jupiter. This illustrates that Jupiter's system was already close to earth being affected by the same phenomenon.

Onward to our moon, by this time our sun has already ignited. The planet that was to be our moon passed through our orbit and headed toward the sun. As it approached the sun, the surface temperatures raised. It would have been like a comet, but the mass was too large. The gases attempted to remain within the gravitational field of the moon. However, its entire atmosphere was lost during its journey to the sun. The surface facing the sun began to boil (evaporate) away. In fact, the entire surface of the moon was boiling as it rotated. This also aided in destroying the atmosphere as the metallic vapors displaced the gaseous vapors near the surface. After the surface cooled, the metallic vapors dewed upon the surface leaving no atmosphere. This also destroyed the presence of hydrogen, helium and water vapor even from the crust as indicated by the samples of moon rocks. However, oxygen is found in the rocks. If the moon formed in the orbit of the earth, hydrogen would be found upon the surface, even water molecules would be present because of the way the earth formed. Most of the lunar transformation occurred after the initial crossing of Mercury's orbit.

Under the moon's surface another phenomenon occurred. Recall that the center of a given planet is plasmatic. A cold planet is then defined as one with its plasmatic core cooled. Since these are heavy metallic substances, we observe other peculiar manifestations. During its comet experience, friction pushed the interior ocean toward the front (boiling side) and "froze" there.

Complex Gravity

The basic concept of gravity is that an object weighing one pound will fall to the earth in exactly the same time as one that weighs 100 pounds. This is not entirely correct. For all practical purposes, this is true. However, consider the rate at which they accelerate: approximately 16 feet per second squared at sea level. A more general

Part 5: Creation of the Macrocosm

formula includes distance as a variable that is an inverse square within the equation. The point is that both equations require a constant. Example: 16 is a constant in the first equation. The mass of an object and the size of the measuring unit derive this constant. Larger masses have greater constants using the same measuring unit. Taking this a step further, two objects will attract each other via gravitational pull. Their attraction is equal to the sum of the two masses multiplied by a constant divided by the distance that they separate squared. Let us get back to the one-pound object and 100-pound object falling toward the earth. Both masses when added to the earth's mass modify the amount of mass in this equation insignificantly. However, there is a difference even though it is beyond our measuring capability.

Actually, there is yet another factor involved in determining the constant other than mass and measuring unit, which is density. In long distances, density plays less of a role. Upon the surface, it is quite significant. The reason exists in the very nature of gravity itself. All gravity stems from a collective of each atom individually via xyzenthium crystals (particles composing submatter). The center of gravity is only the apparent direction established by the collection of atoms. Imagine the entire spherical structure of mass being compressed into a spherical space 1/8 its original size. Let us establish a point of reference upon the surface. Now, examine the gravitational input by the atom the furthest away. Its distance is cut by one-half. The inverse square of 1/2 is 4. Its pull would be four times as strong upon the surface. This does not mean the entire gravitational experience is four times stronger. Note that atom at our point of reference has the same input as it did before the compression. The effect upon the long range is that the amount of gravity felt for the same mass approaches to become the same as the uncompressed mass.

This brings us to our objective concerning the topic of gravity. It has been noted by scientist that objects weigh more at the polar region than at the equatorial belt. This difference is beyond that which can be accounted for by the effect of centrifugal force. Our next observation is that the earth is an ellipsoid in nature. In this the major axis forms at the equator and the minor axis at the rotational axis. At first glance we might think the geometric structure would promote the opposite response. That is we would weigh more at the equator than at the poles because the depth of earth underneath is greater. However, consider the effect of gravity by a cubic mile of ground directly underneath to one 8,000 miles away. Its effect would be 1/64,000 as strong to the pull of gravity generated by the cube directly underneath.

We need to look at another possible answer. That is that magnetic energy may be responsible for the differentiation. If this were to be true, then a magnetic object would have a greater response than one that was non-magnetic. Moreover, the non-magnetic object would weigh more at one pole than the other as the electrons formulating the atomic exterior would be drawn at one pole and repelled at the other. Then we recall the scenario of two objects of different weights thrown off a tower falling at virtually

Chapter 16: Initial Formations

the same rate; we would not be able to detect the difference, perhaps even into the computer age without mathematics.

Adding yet another feature to the equation, the density of the mass underneath is not uniform. The felsic layer provides us with a density of 2.7 to 2.8 times that of water. However, the general density of the earth is 5.5 times that of water. Strangely enough, centrifugal force does have a role. The variation of the axis is greater externally than with the layers internally. The reason for this variation is that of distance. The internal layer is closer to the center generating smaller linear velocity. By this, the inner heavier layer is more spherical than the surface. This gives more padding at the equator of lighter materials than at the poles. The result is that objects at the poles are closer to this heavier mass than those at the equator. The net result is that the closer heavier mass at the poles above the earth's core has more influence than that at the equator. This effect is magnified by the intrinsic principle that distance has upon gravity ($1/d^2$).

Finally, we are going to examine one more shape. We will start with the uniform sphere giving uniform gravity. Let us project this sphere at a distance from an axis greater than its diameter. The resulting image is that of a doughnut. Imagine this doughnut as being uniformly solid as the sphere. At a vertical cross section, we observe two perfect circular areas with equal gravitational fields. The gravity external to the doughnut (ring) is greater than the internal gravity. The reason is the side facing away from the doughnut center adds the gravity of the second circle to its own. The side facing the center subtracts the second circle directly above it. This is because the second circle pulls against the gravity of the circle referenced. Next, we are going to rotate the doughnut on an axis $90°$ to its central axis to form a solid spherical shell. The same relationship applies to the two surfaces. By this, we observe there are two types of gravity expressed, local gravity and general gravity. General gravity is the gravity we measure from the space outside the planet's surface. Local gravity is experienced upon the surface externally or internally and is determined by the shape. This kind of phenomenon occurs in the large end of the planetary sizes even down to the size of the earth. Even the moon is actually above the small end of the spectrum. At the small end this type of local gravity does exist but not as dramatic.

As scientists note, at the center of the earth, the earth's gravity would cancel out giving the effect of no gravity. However, there is a more complex system of gravity. The reason exists in the nature that our planet formulated, as previously mentioned. Viewing the planet from atmosphere to core, we find the following: vapor, solid/liquid, solid, liquid, vapor, lavor, molecuum, lasma and plasma. Note: solid/liquid refers to felsic cap and the ocean. The molecuum transitions occur near the shell surface. If the core were molecuum, then the magnetic condition of the earth would exist in a similar condition as Mars. It is the free moving electrons movement and polarity that gives us the magnetic poles. The location of the molecuum shell, **molecuumsphere**, will become evident as we study the "dividing of the earth."

Part 5: Creation of the Macrocosm

The molecuumsphere of the earth, like the sun, is covered by the liquid-like lavor. Under it is an ocean-like liquid of lasma. The lasma ocean-like liquid surrounds a central region of pure plasma. Most of our gravity experienced upon the surface stems from the molecuumsphere of the earth. The reason is that the nuclei are larger and held close together like a solid, even though they not as rigidly held via microstrong repelling force between nuclei. This is why scientists find that when we dig into the earth that gravity increases instead of decreases, at least upon the surface.

The "**hollow earth**" concept is not unique to our article. However, we do not consider the center of the earth to be void of compressed mass. Unfortunately, our view still is in direct contradiction to the presently accepted conclusions gathered from the data. The first problem is that the mantel is viewed as a solid because we know that the earth has an average density of 5.5 as compared to the surface felsic material of 2.7 times the density of water. Denseness does not necessarily mean solid. We also know that metals expand very little from a solid form to a liquid form. Iron has a density a little under 8.0 times that of water as a solid. In a liquid form, it still is over 7.0 times as dense as water. We also have elemental metals as osmium and iridium that are even denser around 22 times as dense as water or about three times as dense as iron. We do not observe much of these metals upon the surface, as aluminum, silicon and iron is much lighter.

One problem with the solid mantle projection is that the Atlantic Ocean Floor is separating via volcanic activity and internal pressure. With a solid mantle, we would be saying that it is easier for the ocean floor to drag across the solid mantle with the weight of the ocean above it than for lava to push out through the thinner layer of cracked crust onto the surface through the water. Actually, this would not occur as the lava would be chocked by cooling as the solid mantle stops more lava from flowing to the surface.

In a crystalline substance the sound moves from atom to atom forming a nearly perfect straight line. In a liquid, the line is jagged, and in air, it is even more jagged. If we were to measure straight across three inches and draw the two paths of sound by segment we will find that the jagged line adds up to a greater length hence a longer time for the same distance. The jagged line occurs from the movement of atoms (molecules) within the fluid before we introduced the sound. Within this scenario, we observe a thinner mantle and a larger core because of a greater amount of liquid existing toward the crust-like shell than originally perceived. This is only half the puzzle; we will describe the inner core shortly. Therefore, the time elements of sound testing of the earth's interior then needs to be multiplied by a smaller constant, as it takes sound longer to pass through a liquid. The reason is the difference in nature of atomic movement.

When we examine the vapors, we find that their densities to be far less than that of the liquid. However, they have a wide range of compression available to them. The same is true with plasma. Plasma has the added feature of being able to exist in other states of matter.

Chapter 16: Initial Formations

The concept then becomes almost like a balloon. This is a denser substance pressing into a substance that is less dense via different states of matter. By this, we observe a molten ocean under the earth held to a solid shell by gravity. Under this are vapors and plasma. The interior vapors do not escape to the surface because of the inverted gravitation. Within normal conditions, the escaping vapor would be pushed back to the interior surface by the denser liquid. The exception to this condition is that if the shell would be totally cracked internal pressure would override gravitational considerations.

On further examinations of shell gravitational phenomena, we can imagine even thinner crusts. Using the earth as the model, we can imagine other soil compositions that are much thinner that will give a 5.5 average density at 3960 miles to the center. Using osmium, the shell has to be only around 362 miles deep leaving approximately 3,598 miles of absolute vacuum to the center. Using Iron the thickness needs to be considerably more: a shell of 1,273 miles thick leaves a vacuum distance of 2,688 miles to the center. However, the surface crust is between 2.7 and 3.3 times as dense as water and the core is not a vacuum.

Recall the observation that as we dig down into the crust, we find that we weigh even more than at sea level. As noted earlier, this is indicative of moving closer to the denser material, giving this heavier local mass a greater affect upon our mass. Furthermore, the felsic layer puts us further from this mass than it is able to compensate with its density. This is because gravity decreases as in inverse square to distance per particle. Even though there is less mass (by density) directly under us, it is the closeness to a much denser mass that controls the gravitational pull.

Within the earth, the ocean of lava formulates differently than what we may think. When the metallic substances that form the solid crust was still a liquid, it pressed inward upon the internal plasma and vapors creating an immense internal pressure. The vapors and plasma of the different heavier elements behaved just as we see when a variety of gasses is under pressure within the same container. Each element spreads throughout the entire volume, and the heavier metals collected with the other metals as the crust forms. The lava ocean has a variety of metals within its composition. Eventually, this ocean becomes hundreds of miles thick. The metals within this ocean range from about 4 to 22 times that of the density of water. The heavier metals move toward the center that forms between the crusts outer edge and the inner ocean surface. This formulates three distinct ocean levels of lava: The outer magma (submerged lava) ocean away from the earth's center, the elements that we observe forming the lower mantle is primarily iron based silicates. The central magma ocean is of the heavier metallic elements. The inner lava ocean filled with lighter elements than the central magma but heavier than the outer magma level. Within these internal layers exist various elements in island-like pockets. These elements also affect the sound waves. Each island alters the timing of the sound wave slightly different. Note: this is not enough to give the effect of a solid core.

Part 5: Creation of the Macrocosm

Using the same measurements of sound reflections, we can show that the rock core within our planet is nonexistent. Our task includes the examination of plasma under pressure and examining how it affects measurements. From these items of information, a reconstruction of the phenomenon will produce a model of a non-solid core.

As experienced upon the earth's surface, a solid generally takes up less space than does its liquid counterpart. The density within the solid generates its hardness. Secondly, the gaseous expression of the same substance requires a greater volume and is more intangible than a liquid. Similarly, plasma necessitates more spatial volume than a vapor. Thirdly, Sound passes through a solid faster than a liquid, through a liquid faster than through a vapor. The reason is the distance between atoms. This phenomenon continues with external plasma; the atoms are further apart.

Heat extends the domain of the microstrong force (that which prevents the electron from crashing into the atom, as stated earlier) pushing electrons further away from the atomic nucleus. In the plasmatic state of matter, the microstrong force is so great that electrons cannot remain attached via magnetic attraction to the nuclei. At the same time, the gravitational pull is weakening as the distance between atoms increase. Because of these, we observe the different "cooler" states of matter.

Pressure now enters the picture. In solid matter, gravity pulls the atoms together, barred by the magnetic repulse generated by the electrons. In essence, gravity provides pressure between atoms to give the "hardness" of the solid. Our given visualization representing this phenomenon is that of a compacted sponge. Now, let us equate the size of the sponge to the size of microstrong field. As the size of the sponge increases, the size of its compacted form also increases given the same pressure.

Core pressure of the planet provided by the surface material causes these plasmatic "sponges" to compact. Nevertheless, their compacted form is larger than that of a "sponge" of a solid atom. Therefore, we see two phenomena occurring: plasma behaving as a crystallized substance; two, this substance formulates by larger particles (in volume) than a solid. Sound would pass faster through this "crystallized" plasmatic mass than through a solid because there are fewer atoms involved in the transference via volume per atomic space. Therefore, instead of reaching the supposed solid core, the sound traveled to the shell's opposite wall and echoed back.

To illustrate this: imagine a set of cubes that are all equal in size having a dimension of a half-inch per side. Imagine them being lined up to form a distance of 24 inches spaced a half-inch apart upon a frictionless surface having a facial structure of each cube facing squarely in the direction of the alignment. There would be 24 of them and the last half-inch would be vacant.

Imagine a force hitting the first cube causing it to hit the next cube a half of a second later. This is a speed of one inch per second. After hitting the next cube, the first cube stops and the second cube moves at the same rate. This process continues until the last

Chapter 16: Initial Formations

cube in the line moves. Since each cube, 24 of them, only has to move a half-inch, it takes the entire process 12 seconds (24 x 0.5in /1in per sec). This represents sound moving through a solid.

Let us now remove every other cube leaving the first cube in place. This leaves 12 cubes with a space of 1.5 inches between them and 1.5-inch vacant space at the end. The same force hits the first cube. This will take the cube three times as long to reach the next cube as it travels 1.5 inches instead of a half-inch giving a time of 1.5 seconds until impact. The entire process takes 18 seconds (12 x 1.5in /1in per sec). This represents sound traveling through a vapor without any major temperature increase.

Lastly, the same twelve cubes increased in size to 1.5-inch sides (for dramatization of the effect) will represent the pressurized plasma. This gives a half-inch space between cubes and a half-inch space at the end of the two-foot line. The time it takes for the last cube to reach the two-foot mark is 6 seconds (12 x 0.5in /1in per sec).

Using the illustration given, visualize at the two-foot mark an opaque barrier such as a paper wall sitting at a perpendicular angle to the row of cubes. Next, imagine being on the other side of that wall setting the process in motion with an electronic switch. We think that we are examining the process in which the cubes are a half-inch measurement when they are 1.5 inches in measurement. Since the process took 6 seconds instead of 12, we conclude that the length encompassed by the process equates to 12 inches instead of the actual 24 inches.

While the illustration is somewhat simplistic, it shows how scientists conclude that a rock core is reached rather than contacting the earth's shell opposite wall. Other factors to include in the model are variance in cube size and in distance of their spacing. In general, the cube size would start out small and grow larger as it approaches the earth's center via heat. Secondly, leap in sizes occurs as we pass through the different layers. The phenomenon that generates sudden change in cube size generates sudden changes in spacing, as different elements have different properties.

As stated earlier, the moon's soil sample shows the presence of oxygen, but almost no hydrogen. The presence of oxygen is indicative of plasmatic conditions existing within the solar disc region of a sun, but does not necessarily mean the moon formed within our solar system. Within this scenario there are two possibilities could occur. One: the moon formed near the solar disc of our star, just beneath the asteroid belt of our star. Two: the formation occurred in a levanastorm, moon and planet storms forming in one of the separated 'arms' of the plasmatic hurricane-like storm, form our solar system. Moreover, from examining the surface, we observe comet-like boiled features, the presence of oxygen, and a relatively small mass having a density 3.3 times that of water. This scenario is, again, indicated by the lack of hydrogen within the surface material. It is highly improbable that the moon formed under the asteroid belt. We can calculate, using Mars and the earth as a guide via distance and density, the moon formulated

Part 5: Creation of the Macrocosm

within the asteroid belt. This is a minimal distance as the levanastorms are separated arms that extend the heavier solar disc material far beyond the governing conditions that formulate the solar disc phenomenon.

Unlike the earth or sun, the moon has very little of the heavier elements within its plasmatic core. After the crust developed, the internal pressure was not as strong as observed on other planets for the moon had less mass in ratio to the coolness. However, the pressure was still immense. Just as different gasses inside a pressurized planetary core will occupy the entire core, lunar plasmatic elements respond similarly. This means that the heavier generally metallic substances will touch the shell as well as the lighter elements. Upon touching the cooler shell, the metallic elements collect into a liquid. The gaseous elements, naturally, does not liquefy in the cooler temperatures.

As in all planets, both the liquid condensation and bubbles of vapor consolidated independently. The drops of liquid merged into larger drops and the bubbles of vapor merged into larger bubbles. The liquid eventually merged into a shallow ocean of material heavier than the surface, because the internal elements are heavier. This internal ocean pushes the vapor bubbles back toward the plasma. As the moon was moving toward the sun, the interior plasma of the moon had already turned into liquid and the plasma into vapor. Previously, the moon traversed through deep space, encountering temperatures that were extremely cold, and the moon had relatively little mass to sustain heat. The observation is: even though the vapor provided pressure; its magnitude is far under the immensity of the plasmatic condition. This caused the heated surface at one of the polar locations to sink somewhat toward the core.

After friction against the motion toward the sun became a factor, the liquid ocean-like shell under the solid shell responded independently to the external lunar shell. The liquid gel-like ocean began to migrate toward the 'front' of the moon. Despite that the surface of the moon was losing material by external friction; this migration made the front of the moon heavier. The reason is that the composition of the internal "ocean" is of heavier elements and of more mass than the mass lost from the surface.

The gasses within the "hollow" moon are also of the heavier elements. They would be the noble gases of krypton, xenon and even the lighter argon. The oxygen found on the surface probably would not reach that far down, as it is an external gas to the metallic plasmatic storm of the moon. Radon may exist in small quantities, as this is a heavier noble gas. Note also that these heavier noble gasses are also found in the atmosphere of Mars agreeing with the plasmatic core eruption on that planet.

Lastly, as we examine the "seas" of the moon, we find that the largest sea is off to one side trailing by smaller "seas." Another observation is that these seas exist only on one side of the moon. The reason for this coincides with the idea that the moon had a rotation and behaved similar to a comet. The trailing "seas" are the historical evaporations of the lunar "felsic" material as the moon slows down on its rotational

Chapter 16: Initial Formations

axis. Finally, the axis slowed down to a stop leaving the region of the largest "sea" facing the sun evaporating the "felsic" material for the longest period. After the moon stop rotating, rotational movement of the nonmagnetic planetoid still occurs. If we examine the moon's surface, we find that the large sea is off to one side instead of facing directly toward the earth. This is because of the attempted solar gyro movement of the moon as it moves at its only perihelion around the sun. During this time, the moon was able to shift the imbalanced interior thick "ocean" slightly beyond the location marked by the largest sea. Thereafter, the interior ocean freezes, permanently holding the imbalance.

After the moon leaves its close encounter with the sun, it passes back beyond the orbit of Mercury. Elohim places it in orbit around the earth. The side that was always facing the sun is now facing the earth. During our journeys to the moon, we found that the gravity on the inner side of the moon to be slightly greater. Because of its off centered gravitational centroid, this side pulls toward the earth during entry into an earthly orbit.

The moon's original orbit was farther away than that of today's orbit. The reason for the original orbit being further from earth is twofold. One: this kept the tides gentle. Two: this created even 30-day moon cycles per year. Even so, the earth was on a different path as well creating a 360-day year. This will be described in more detail later.

If Elohim brought the sun and moon into the equation before creating plants, the coarse granite-like structure between organic and inorganic materials would have been destroyed. Remember that the felsic cap is very shallow in comparison to the surface of the earth today. Secondly, recall that the earth and moon "revolve" around each other as manifested in Pluto and Charon, but not as dramatic. The result of this slight wobble by the earth forms tides. If the moon was brought in a Creation Day before the plants form, tides caused by the moon for a thousand years would have eroded most the "top soil" destroying the separated pockets organic chemicals. Along with this, if the sun had been ignited, the remaining organic material would have baked upon the surface, imprisoning the elements into a solid pattern. Therefore: After the felsic cap gradually emerges out of the sea, it is important to put plant life upon the surface to prevent erosion and to provide shade from the ignited sun. Some may argue that HE can do anything, even form life from the solidified mass. While it is true, HE could do that. However, the primary reason is that HE would be working against HIMSELF, making the task more complex. As we shell observe shortly, even the relationship of the rotational axis of the earth to the sun formulate the yearly tides.

The Process of Creating Life Continues

After establishing daylight, Elohim continues creating life on earth in the next day. The birds and sea creatures were created within Day Five. Naturally, the Hydrocarbon pockets on the felsic cap out of the water formed the birds, and that which existed

Part 5: Creation of the Macrocosm

under water was used to form the various sea creatures. The inorganic material surrounding the organic (carbon compounds) material provide a protective shell. Moreover, these "edge" inorganic elements provide more input material for formation.

The riddle, "Which came first, the chicken or the egg?" has an answer. The egg came first, but it was not laid, nor was it another creature preceding the chicken. Instead, it formulates within the felsic cap surface. The heat on the surface incubated the egg. Once again, life is not intrinsic to the chemical compounds forming the chicken.

On the sixth day, Elohim created the animals and lastly natural man. The dinosaurs count as being part of the animal kingdom formed before natural man. Another peculiar observation is that natural man was created as a multitude male and female, just as the rest of natural creation. This brings us to the seemingly paradoxical nature in Scripture concerning the creation of the Human Race.

Adam and Eve

By examining the first two chapters of Genesis closely and archeological data, we find that the traditional interpretation of the accounts contradicts itself on several accounts about the Torah and with the archeological data. The Creation of the "singular" Adam occurred at least before the animals (Gen. 2:18-19). There are even hints that he was formed before vegetation (Gen. 2:4-9). Adam's formulation was not as a natural man, but as a supernatural human above all the elements and natural laws. In fact, the Garden of Eden was not only a physical place, but it was in union with the dimensions of Heaven was well. The creation of Eve occurred on the Sixth Day of Creation. Adam was formed earlier, on the Third Day, to be a living human witness to the creation of all life on the planet. As we shall see in the last chapter concerning the logic of creation that Eve was not some afterthought as some have interpreted, but has a primary role in symbolizing the relationship of the creator to the creation.

One major erroneous conclusion is that Elohim required incestuous relationships by the children of Adam and Eve to populate the planet. As stated in the Torah, Elohim detest incestuous relationships and sexual activity with the animals. It is also written that Elohim changes not. Then, was Adam looking for a wife from animals? Everyone wants to know, where did Cain got his wife, and who are the people of Nod? But, the real question is where did Seth get his wife without sinning with his sister? We could say from Nod, but as we examine even more scriptures from the Torah, we learn differently. While this book is not first in discussing dual-creation, it is in Seth's issue of finding a wife.

The idea is that there are two kinds of human beings formed upon the planet. While this may seem wild at first, consider the formation of the Neanderthal and Cro-Magnon races. The Neanderthal disappears for no apparent reason, as far as present day scientists have been able to determine. Secondly, there are no transition stages from

Chapter 16: Initial Formations

Neanderthal to Cro-Magnon. Thirdly, the skeletal remains of the Neanderthal end where the skeletal of Cro-Magnon remains begin.

The **Neanderthal** set of humans form in "Genesis" 1, and the **Cro-Magnon** in the second chapter. All the human races existing today are all from the Cro-Magnon race. There are some interpretations that conclude that Ham's wife may have some Neanderthal blood, because of the reoccurrence of **Nephilim** (Giants) after the flood. Again, there are other solutions more congruent with the Torah. However, we are getting ahead of ourselves. Now, we will expound the answer concerning Seth.

As stated earlier, the bones from the Cro-Magnon are not found much later, yet according to scriptures, he was formed first. The reason is that the Cro-Magnon race did not die until their lifespan became 900 plus years after the fall, and they started out with just two beings. The Neanderthal started as a small multitude, and after the fall had shorter lifespans between 100 to 150 years. By this, the bones of the Neanderthal were collecting in the earthly strata long before those of the Cro-Magnon.

When Cain went into the land of Nod, or the land of wanderers, he went east. Note: the Garden of Eden was east in Eden, and Nod was east out of Eden. To be east, is to be toward Havilah. An interesting item concerning the names of these two places is that they describe the condition of life.

Eden is composed of the Hebrew letters: Ayin, Dalet(h) and Nun. The picture language is: Eye, Door and Fish. The Eye and Door represents eternity or that which is eternal. It is interesting to note that on back of a dollar bill is the "all seeing eye" over a pyramid. The Dalet(h) was also represented as a triangle, and the Ayin as an eye. Back to Eden, the Nun often means life. By this, Eden means eternal life, which has been translated later as pleasure. But, it means that life was eternal there before the fall. This is the location and condition in which the Cro-Magnon lived.

Nod (sounds like node, in Hebrew) is composed of Nun, Vav and Dalet(h). The picture language is that of a Fish, Nail and a Door. Since a fish represent life, and a nail means to attach or add, we see life being added. The door is a path or way. The result is the "way to add life." This implies birth. Where there is a beginning, there is an end. So consequently, this also implies death. On that note, before and after finite existence is eternity which will always be. He went into the land in which the Neanderthal lived.

We still have not answered: Where did Seth get his wife; since Elohim created Eve from Adam, as only a Cro-Magnon woman would be a suitable mate for Adam? If Seth went into the Neanderthal race for a wife, the Cro-Magnon race would dissolve like it did with Cain. The question then becomes: Where did other Cro-Magnon people come from, if they were not the children of Adam and Eve?

The answer is found by examining the life of Adam and Eve before the fall. Elohim did not create two races of humans just because HE thought it would be nice. HE has a

Part 5: Creation of the Macrocosm

purpose for everything. HE did not create two classes of people without providing the "lower class" a way to become equal with the "upper class." Perhaps the best example is seen today. The people of the "upper class" are those who have been regenerated by the Holy Spirit. The people of the "lower class" are those who do not. But, the lower class is not doomed to be permanently outside, as they can become grafted in by their own choosing process.

Along with this, there is a future transformation of the human form that will genetically alter formulating an immortal human body. Again, this phenomenon can only occur in those who are already transformed internally by being a vessel for the Ruach HaKodesh (Holy Spirit) which anyone can have by the choice they make. The same was true before the fall with the Neanderthal. The difference was that there was no sin nature to impede the process. The result was a genetic alteration at the point of accepting the message and receiving the Ruach HaKodesh. The Neanderthal was transformed into a Cro-Magnon in its full nature being above the natural instead of being natural.

Obviously, there was a role that Adam and Eve had in interceding for the individual. If this were not so, then sin would not have been imputed to all peoples when Adam sinned. Satan thought he had won, preventing the Creator to reform that which was destroyed in the rebellion in Heaven. This will be explained in the last chapter of this book. At any rate, Elohim knew all this was going to happen before HE created the physical universe. Even so, HE created it anyway knowing that in the end HE will attain the desired results.

Therefore, Seth got his wife from one of the many Neanderthal converts. This opens another hole in logic: Could not Adam just "transform" a Neanderthal woman and take her for a wife? There is an answer to that question. Eve was to be equal in nature to him. He was not to say that he made her into something that she is. She was to be as he was, a Cro-Magnon without choice. This attribute is found later in Torah as the First Born of the Tribes of Israel had no choice but to belong to Elohim.

Another point to make, the bones of the Cro-Magnon is from a fallen human race. What was the nature of Adam and Eve before the fall? Scriptures state that they were naked and not ashamed; but after the fall, they saw that they were naked and were ashamed. They were not blind or stupid before the fall. Something happened to make them see their nakedness as something to be covered. Some say it was the sexual act, but that does not account for the shame for they were husband and wife. Elohim created the two sexes, and it is good within its proper design. All we can truly say was that it was a sin of disobedience. It was not until Adam disobeyed that, "their eyes were opened." This does not imply that they were blind or unaware of their existence, but became aware by the means of losing a "covering."

They lost something that altered their nakedness. Afterward, the external and the internal were different as life and death and as, good and evil. Consider for a moment

Chapter 16: Initial Formations

that Adam and Eve's had a glowing appearance. This glow did not exist by some chemical composition, but it was by the linkage with the external infinite manifestation of Elohim. This is not as far-fetched as we might think. Consider Moshe (Moses) on Mount Sinai. After being in the presence with Elohim, his face radiated light; so much so, the he had to put a veil over his head by request of the Israelites. When Yeshua (Jesus) was transfigured, HIS Skin radiated as well. In the New Earth of Revelation, the inhabitants of New Jerusalem need no light from the sun as Elohim is their light. Adam's punishment of being cut off from Elohim literally means having no "physical" connection with HIS Infinite Nature.

Unfortunately, we, the people of today have very little conceptual knowledge of the nature that was lost by Adam and Eve. Sharing Elohim's infinite nature, HIS Shekina Glory is truly incomprehensible to our present finite (natural) minds. While it is hard to prove scientifically, there is data indicating so. Scientists have determined that there are at least ten dimensions instead of just four. These six dimensions exist in a defunct-like state in that their manifestation occupies almost no space. These indicate that there is something that has lost total manifestation. Consider proving the existence of red and the task of describing its appearance to a person born blind.

Our fallen minds are so decayed that we need books to hold what we think as being vast knowledge. We often think the information within the Torah did not exist until it was written. As stated earlier, Elohim does not change, and Adam talked with Elohim! By deductive reasoning, we find Noah's father knew Adam, and Noah knew Abraham's father. Even more peculiar, Noah was alive when Abraham has born. Knowledge of the fall was not a long string of hearsay to Abraham extending through generation after generation for thousands of years through a multitude of people. The Torah was written because humanity has lost the ability and had no desire to retain it in their minds. We often unwittingly pride ourselves in our machines doing things that we lost the ability to accomplish via our decaying physiology because of sin; not that we should be ashamed of creating machines, we are given this knowledge by our Creator.

As written in the Torah, they ate the fruit of the Knowledge of Good and Evil. Having a physical presence requires physical attributes of the fruit within the Garden of Eden. So, they actually ate some kind of fruit. The physical nature of the fruit is not as important as its spiritual linkage. The intoxicant was the glory of selfishness, self-exaltation. It feels "good" to get that which an individual desires. Not that it in itself is evil, but placing it above the will of the Creator is "evil." Hence, we get: the knowledge of good and evil. Notice, that both notions of good and evil exists within the same action.

Just as in Sodom and Gomorrah, the end result is a sexual sin. Adam and Eve had no children before "eating the apple." The first resulting manifestation of this sin was found in the sexual arena. Eve wanted to have children "like all the other women." Adam wanted to have a family to help in the field to build a better "garden." The

Part 5: Creation of the Macrocosm

solution then became obvious to them. The idea is not that sex is evil, as it was the intended function between a man and woman. The evil was that instead of bringing in the Neanderthal into the Kingdom of G-d, they were going to create the Cro-Magnon species themselves leaving the Neanderthal outside the loop. Note: their (Adam and Eve's) reasoning is found in the manner of curse that was laid upon them by Elohim. These scriptures can be found in the Torah B'reshyit (Genesis) 3:16-19.

Before we make conclusions against marriage, remember that Elohim instituted marriage. Secondly, Adam and Eve did have a love relationship before the fall. She was not just some kind of emotionally uninvolved co-worker. Before the fall, the abstinence from a sexual relationship was not a struggle with hormone impulses as it would be today, but a natural state of being. Remember, they were above the elements. After the fall, they became like the rest of creation needing sex to reproduce.

Afterward, the transformation process stopped. A Believing Neanderthal remains a Neanderthal for the remainder of their lives, but saved from the final punishment after death. However, there was sin between the Cro-Magnon race and Neanderthal. This is the mixing of genes. It says that the "Sons of G-d" found the "Daughters of Men" to be fair and took of them wives as they choose B'reshyit (Genesis) Chapter 6. For this, Elohim numbered the days of man to be 120 years, via the flood. It also states that Noah was the only one left that "kept his pedigree pure" or "perfect in his generations."

The primary evil in this was the idea of destroying the linage in which the Messiah was to be born. If there is no pure Cro-Magnon man left, the evil done by Adam and Eve could not be undone. The fall is directly linked to the nature of the pure genetic code Cro-Magnon. If the Neanderthal genetic code was mingled, the shedding of the blood would be an incomplete ratification of sin as the Neanderthal blood had no direct linkage or responsibility to the spiritual nature. Secondly, the mingling further destroyed the initial purpose of the two races. Here, the objective was to destroy any separation between that which was set aside for Elohim and that which is common to the world. Ultimately, the result would be the total obliteration of the Cro-Magnon race as it becomes swallowed up in the Neanderthal race as there were more Neanderthals than Cro-Magnon. After ten generations, traits become lost.

We are a little ahead of ourselves in the scenario. We mentioned the previous information, as there is a need to know that there is a relationship between spiritual realm and the physical realm, and Elohim establishes this relationship. There is much more to address concerning Adam and Eve and their impact upon our world. However, most of this topic is beyond the scope of describing creation and its resulting physical nature. There already exists much biological information concerning creation. Biology is another topic in itself with the entire system of DNA information needed to form all the different biological "houses" for life on earth. Then there is the process of insertion of life into these forms. The latter exists as a mystery, even to this day. Scientists have only been able to alter life, not create it. For our purposes, we will limit

Chapter 16: Initial Formations

ourselves to the knowledge that chemistry within itself cannot and does not constitute life. If we think about it, when we die, what element or elements disappear? The answer is naturally none. Actually, the body continues to function somewhat after death. Example, hair and nails continue to grow.

Lastly, we need to have a time element to the fall. It is generally accepted that the fall occurred about 6,000 years ago. However, the length of time after being created to the fall has been somewhat obscured. Because, Rosh Hashanah occurs on the first of Tishrei commemorating the Beginning of Creation or the First Day and Yom Kippur, the Day of Atonement, occurs on the tenth day of Tishrei, it has been taught that Adam sinned on the third day after the Creation Week. There is yet another factor involved. That factor is a thousand years equals one day in the sight of Elohim. This equates to a 3,000 year period between the Creation Week and the fall. Moreover, counting from the creation of the Neanderthal, Adam and Eve both lived about 5,000 years afterward. Yet in scriptures, it is accounted for Adam's life that he lived 930 years. The reason for this apparent contradiction is that before the fall a thousand years was counted as only one day in his life. Then counting from the Third Day of Creation, Adam lived about 7,000 years before he sinned. Recall that Elohim said that in the day that they eat the forbidden fruit that they would die, no one after that lived to be 1,000 years old or older. The account of the age of Adam is a verification that he died within the day he ate of the fruit. By this, the extent of human history, excluding Adam, is about 10,500 years. However, after the fall the countdown of 6,000 years has been working. By this, we know that we are on the brink of the "Last Days."

Finishing the logic of time, 4,000 years after the fall Yeshua (Elohim in human form) arrives on the earth and eradicates sin. Looking back on the Festival schedule, we find four days after Yom Kippur, we wave the Feast of Sukkot or the Feast of Booths, Tabernacles or temporary dwellings. Recall that we are using one day for 1,000 years. Moreover, Yeshua was born on the first day of Sukkot and circumcised on the eight day (last day) of Sukkot. The Eighth Day represents the Kingdom of the New Earth and New Heaven after the coming Millennial Reign. By the act of dying for our sins, HE restored the original plan for humankind to rebuild HIS Kingdom. HE regained all that was lost in the old world and all of those up through Avraham (Abraham), Moshe (Moses) and to the point in time of HIS Giving of HIS Life, and continues to move toward HIS Goal in this world (see Appendix: Life Continuum for augmentations of life un both space and time).

The Feast of Sukkot has many manifestations; the first one celebrated by Israel commemorates their journey through the wilderness as commanded by HaShem. The Prophetic Feast was personalized to Israel as a nation. One manifestation of the prophecy of Sukkot is that the earth after the fall would be a place of temporary dwellings for fallen man for a week or 7,000 years counting the Millennial Reign. From this, we can realize that this is the same amount of time originally intended with the

Part 5: Creation of the Macrocosm

Neanderthal conversion. On the Eighth Day, just as Yeshua was circumcised, Elohim will circumcise the flesh of the world, discarding those "extraneous" to HIS Original Design and recreate Heaven and Earth. By this, we know that the celebration of these Feasts declares to the world the history of the human race. This pattern of history is repeated within the history of Israel as a sign that HE is the one that set the nation apart, not some other "god." Moreover, it declares to the world HIS Prophetic Pattern of future events. Judgment has been passed upon the world because of the sin of Adam and Eve forming the first Yom Kippur beyond the "Fall."

Chapter 16 Quiz

1. What caused Dry Land to Appear?

    A. Evaporation of Water
    B. Violent Earthquakes
    C. Axial Shift by the Earth
    D. Venus Passed by closely

2. How was the Sun Ignited?

    A. Crashing Planetoids
    B. Gravitational Imploding
    C. Cooling Down
    D. Friction

3. Which came first the Chicken or the Egg?

    A. The Chicken
    B. The Egg
    C. Neither
    D. Both

4. The Moon came from?

    A. Beyond Mars
    B. The Sun
    C. The Earth
    D. Collecting Meteors

5. Adam and Eve were:

    A. Neanderthals
    B. Aliens
    C. Apes
    D. Cro-Magnon

6. (T/F) Dry land appears as a natural progression of Earth's formation.

7. (T/F) The chicken came before the initial egg.

8. (T/F) The sun ignites because of pressure,

9. (T/F) Complex gravity occurs only in the sun.

10. (T/F) Adam and Eve were Neanderthals.

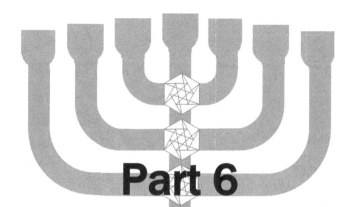

Part 6

Post Creation Activity

Chapter 17

Old World Phenomena

The Felsic Cap was circular. Instead of two plate tectonic shields, there was only one. This shield is called **Kadummagen** (Ka-doom' ma-gan). This word is an irregular combining of the Hebrew words for ancient (**kadum**) and shield (**magan**). Hebrew grammar rules generally would not combine the words in this manner, as the adjective follows the noun. There was only one ocean. This name of this primordial ocean is the **Sevievah** (Se-vee'vah) **Ocean**. Sevievah means surrounding literally in Hebrew. The Torah states that the seas were gathered into one place. This gives only one ocean surrounding a "central" landmass.

In relationship to the land positions today, all of the landmass existed primarily under the equator. Its center is off the rotational axis by about 30 degrees. The North Pole existed at the rotational axis centered in the land of Havilah as mentioned earlier. If the entire felsic cap existed above water, the coastline would extend to the Equator down to about seven latitudinal degrees above the Antarctic Circle. The center of the landmass existed within the triangular region known as the Sinai Peninsula. It is interesting that the land promised to Israel lies just east of this region, because this also fits the description of where the Garden of Eden once was, eastern Eden.

Before continuing on, let us examine the sense of direction. We assume that the top of the world exist within the Northern Hemisphere. Actually, the Southern Hemisphere is the top, as it was the external face in the multigalactic shell that forms our galaxy. We say that the earth moves counter-clockwise in its orbit. From this standpoint, we move clockwise. Another interesting point is that the older languages read from left to right and the newer languages read from right to left.

The sense of direction derived from three different observable occurrences. One: it is the placement of human beings within the landmass. Humankind placement was in the central region of the landmass, which encompassed Eden. The mammal kingdom existed external and overlapped into the kingdom of humankind. The mammal animal kingdom forms a doughnut surrounding the kingdom of humankind. External to this was the kingdom of dinosaurs on out to the water. The crescent continental shelf contained tropical sea life. Then outside that region was deep-sea life. Two: it is by the direction of the sun. This gave our direction of east, determined by the sun's ascending from and descending upon the horizon. The result is the land of Havilah existed due east of Eden as Havilah becomes the center of the rotational axis. The sun rise and sets

Part 6: Post Creation Activity

nearly due east viewed from the Garden of Eden. Three: it is the direction of the apparent circuit of the sun in the sky, it moved counterclockwise in the sky. The words used in the Torah put north and south in inverse positions.

East in Hebrew is Kedem, also implies meeting, disappoint and prevent. At first glance, these three words seem unrelated. The image is that the sun is heading toward the eastern horizon. It meets the horizon, but it cannot go below the horizon. Instead, its path leads it back up into the sky. In this, we could say the sun was prevented from going below the horizon, or the idea of going below the horizon was disappointed. The reason for this scenario is the same reason that one day of the year at the Arctic Circle; the sun does not go down.

West is Yam or sea. The reason for this is that if we travel away westward from Eden, we will encounter the sea sooner than in any other direction. Even though the cap is nearly a perfect circle (not that there were not variations locally in the coastline), sea level is not. The reason for the variation is that at the equator, centrifugal force is the greatest. The water is more flexible than the solid surface underneath. Then we have the tides, which were slightly less than today as the moon was slightly further away. Even so, a good percentage (not over a forth) of the cap was underwater. The reason for such a high estimate is that the outer regions of a circle involve more area than the central area. This region that was underwater covers the west coastal regions of the Americas.

North in Hebrew is Tzafon. The root also suggests peering into a distance. Putting it together, the hottest region is that toward the equator and that the sun travels in its circuit first, then to the zenith in the western sky, and then travels the greatest distance to reach the last directional reference before meeting the eastern horizon again.

South in Hebrew is Negev also implies parched or scorched. The hottest region is that toward the equator and that the sun travels in its circuit first, then to the zenith in the western sky, and then travels the greatest distance to reach the last directional reference before meeting the eastern horizon again. After the "division of the earth," the relationship flipped being above instead of below the equator.

Putting it together, the hottest region is that toward the equator and that the sun travels in its circuit first, then to the zenith in the western sky, and then travels the greatest distance to reach the last directional reference before meeting the eastern horizon again. After the "division of the earth," the relationship flipped being above instead of below the equator. The reason for this disorientation is that despite the flip by landmass the sun moves upward from the same direction.

Technically, east is south by our orientation. Consider the Summer Solstice near the Arctic Circle in our Northern Hemisphere. For convenience, we will pick Fairbanks, Alaska. At this

Chapter 17: The Old World

particular time of year, the sun rises slightly east of true north and at noon is reaches nearly 23 degrees above the horizon at due south and sets nearly due north again; except this time, it is slightly west. If we were on the Arctic Circle the setting and rising of the sum would be at true north. Similarly, this is the occurrence in the pre-deluge.

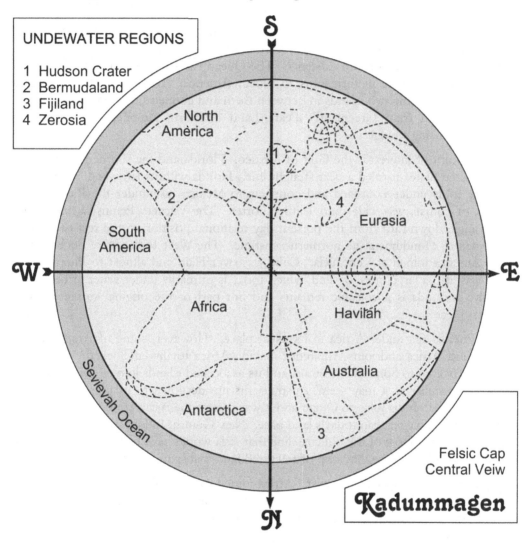

North America formed part of the "southern" coastline of this single landmass. Originally, the west coastline of North America down to the tip of South America was a straight coastline spherically speaking. We could lay the entire felsic cap centered upon the southern rotational axis and the coastline will sit perfectly upon the 23.5th latitude south. Alaska's coastline ends at the Kamchatka Peninsula where the Aleutian Islands point. Panama ends at South America; connected a little more with its eastern coastline,

Part 6: Post Creation Activity

and it was not as long longitudinally as seen today. South America's western coastline extended about 6.5° above the equator. The Baja Peninsula joins into Mexico even to the point seen in the Mexican Coastline south of the peninsula. Even the Gulf of Alaska formed a straight coastline in its connection to Russia. While the coastline was not perfectly smooth, there was not much deviance from it's the circular shape.

The Hudson Bay formed as a crater, which does not yet exist until after the flood. Baffin Island connects the mainland of North America to Greenland just a little above where the crater will come into existence. The Queen Elizabeth Islands join Greenland to Banks Island, which in turn joins to North America. Baffin Island connects to Victoria Island via the two islands in between them and connects to the mainland. The coastline formed from Greenland's Peary Land to North America formed a high latitude arc making a bulge.

The next point of interest is the Gulf of Mexico. Florida and the southern East Coast of the United States formed a near straight line. Florida with its large continental shelf originally folds under Louisiana and connects to Mexico just under the Rio Grande. Mexico, of course, was wider and a little shorter. The Yucatan Peninsula was at 90-degree angle downward from the present day relational position to the rest of Mexico and joined to Honduras at its northern coastline. The West Indies were once part of South America joined at Venezuela. Cuba was over Haiti and almost to Puerto Rico. There was also a large piece of land, which today is primarily under water at Bermuda. Actually, Bermuda is a volcanic remains and not part of the original surface of this landmass piece.

South America fits under Africa in its usual place. However, some of Argentina fits under South Africa and joins Antarctica making Africa totally landlocked. Antarctica joins to Africa up to Somalia. Australia and its associated islands join to Antarctica and India. As peculiar as it may seem, Australia fits upside down to its present relational position to Asia. India joins to Oman and it was part of the Asian continent. However, India was much bigger than today's land base. New Guinea, Indonesia, the Philippines, and Japan created most of the old coastline that Asia would have had up to Kamchatka. These islands form a long thin strip of fractured land that spans nearly 120°.

The Mediterranean Sea did not exist. Africa joined to Europe. Italy and some of the major islands existed in the Libyan Gulf. Spain was originally at a near 90° angle to its present position to France. In fact, Spain joins to the French coastline. England and Ireland joined to Denmark, and Denmark under Norway. Norway connected to southwest coastline of Greenland. The North America mainland connects to Ireland via Newfoundland above France and Spain plus the Northwest coastline of Africa. Most of the western coastline of northern Africa (above the joining of South America) connects to **Bermudaland,** the land in which Bermuda now exists. We might equate the fall of Atlantis to this parcel of land since all that remain of this region is volcanic islands. The story of Atlantis seems to be more of an external account about the flood

Chapter 17: The Old World

of Noah than about a post flood culture. Consider, the elements; great technology, great wickedness and great destruction, especially, by being underwater.

There are three more points of interest. There is another land that in under water in the Pacific associated with the Fiji Islands, hence, **Fijiland**. **Zerosia** is another sunken land between Russia and Greenland. Finally, Iceland also did not exist before the flood. To explain its existence will put us far ahead of this discussion, but we will discuss it later.

Cursing of the Ground

After the fall of Adam, Elohim cursed the ground. Two primary phenomena occurred: the transformation of the nature of some plants and the alteration of the ground. The transformation of plants primarily was the supernatural formation of thistles and weeds. The water table of Kadummagen lowers. An interesting item about weeds is that they generally have a deeper root system. Human beings now have to till the ground to plant the seeds deep enough into the soil to get water. Remember that it still has not yet rained (water).

The lowering of the water table was achieved by raising the altitude of the felsic cap above the water table. This alteration does not naturally occur; it is a result by Elohim's intervention. The mafic layer under the felsic cap experienced cooler temperatures via the rotational relationship to the sun. The South Pole no longer holds itself toward the sun by solar magnetic energy. The sun's magnetic energy destabilizes to become as the sun is seen today, via magnetic disturbances of passing stars brought by Elohim. When the solar poles flipped, the magnetic optimum was lost by the earth's magnetic system. Winter caused the Kadummagen to cool, which in turn cooled the mafic layer under it. As the mafic layer cooled, it contracted pushing the felsic cap upward.

Records of this event appear in the several times that the earth tried to maintain unity with sun by flipping its axis to match the sun's polarity. As the solar poles gradually destabilize before flipping, the weakened magnetic energy was unable to hold the earthly poles in its centric pattern around the sun. The plasma underneath the surface had the capability to respond without destroying the surface, as the shell magnet was not fully developed. The earth became unable to maintain its magnetic interactions, as the crust cooled and continuously thickened its barrier. Finally, the core stabilized via the shell magnetic formation.

The logic of the contraction of the mafic layer raising the mafic plain can be seen by using a waterbed. Imagine a waterbed about 3/4 full of water. After putting a small dog on this bed, the surface under the dog will pull toward the ground via gravity. By grabbing the edge of the waterbed and twisting the material, we decrease the surface area enclosing the water. As this happens, the dog becomes elevated. Thus, we have decreased the surface area while the internal volume of water remains unchanged. The same happened with the land of Kadummagen.

Part 6: Post Creation Activity

Resulting Solar Features

The lighter materials iron and above minus the elemental gases form an ocean of fire. The atmospheric gases compose the corona of the sun. Unlike the atmosphere of a given planet, the sun's atmosphere behaves similar to water hitting a frying pan. On a hot day on earth, we observe heat waves rising upwards. The same occurs on the sun's entire atmosphere but with much more force. The fiery ocean appears as constantly boiling liquid cells. The lavor liquids themselves exist by "coolness." The dark edges of these cells are cooler material falling back toward the ocean bottom. The central region is slightly higher as the density of cooler elements pulls down the edges.

The cooling lavor does not thin as it cools within this process. The primary reason is that the lavor never cools enough to become vapor. Secondly, the electrons that are normally present in the cooling of plasma are not there. The reason for the lack of electrons upon the surface is that they trapped underneath the moleucuum "ocean floor" of the sun. This moleucuum spherical floor, or the molecuumsphere of the sun, is relatively solid. This condition causes the cooling lavor to grab electrons from adjacent atoms causing a "plasmatic" bonding (soft bond). The result is that larger molecules form, which is heavier than single atoms; solar gravity then draws them downward. As the molecules reach the "ocean floor" of the sun, they heat up and cause the bonds to break between the elements. The result is that of formulating lighter material than the surrounding ocean.

Under the fiery ocean, there is another process occurring. Since the elements of the ocean floor are radioactive, they give off the three kinds of radiation. The most penetrating are the gamma rays, which aid in heating the surface. Beta particles and alpha particles emit from the floor material. The beta particles are electrons and are able to travel further through the ocean before being stopped than the alpha particles. However, this difference of distance is minor when compared to the thickness of the fiery ocean. The alpha and beta particles form plasma under the ocean. This plasma is comprised primarily of helium and electrons. The new plasma starts as a thin layer of tiny bubbles magnetically held to the floor's surface. As it gathers more plasma, the bubbles grow and collect into larger bubbles. During the transitional phase between polar flips these bubbles release via weaken magnetic energy. The bubbles then are too big to remain attached to the ocean floor without the magnetic hold. Gravity then presses in at the walls at its base. The sides fall inward squeezing the bubble interior upward. As the altitude of the bubble increases, its lateral diameter decreases. Eventually, the surface ocean above the bubble becomes a thin layer. When it burst, it exposes all the hidden plasmatic accumulation. A magnetic storm ensues as the negative and positive charged plasma separates toward the polar magnetic poles. The result is the appearance of a region having an inverse polarity even though initially this plasma was neutral in nature during its formation.

Chapter 17: The Old World

The destroyed bubble exposes the molecuum forming the ocean floor of the sun. It appears as a dark spot called the umbra. Recall that the molecuum primarily has electrons at the outermost shell level of the atom. These form bonds with other atoms forming a solid-like substance, which is as one plasmatic vapor molecule. The bonding electrons, relatively speaking, rarely get cool enough to release photons. Their photons are long red to infrared in size as juttorial energy dominates the process. The penumbra is the "gray" material on the sides. This is a mixture of plasmatic wall of the bubble and regular ocean material. Scientists have noted that this material is hotter than the surrounding fiery ocean. The reason for this is that the plasmatic-bubble wall heats up this region; the electrons release stored heat beyond that exists by lavor photons as they cool and collect to atoms regained back into the standard fiery ocean.

Solar flares are similar to sunspots in the process that formulates them. The difference is where they form. **Sunspots** form on the top of the ocean floor, while solar flares formulate within the "soft molecuum crust" of the ocean floor. They start out as small pockets within the plasmatic solid; than they collect into bigger pockets. This occurs, as these smaller pockets push upward by gravity's influence upon the surrounding molecuum. As they become bigger, they accelerate due to the increase of magnitude of the "heavy liquid" pushing it. These pockets continue to gather the various smaller pockets in front of their path. Eventually, the pocket reaches the top of the ocean floor at its maximum speed. As the heavy molecuum crashes back together on the ocean floor surface, the pocket is propelled through the fiery ocean and into the atmosphere. In the process of displacing the fiery ocean in its path, the pocket pulls some the ocean with it into the solar atmosphere.

Magnetic polarity of the sun completes a flipping cycle every 22 years. Every 11 years the polarity of magnetic energy inverts. Therefore, the second inversion brings the polarity back giving a 22-year cycle. Another interesting observation made by the scientists, the magnetic flip process starts at 40° north and south of the equator and not at the poles. Bands of inverted polarity begin at these locations. These bands grow toward the solar equator at a rate of about three to four miles per hour. As they approach the equator, their magnitude becomes nonexistent. At this point, the process repeats begins again at 40° north and south of the equator with the inverse polarity. Hence, the total distance between the two magnetic phenomena is 80° instead of 180°.

Plasmatic reservoirs of neutral plasma exist at the solar poles. Neutral plasma is not neutral in the sense that they have no charge, but in the sense that their net charge is zero. This does not lessen the magnetic magnitude of the plasma within these regions, when we add a positive 1,000,000 to a negative 1,000,000 the answer is still equals zero.

Utilizing this information, we are ready to examine the nature of these phenomena. Let us start at the 40° north and south of the equator. We know that the new inverted magnetic polarity is the strongest at this new band. The image is that of plasma rotating just under the thick shell of the fiery ocean's heavy liquid floor. The centrifugal force of

Part 6: Post Creation Activity

this plasma pushes the plasma downward to the equator via the curvature of the liquid shell. Two processes occur simultaneously. While the rotating plasma moves toward the equator, its thickness diminishes and the plasmatic charges begin to merge with the approaching opposite charged plasma from the other side of the equator internally.

Thickness of the charged plasma diminishes because of the alteration in shape. Imagine a rectangle that is four units at the base and one unit high giving an area of four square units. As the base increases in length, the height must decrease in order to maintain the same size in area. If the base becomes eight units wide, the height has to reduce to a half unit high to maintain four square units.

Merging of the two plasmatic charges at the equator generates neutral plasma. Actually, the scenario behaves quite different than the observable phenomenon. Unlike the normal movement of charged plasma only having electrons to move maintaining the polarity, we have two bodies of charged particles moving over a relatively neutral territory.

Imagine two sets of springs moving back and forth in opposite directions from each other. Each spring has a leading edge and a trailing edge. They each have their own momentum. We will start each spring at their optimum located today at $40°$ north and south of the equator. At these locations, each spring is compressed at its maximum producing the maximum display of charge. The springs then "jump" toward each other as they are drawn by their opposite charges. As the springs jump, they decompress. This relates to the lateral growth of mass as we move away from their polar concentrated regions to the vast equatorial regions of the sun. The two masses pass through each other via momentum and weak magnetic pull via lack of density. As the two spring pass through each other the net charge is zero. After passing through, the "elongated" spring approaches the opposite side of the sun. The most compressed region of the spring is not always the leading edge. The manifested compression moves backward on the spring after reaching the leading edge of the opposite charge via magnetic friction. The magnetic stretching of the plasma weakens the manifested charge of the assembly. The leading edge of the spring reaches the polar location and stops, allowing the rest to arrive; thereby, the spring compresses again. Then, the process starts all over again.

Adding yet another factor, the charged plasma at the core travels faster than the plasma upon the surface. There are two major reasons for this occurrence: one, shorter distance. Two, the plasma is hotter generating less friction. The cross section view of the movement is like a line segment that moves laterally to its shape, and as it travels both ends of the line bends backward forming a u-like v. Returning to a three dimensional view, a cone-like object formulates. By this, the core charged plasma hits the neutral plasma pool first and splashes toward the surface forming the observable magnetic "polar ring" around both of the magnetic poles, which are in union with the rotational poles, at an observable distance.

Chapter 17: The Old World

Perhaps the most unusual aspect of this phenomenon is the movement of positive charged plasma. Positive plasma forms from "free" protons. While this is not unusual in itself, consider the difference in mass of a proton to an electron. If it is a nucleus of any size, it has more neutrons than protons. The result would be that positive plasma generates more friction in movement. However, consider hydrogen. In general, it has only one proton with a charge equal to an electron. Secondly, the magnetic charges of these particles have a greater impact upon each other than gravitational and inertial considerations. Recall that Hydrogen-1 forms also within large nuclei near, next to, and the neutron snow inclusively, which are centrally located within the celestial formations.

The next item to examine is the nature of these neutral plasmatic reservoirs. The plasma stops and collects at 40° north and south of the equator as it collides into the reservoirs of the polar neutral plasma. Recall that the two plasmatic masses move through each other. Some of the electrons join with the nucleus moving in the opposite direction as they are caught in a gravitational aided magnetic pull. These electrons do not truly join with the nucleus to finish the atomic structure, but merely tag along. These do not return with the charged masses. They stay at the polar region in which the positive charged particle was moving toward. The chance of this phenomenon to occur is very slim, but the number of charged particles is extremely large. In essence, these reservoirs are growing at an extremely slow rate. This indicates a starting point occurred.

There is yet another item to discuss about the neutral plasma pools. The charged plasma travels through thousands of miles of neutral plasma only to splash against a pool of neutral plasma. The plasmatic polar pools must have another attribute that stops the charged plasma, and they do. The neutral plasma that the charged plasma passes through is large with larger "holes" between particles. The neutral plasma at the pools formulates not only of the neutral plasma of the original mass, but it contains the neutral plasma by the small neutrally charged particles that fill in or saturate the "holes."

Solar Flipping Deceleration

Flipping of the solar magnetic axis initializes at a more frequent rate. It kept the earth from realigning with the solar magnetic axis. Before the earth could respond and flip magnetically, the earth's solar centric magnetic alignment was disturbed. It was not a true gyro centric alignment as it was still affected by the magnetic energy of the sun. However, the strength of the solar centric alignment weakens at each flip.

There are two reasons that the hold weakens. One, the earth cools generating a thicker crust under the felsic cap. Two, new neutral plasma formulates on the sun at each flip. The reason for the formation of neutral plasma is that the two charged masses mingle during the flip as described earlier. This subtracts from the magnetic charge emitted from the sun as the amount of magnetic imbalance or polarized plasma among the

Part 6: Post Creation Activity

plasmatic elements lessens. The "new" neutral plasma initially moves with the charged plasma via momentum of the positive charged particle's mass. Separation occurs after reaching the neutral plasma polar pool as it no longer able to respond to the draw of the next magnetic flip.

The rate of the magnetic flip also decreases after each flip. The reason for this decrease is that there is less polarized plasma to cause the flip. This implies that the neutral plasma "pools" at the polar regions increase in size. As the charged mass decreases, it slows the flipping process, even though it has less distance to travel. Instead of flipping once a year (for an example), it is flipping about once every 11 years. Some scientists have concluded that there is a large variation of cycles for 7 to 27 years.

Since we know that the average radius of the sun is about 432,000 miles and the phenomenon initiates as a ring 40° above the equator, we can calculate the velocity that the ring travels toward the equator. This angle equals exactly 1/9 of a circle. The equation for the circumference is $2\pi r$. We divide the result by 9. Then we divide by 11 years, the time it takes to reach the equator. To find the miles per hour, we need to divide by 365 days per year, divide again by 24 hours per day. The result is about 3.1 miles per hour.

At the once a year cycle, the velocity was much faster. Instead of moving 40° is traversed nearly 90° to reach the equator. Instead of recalculating, we can multiply our previous answer by 11. This brings us back to one year movement at 40°. Since 90° is 1/4 of a circle, we multiply by 9/4 giving about 77 miles per hour. This also points to an end in which the velocity will become zero.

However, we also need to examine a counter process; this is the fact that as the neutral plasma polar reservoirs grow the distance between them shrinks. By this, the time between flips will decrease to zero as the velocity of charged plasma moves to zero. In graphing the phenomenon, we get an image similar to the initial arc of a sine wave (like an upside-down cup). The peak of the sine wave occurs at the angle of 45° above the equator. Since the phenomenon occurs today at the 40° parallel above the solar equator, we are already past the peak or the slowest time lapse of the solar axial flip of the magnetic axis. However, it will be thousands of years before the process ends and the magnetic activity ceases to exist. However, Elohim will create a New Heaven and a New Earth long before these dates occur.

The "magnetic polar cave in" by the sun, is not limited just to our solar system. The nature of the entire physical universe altered. It is not only the flipping of the axis, but the development of the polar neutral plasma pools that depicts the solar decaying process. The scriptures say that all creation travails together waiting for the redemption by Elohim. The cursing of the ground marks the beginning of the decay of the entire physical universe within our finite space. After the Millennial Kingdom, Elohim recreates the entire universe again.

Chapter 17: The Old World

There is another interesting detail that we need to examine. When Elohim alters natural phenomenon, it operates as if it has always been part of the natural process. Consider a supernatural healing. Once it occurs, we cannot find any indication of the previous injury. The opposite is also true. This applies with the reservoir building and the polarity flipping processes seen in our sun.

This alteration was not a matter of initiating a new phenomenon, but to remove a supernatural sustaining force. The image is like a novelty device, in which, spheres are dangling from strings forming a horizontal line a distance above the surface. To activate the device, pull out an end sphere and let it fall toward the lineup of spheres. The sphere hits the lineup of spheres and displaces sphere at the opposite end. It swings out and falls back to the lineup. In essence, it moves back and forth gradually losing kinetic energy. Before the cursing of the ground, it is as Elohim holding the end sphere away from the lineup. Afterward, Elohim lets go and lets the process begin.

Pre-Flood Atmospheric Conditions

A canopy image of the water vapor is not entirely accurate. While there was water vapor above the ozone, it was not a uniform cover. Recall the plasmatic storm in which our planet formed. Just as Saturn has rings, the earth had a similar phenomenon. Unlike the rings of Saturn, the water vapor formed a thicker layer and connected to the lower atmosphere via ozone. A side view would be similar to the image of the solar plasma described during the Fourth Day of Creation. The water vapor did cover the Polar Regions, but was much thinner. The polar water vapor existed as gravity pulled the water vapor into the eye of the planetary storm. If the canopy were uniform at 2.5 atmospheres, the height of the water formed during the flood would be only about 57 feet high. True, it is high enough for the account given in the Torah about covering the highest mountain of 22 to 30 feet. However, the base from which this measurement was taken was not given. The reason is the nearly non-existent slope of the cap. It would in general appear perfectly flat excuse the bumps. As an average, we could travel approximately 9 miles toward the central point (Mount Sinai) and gain elevation of one foot. We calculated this figure from the following data: There is 4,595 miles to center of cap at 500 feet high. We divide the distance to the center of the cap by the central elevation. This is a purely linear answer, variation occurs via a slight curvature.

Scientists have determined that if all the ice upon the earth were to melt the water level would raise 400 feet. Furthermore, they have estimated that the water level has risen 300 feet after the ice age. This will give a total of 700 feet of water. However, much of our land features would still be above water. Since we are concerned with total coverage occurring, the water level is somewhat lower. For there to be 10 atmospheres of air pressure before the Flood, the amount of water needs to be about 611 feet. We get this figure from today's air pressure being 14.7 pounds per square inch. Multiply this by 10 to represent 10 atmospheres, we get 147 pounds per square inch. Then we

341

Part 6: Post Creation Activity

subtract todays atmosphere (147 – 14.7), we get 132.3 pounds per square inch. We need to double this number because we are looking at the variation of pressure from the earth's surface being the maximum value to the edge of the atmosphere being zero; this gives a value of 264.6. Divide this by 8.34, as there are 8.32 pounds of water per a gallon giving a pillar of about 31.73 gallons of water exerting pressure upon a one square inch base. We multiply by 231 cubic inches per gallon giving 7,328.8 cubic inches of water standing upon one square inch. We divide this number by 12, which give us the answer of 610.7 feet of water.

Our reasoning for the 10 atmospheres will be expounded in detail as we look into the Ice Ages. It is however, interesting that scientific data correlates with the need of 10 atmospheres. It also correlates to the flatness of the initial felsic cap which is also of necessity considering the manner in which the cap was formed. That manner is the fact that the entire surface was molten simultaneously.

The weather changed slightly during this period of time. Even though the two primary air masses remained separated, seasons developed within the southern air mass. While there was yet no snow for winter, there was a definite coldness to it. The coldness, however, did not go below freezing.

The Flood

Rain did not exist upon the earth before the flood because of the denser atmosphere and lack of water vapor in the troposphere. Recall that the ozone separated the water vapor from the troposphere. Instead, there was a mist that formed near the ground. The image is similar to carbonated water. Carbon dioxide exists in magnetic balance, while water is bipolar in nature having both negative and positive magnetic charge sides to the molecule. In the case of the atmosphere the situation is reversed. Instead of the bipolar molecules of water vapor dominating, the magnetically balanced molecules dominate being oxygen gas (O_2), nitrogen gas (N_2) and argon. Moreover, these gases are heavier than water vapor. Their weights in atomic units or nucleon counts are about 32 to 36 for oxygen, about 28 to 30 for nitrogen and the noble gas of argon has 38. Compare this to water vapor having about 18 to 20 nucleon count per molecule depending upon the size of the oxygen atom. The ozone (O_3) lightness stems from its single bonding formations between any two atoms. The oxygen atoms are not as deeply embedded creating a larger volume lessening the molecular density.

As a result, the pockets of water vapor become compressed by the surrounding atmosphere. This pressure brings the water molecules closer together causing them to interact magnetically. Being closer together they absorb more photons. The positive pull by the hydrogen atoms on oxygen of neighboring water molecules causes some electrons to be pulled beyond their normal orbital causing them to release various photons. This interaction causes the transparent water vapor to become opaque clouds.

Chapter 17: The Old World

The next phenomenon to occur is that the water vapor pocket becomes denser or heavier attempting to match the density of the surrounding atmosphere. As more water vapor rises from the ground, it rises faster than the collected and condensed "older water vapor" and becomes collected into the pocket. Eventually, the water molecules begin to collect via magnetic energy into water droplets. When these water droplets become heavy enough they fall to earth.

Back to the time before the flood, because of the thicker atmosphere this condensation process occurs at a faster rate. Therefore, the mist appears as a very low fog that waters the earth; remember also that the felsic cap is relatively flat giving nearly equal pressure anywhere upon the surface.

Elohim could have just materialized the rain and dematerialized water; however, this is punishment for a people who ignored HIM and worship themselves (leaning on their understanding of nature) and creation. This corrupted even the animal kingdom. Therefore, nature was used by Elohim to destroy the Old World. Here, we observe a planet moving very close to the earth. This planet known in today's culture as Venus. As stated earlier, Venus came from another solar system. The star that pulled Venus from this system is known today as Jupiter as indicated earlier.

Venus provided the necessary influences to cause it to rain. The gravitational influences of the planet caused volcanic activity. This is important in that it released huge quantities of dust particles in the air adding more seed for changing water vapor into rain. The planet tore the stillness of the earth's atmosphere. The activity occurring in the earth's atmosphere was much more violent than that of the surface material. The primary reason is that the atmosphere is loosely fitted molecules free to move at the slightest stir. The water vapor pulls toward Venus. This altered the perfect balance between centrifugal force and gravity. The atoms facing Venus in general slowed down causing turbulence between the various stratums of atmosphere. In this, the water vapor pulls into the troposphere from above the ozone with the ozone. Because of the thickness of the atmosphere, rain formed very quickly.

As noted by scientists, Venus has 93 times the atmospheric pressure that of earth. The Venusian's atmosphere should have partially sucked into the earth's atmosphere at the distance that Venus had to come toward the earth. However, Venus did not have near the atmosphere at that time. As scientists note, its atmosphere has a composition of 96% carbon dioxide and freezes around -108 degrees Fahrenheit at one atmosphere. However, the atmospheric pressure of Venus is approximately 93 times that. At 70 atmospheres, carbon dioxide freezes at room temperature. Since the planet came from outside our solar system, the coldness of deep space was sufficient to freeze all the carbon dioxide. The sulfur dioxide in the atmosphere also turns to liquid at temperatures just under the freezing of water. Now, 1/96 of 93 atmospheres is less than 1 atmosphere. This means that, at most, the Venusians atmospheric pressure was slightly below to our present day atmosphere. In fact, it is possible that the earth lost

343

Part 6: Post Creation Activity

some atmosphere to Venus. After it passed the earth, all of the frozen gases melted giving the atmosphere seen on Venus today.

Atmospheric conditions on Venus were warming as it passed the earth. The estimation of temperature is approximately -200^0 F (-130^0 C) as Venus has internal heat warming its atmosphere. This is not cold enough to liquefy oxygen, but far beyond the temperature to freeze carbon dioxide. By this, oxygen was the only element found upon its surface as it entered through the outermost asteroid belt of our solar system. As indicated earlier, sulfur dust becomes entrapped from this region as the Venusians atmosphere. Sulfur meteorites of various sized burned up as they fell into the Venusians' atmosphere. As they burn up, the oxygen combines with the sulfur forming sulfuric trioxide (SO_3). While volcanic activity upon Venus could also account to some of the sulfur content, it is not. The reason is that the masses of Venus and Earth are close, and it would have generated a similar phenomenon upon earth within its outer atmosphere. The water vapor captured from the earth was the final ingredient to formulate sulfuric acid. The sulfur dust collected oxygen before Venus reached the Earth giving it the ability to use the water vapor to generate the sulfuric acid (H_2SO_4) outer atmosphere.

The amount of rainfall was near 500 feet altogether as mentioned earlier. The water covered the entire Kadummagen with an average of 300 feet. Note: the higher central region forms only 1/3 of the total cap area. The only possible structures to remain above water were trees and tall buildings. However, the interaction destroyed these structures as well. Venus came very close to our planet, just outside the orbit of the moon. This generated a huge tidal wave that crushed across the entire Kadummagen. This destructive force also deposited a large field of sediment. The shape of this field is like a crescent the size of the felsic cap. The thickest part of the crescent centers about the central to northern region of North America. We have not forgotten the continental shield of Laurasia; this is not the end of the destruction that occurred during the flood.

A large meteor smashed into the earth after Venus passed the earth's orbit. We might think it was a small moon of Venus, and be satisfied with the concept. However, more probable is that it was a large planetoid within the asteroid belt caught by Venus as it moved toward the earth. It smashed into the earth later forming today's Hudson Bay. We call this planetoid **Hudson Comet** from its interaction with the earth, as we shall observe. However, toady's felsic location is far from where it struck. During the time it struck, the crescent of sedimentary deposit already existed and water still covered the earth. The planetoid plunged into the earth near the middle of the thickest region of sediment. In doing so, it removed all the sedimentary material from the impact area outward for several hundreds of miles. As the result, the earth has two shields, Laurasia and Gondwana, **Laurasia** being the shield formed around the **Hudson Crater**. Gondwana is the larger shield opposite of the original sediment crescent laid

Chapter 17: The Old World

by the passing of Venus. The sizes of these shields shrink when other phenomenon adds sedimentary material.

Hudson Comet Impacts upon Earth

It appears that the planetoid had a diameter of approximately 160 miles. At this diameter, Hudson Comet is slightly greater than 1/50th that of the earth; however, this makes the volume to be about 1/125,000 of that of the earth. In the Torah, it does not mention that this event occurred. The primary reasons are the following: There was no living witness except Elohim to the crash and was insignificant to the catastrophe occurring at that time (death was already assigned to all flesh). Hudson Comet did not just crash into the surface and embed within the crust; it bore into the asthenosphere into the molten liquid where it finally most of the material was absorbed. This is like having a sphere of on inch sphere crashing into a 50 inch sphere. The Earth would have ended up like Mars except for two reasons. One: the entire surface was covered with water; and two, the meteor was not a solid rock to the core. When we look at the Hudson Bay, we are not looking at the crater; we are looking at the puncture "wound." The crater is extremely huge, as we shall observe shortly.

Hudson Comet, like most spherical planetoids, is hollow spherical shells filled primarily with a thin atmosphere of krypton and xenon. The core is as foam, in our case, scientists found iridium splattered across the surface of the Earth. Upon impact, the interior dust and vapors push out via falling outer shell surface. Being heavier, this dust settles before the earthly sedimentary material. Its original rotation obscured by the rotation generated via friction of its trajectory. The earth absorbed most of its rotational momentum and twisted slightly.

When Hudson Comet smashed into the surface, it created a crater, radial fracturing, and an elliptical fracture ring. The fracture ring was about two thousand miles at the minor axis and three thousand miles at the major axis. The downward collapsing force of the surrounding landmass at the point of impact caused the fracturing. This force occurred by the sudden impact upon a relatively thin floating surface. By throwing a rock upon a large puddle covered with thin ice, we can see this kind of fracturing. Our image emerges because the crust rests on a soft pliable ingot (white hot metal) surface known as the asthenosphere. Yet, asthenosphere for all practical purposes is as a liquid in the view of the forces involved and rests upon an ocean of molten liquid metal. The impact crater exists at one of the focal points of the ellipse. Hudson Comet appears to hit the earth at the latitude of about 43° latitude and by an angle about 39° west from being due "south" from Havilah and moving faster than the Kadummagen in the direction of its rotation. This means that it did not fall to the surface at a 90° angle. This angle is about 42° from the surface putting the base of the angle in the direction of rotation. This is indicative of a larger object hitting the earth with an initial velocity comparable to its acceleration to the earth; otherwise it would be a direct hit via gravity. Since the

Part 6: Post Creation Activity

planetoid's core was hollow, the planetoid flattened to about a diameter of about 250 miles. Despite this hollowness, its crust was thicker than the earth's; this is seen by its circular puncture into the earth's crust. The twist of this puncture is counterclockwise in its formation. At this point, all of these fractures lay dormant within the crust providing little change in the landscape, with the exception of the puncture.

Other factors influenced the fracturing of the felsic cap by this puncture. The ellipse was disrupted from forming a true ellipse. The planetoid hits the landmass quite a distance from its center. Secondly, the rotation of Hudson Comet itself alters the shape. It appears to be rotating in the opposite direction of the earth. This becomes evident as the large swirl ending with the multi-fracturing forms little plates that will later form islands west of Greenland. Not as obvious, we see it partially today as Denmark's formation as well.

The result of the impact occurring at a distance from the pole was the resistance of the fracturing was weaker on the side toward the water creating more cracks thereby releasing the energy of the twist creating the small islands west of Greenland mentioned earlier. There is a backscattering effect of this release outside the fracturing ring. These fractures will later form Kamchatka and the separation of Alaska from Russia. The fracture line behind the focus of the ellipse only slightly cracks and later causes many large sized lakes in North America to form.

Yet in another sector, there formed a temporary upward lift at the ring's rim. This existed as an arc opposite from the weaken arc by the coast. This effect compounds the northerly direction of the impact. As the interior fractured plate pushes upward at this edge, the backside of the brim, approximately 400 miles thick, cracked. This forms another fracture line forming the northern coastline of Africa and the eastern American coastline. This fracture extended unto the center of the Kadummagen and onward forming the southern border of Turkey.

Later another fracture phenomenon happens because of the impact of Hudson Comet. A shock wave traveled around the world and collided on the opposite side of the planet. Its collision point was at the bottom edge of the Saudi Arabian Peninsula and shattered southward. This occurred at a higher location because the shock waves are circular in nature and traveled about a circle, which was less than a great circle (not a circle like the equator). It created the fracture that later determines the extent of the fractures that separated the following landmasses: Australia from Antarctica, Australia from India, India and Africa from the Saudi Arabian Peninsula, Antarctica from Africa.

As noted earlier, scientists also have found iridium associated with the impact. The iridium came from Hudson Comet and not the earth. Iridium (77 protons with at least 77 neutrons) is not a surface element material of a spherical planetoid or even planet or star, but depicts fractures of the spherical shell-like mass allowing its inner core material to gush out, even though it has a thicker shell in ratio to the earths. Water from the

Chapter 17: The Old World

flood quickly cooled the material down. A side note, this impact has been blamed for the disappearance of dinosaurs. The comet impact is only secondary event beyond the occurrence of the flood.

Interior to Hudson Comet, dust particles are solidified plasmatic mixture of the heavier elements. The image is as thinly constructed foam. The bubbles, naturally, are the heavier noble gases of xenon and krypton, primarily. The substance of the foam was not only iridium, but appears to be the only heavy element of any massive quantity. The foam structure, itself, is very flimsy and was easily crushed into powder by the collapsing spherical shell of the planetoid. Sometimes, in cold planetoids, the foam collapses by its own weight and becomes dust; especially

Post Flood Activity

After the volcanoes quieted and the rains stopped, the earth was covered totally with water, except Noah's Ark. The water would have remained forever without the intervention of Elohim. The atmosphere has already done all it could do to become stable once again. Water vapor could rise, but the vapor then collects. It forms rain and fall again to the ocean. Therefore, Elohim sends a wind to blow upon the earth. This wind could not have occurred of itself. The nature of this wind was more than just creating water vapor from the surface of the all-engulfing ocean. This wind accelerated the water vapor so that much of the vapor may again be restored to its original position. The basic idea is that the ozone layer removes from the central equatorial atmospheric region creating a low-pressure region causing faster moving atmosphere encountering the Troposphere. Both cause the water to vaporize rapidly. Again, this process was generated purely by supernatural means. As stated earlier, the condition would have remained permanent without external intervention by Elohim.

As noted in the Torah, a bow was set in the clouds. In order for there to be clouds, there still had to be water vapor in a thinner Troposphere. The landmass was still in one piece, except for one large crater. However, the land formations have changed. There are now mountains higher than ever before via sedimentary deposits and volcanic activity. The ocean level is still higher than what it was before.

The ratio between fresh water and salt water generated by the flood is much less than anticipated by scientists trying to cover Mount Everest with water. Instead of trying to account for thousands of feet even miles (5.4978) of water, which proved improbable, we are dealing with only a few hundred feet. This has been an obstacle for many years. It gives a ratio of about 1 part fresh water to 13 parts salt water or 1:13 instead of 11:3. Another interesting factor, all the fresh water stayed primarily upon the surface of the ocean as salt water is heavier. Now, we are ready to examine the process by which the continents became divided and the manifestations of high altitude mountains.

Chapter 17 Quiz

1. Where is Eden?
   a. Center of the Felsic Cap
   b. East
   c. South
   d. None of the Above

2. Elohim used what planet to generate Noah's Flood?
   a. Mercury
   b. Jupiter
   c. Mars
   d. Venus

3. What existed at the core of the Hudson?
   a. Water
   b. Nothing
   c. Iridium
   d. Osmium

4. Within Today's Orientation, Which way is East before the Flood?
   a. North
   b. South
   c. West
   d. To the Moon Rise

5. Where does the Name, Kadummagen Come from?
   a. Hebrew
   b. Words for Ancient and Shield
   c. Both a and b
   d. Neither a or b

6. (T/F) Israel and the Garden of Eden exist in the same location.

7. (T/F) Solar changes occurred by Elohim's cursing of the ground.

8. (T/F) Noah's Flood occurred because Venus came close to the Earth

9. (T/F) Atmospheric pressure decreased because of the flood.

10. (T/F) Hudson Comet created the division between Europe and Africa.

Chapter 18

Peleg's Earthquake Overview

Our first task is to examine the events leading to Peleg's Earthquake. After the flood, people began to repopulate the Earth. Instead of moving all over the Earth as Elohim commanded, they chose to stay in one group in the land of Shinar (in Iraq today). To cement their position, they decided to build a tower as a monument to this cause. Elohim destroyed this tower, and confused the languages of the people so that they could not understand each other. It is for this reason that the tower became known as the Tower of Babel. However, not all were involved in the Tower of Babel. Elohim always has people that are reverent towards HIM. The language of the uninvolved people were not confused, these people were the Hebrews. Even though the Hebrew today is somewhat altered in pronunciation, the character format of the words still have the same picture language of the old language represented. The need of using written language for something beyond a memorial and business is a continuation of this decay.

They were scattered by families. As far as races are concerned, it seems that family roots have very little to do with the color of skin. The image is that all people were a brown (or red like Indians). Imagine the whole felsic cap filled with this color. Now, imagine a line cutting the cap in half separating the northern half from the southern half. Each area has a center. Approaching the center in the northern half, people get whiter. Inversely, people get darker in the southern half. From these central points, the variation strengthens. This leaves "red" or "brown" across the middle and around the circumference of the cap. Scientists call this phenomenon, genetic drift. The required observation is that most of the families of Ham did go south and east and most of Japheth's families went north and west. Shem has stayed primarily within the central region. Because of this, we could say the descendants of Ham generally became black and the descendants of Japheth generally became white. It is from the word, Shem, that we get the word S(h)emitic.

After the people dispersed throughout the altered Kadummagen, another phenomenon occurred. Elohim divided the Earth. Before describing the occurrence, we must first examine the forces involved.

Analysis of Magnetic Plasma

Recall that the initial magnetic flip of the Earth during creation moved the rotational axis with the magnetic axis. In this phenomenon, this movement did not occur. If this

Part 6: Creation of the Macrocosm

were to occur, then the Earth would be rotating in the opposite direction than the other planets within our solar system with the exception of Venus. To understand this difference in behavior, we need to examine the nature of plasmatic magnetism. Neutral plasma is a mixture of positive and negative ions which are disassociated electrons (negative ions) and nuclei (positive ions). Polarized plasma occurs when the distribution of positive ions and negative ions exists at an imbalance from the standpoint of the entire mass. Unlike their equal opposite magnetic charges, they are not equal in mass. The negative ions formulated by electrons have very little mass when compared to the mass of the positive ions formulated by nuclei.

Within the initial set of magnetic flips during creation, the magnetic response affected both the positive and negative ions. The result is that the rotational axis of the mass flipped with the magnetic axis. However, this is not the case during this interaction. The negative ions moved without moving the positive ions. The image is that the concentration of negative ions is generally focused at one side of the rotational axis and fades in concentration as we move toward the opposite axis. The positive ions exist more evenly throughout the entire mass. Because the negative ion concentration fades as we move to the opposite pole, the opposite pole is said to be positive even though the distribution of cations (positive ions) are fairly uniform.

During the magnetic flip within this scenario, only the negative ions move. By this, the rotational axis remains unaffected by the phenomenon. The image is that these negative ions move without changing their distribution pattern. Obviously, there is some loss to the pattern via bulk blockage, but it is surprisingly stable.

The primary reason that the positive ions do not move is the effect of cooling. As the plasma cools, the positive ions (nuclei) collect electrons. This weakens their magnetic strength and is too massive to move, while the electrons retain their magnetic strength with little mass and moves easily. This gives us the resulting phenomenon, as seen.

Earth's Magnetic Polar Phenomena Set

The first set of phenomena to discuss is concerning the planet's magnetic reaction. The first of these is the flipping of the magnetic poles. To start out simply, imagine two spherical magnets in space. Strengths of the magnets are equal, and their masses are equal. Imagine them both starting with their poles directly match in vertical alignment, north with north and south with south. Naturally, they magnetically flip so that the north of one will align with the south of the opposite magnet. If one magnet was held in place, the other would simply flip to its magnetic polarity, but both magnets are floating free.

Let us examine the nature of their flip. They take the path of the least resistance. The relationship between these paths, which the magnetic poles take on their journey, can be described as parallel. Placing a plane between these two spheres, their paths will be

Chapter 18: Peleg's Earthquake Overview

parallel to it not perpendicular as we might think. For the latter to occur, one set of magnetic polar interaction between planets needs to push away. This means that the opposite set has to pull together. This cannot occur, as both sets have equal strength and are interacting simultaneously with the equal interaction occurring with the poles in the opposite hemispheres. By this, they flip sideways parallel to the plane perpendicular to the line projected between the two centers of the two spheres and they continue to flip until they reach an optimum position.

For a practical exercise of this phenomenon, take two poles of equal dimensions. Drill a hole in the center of each pole, and attach a small eye bolt at each end of each pole. Next, we will attach each pole on a separate parallel plane (i.e. two walls facing each other) using the central holes that are drilled into the poles. We also need to be certain that the poles are attached directly across from each other and that they are able to spin. Then we place the two poles so that they are both vertical in position. Our next task is to take two strings and place it through the eyes of the bolts. Take one string and pull it through the top eye bolt of one pole and through the bottom of the opposite pole. Do this for both poles. Now, grab the ends of the strings and pull on them toward away from their respective "wall." The result is that they flip until the string cannot be tighten anymore without breaking the poles.

The fastening of the poles to opposite planes represents the inertia of the planet. The connecting of the string to opposite eyes on the pole represents the magnetic attraction of both planetary poles facing the same direction. Naturally, the pulling of the strings represents the pulling force of the magnetic energy between the two.

Next, let us analyze the resulting positions. Since both magnets are of equal strength, each magnet moves an equal distance. The result is that both magnets moved $90°$ from their original positions. At this point, they both are at their closest possible distance without moving the spheres together.

Now imagine two magnets of unequal strength with equal masses (although this is highly unlikely). Giving the same situation as in our first analysis, the stronger magnet will turn only $60°$ while the weaker one will turn $120°$. Note the addition of flips of the pair of objects in both cases equal $180°$. In order for one magnet to turn $180°$ and the other remaining stationary, the stronger magnet needs to be infinite in magnitude. The changes in mass in this phenomenon alter the figures somewhat in favor of the larger object. The reason being as mass increases so does inertia. Another variation is possible because magnetic energy does not directly relate to a structure of mass.

Another factor to present to this picture, these magnets are not stationary. They have a momentum that attempts to hold them in course of some original velocity, usually at different speeds. This creates a phenomenon concerning the nature of the magnetic hold that existed between the two spheres. The term, "magnetic hold", can be misleading as the two magnets were initially held in place by mutual repelling rather than

Part 6: Creation of the Macrocosm

by mutual attraction. To see this, we must visualize two different planes cutting through the two spheres, one per sphere. Each plane contains the magnetic polar axis of the sphere that they intersect. The line connecting the centers of the two spheres is perpendicular to the two planes. Magnetic pressure holds the two planes parallel as the magnetic energy causes the axis to flip. As the two spheres pass each other, the planes rotate keeping parallel to each other. Again, it is by magnetic tension, which we previously illustrated with the eye-bolt poles and strings. By this, there is a lateral shift by the magnetic planes as the magnetic poles move to attain their optimum

Adding another factor, the spherical magnets have a rotation prior to the flipping rotation. The normal rotational axis connects to the magnetic poles during the process of planetary formation. This fact in itself does not alter the behavior of the magnetic interactions. However, after the planet cools, the magnetic poles and the rotational axis are not rotationally dependent. The magnetic poles can separate from the rotational axis without affecting the rotational axis. This becomes possible as the solid crust contains "hard" frozen electric magnets while the core has "soft" plasmatic magnetic forces holding the magnetic charge. Hard magnets are solid and cannot shift without moving the solid mass containing the charge. Soft magnets, as previously described, can shift through solid material without moving solid masses. This we observe in electrical current as moving electrons that do not move the wire as they pass through.

Unlike magnets used in industry, the ratio between the force of magnetism and structural integrity is reversed. In other words, the magnetic forces are stronger than the strength of the crust of the Earth. The result is that, after reaching the optimum alignment between the two magnetic poles, there is **polar compression**. Returning to our example of two affixed poles at the center with strings attach the pole ends to the opposite pole at it ends. At the end of the flip, the two strings are parallel bringing the poles parallel. Now, the strings continue to pull by shortening, this pulls the end of the poles toward each other. Again, there are variations to this phenomenon. Suppose, for a moment that one pole was more solid than the other, the weaker one would bend with little affect to the stronger pole. This illustrates the way the Earth's magnetic poles compressed when a stronger external magnetic field interacted with Earth's.

Lastly, there is also a decompression of the poles. Using our previous illustration, we can imagine the weaker pole as being somewhat elastic and able to spring back a little toward its original position. This is the last magnetic phenomenon associated with these phenomena as the Earth moves beyond the strength of its magnetic field. The primary cause for the decompression is that magnet of the Earth attempts to restore itself to the 180° position between poles. However, this attempt is short lived; the friction of moving surface layers with the molecuumsphere stops the process from completion. They become "satisfied" after reaching equilibrium within the demand of friction and magnetic energy. This magnetic imbalance is ultimately restored by the soft magnetic affect upon the hard magnets via electron movement,

Chapter 18: Peleg's Earthquake Overview

Earth's Shell Phenomena Set

As stated earlier, there is a thin layer between the mafic and ultramafic layer that scientists call the Moho. Within this layer there exists gravel. The gravel provides a lubricant between the mafic layer and the ultramafic material allowing the mafic plates to move upon the surface. However, there is much more to be said about the Moho.

The Moho formulates during the magnetic flip as the ultramafic layer moves in response to an external and internal magnetic stimulus. The mafic layer has magnets as the ultramafic layer, but they are much weaker. Ultramafic magnets strength comes from an iron-like electric magnet scenario, and they are locked somewhat into position within the shell like a permanent one. The mafic magnets are locked into their shell, but their formation disconnects as it is further from the magnetic charge of the molecuumsphere.

Another interesting factor is that the magnetic material is discontinuous as the heat from the asthenosphere destroys the magnetic properties of matter. The heat does not destroy magnetic energy; it destroys the ability of "cooler" (non-plasmatic) materials to replicate a magnetic field. Even so, the ultramafic layer does become a strong magnet. Originally, it continues into the mafic magnet. The mafic magnets also move via inertia and friction during this phenomenon. Most of the magnetic energy moves with the molecuumsphere holding the ultramafic magnet to its course minus the mafic magnet.

However, the mafic magnets still exert force and the external magnetic stimulus affects its polar positions. Today's scientists have identified the phenomena located in the "Bermuda Triangle" and near the Mariana Fault as magnetic poles, which are loosely 180° apart.[3] Unlike the suggested reasoning by scientists, these mafic magnetic poles initially formed before the magnetic and rotational poles moved to form dry land. Their movement later became controlled by movement of their independent mafic plates.

The second phenomenon to introduce is the **Felsic Moho**. The felsic Moho is not a thin layer as the Moho, but it is indeed a separation between two rock layers. As discussed earlier, there are two major reasons that contribute to its existence: The compositional differences that resulted in a different contraction rate in response to coolness. An enhancement occurs by the later tightening of the mafic layer as it contracted by its cooling during creation. The second is inertia of the cap to the movement of the crust underneath. The net result is that the felsic layer is able to act independent to the mafic and ultramafic layers.

There is yet another Moho-like formation to discuss. This formation can be described as a soft Moho. It is the separation between the asthenosphere under the ultramafic material and the molecuumsphere. The bottom edge of the asthenosphere is liquid magma. The surface above the molecuumsphere is also liquid in nature being lavor. Vapor forms between these layers as the elements pass from being lavor to liquid. However, this kind of transition is not permanent as the elements also move from liquid

Part 6: Creation of the Macrocosm

back into the lavor state passing again through the state of being vapor. This interface between the molecuumsphere and the asthenosphere is the Moho of the asthenosphere or the **asthenomoho** (as-thee-no-mo-ho). As we observe, the asthenomoho is not like the one between the mafic and ultramafic layer. Instead of using gravel as a lubricant, the asthenomoho uses liquid and vapor as a lubricant. Friction within the asthenomoho is seen as almost nil, but not nonexistent, in comparison to the other Moho formations. If the asthenomoho was truly frictionless, the surface of the Earth would become motionless because of external friction. The molecuumsphere, with its internal mass, controls the rotation of the planet. It is the steady application of the frictional force provided by the molecuumsphere that returns the surface back to a steady rotational axis as seen in today's world.

This gives four layers in the crust not two. Three, which are continuous and the top discontinuous layer moving independently during the Peleg Earthquake. However, it is the upper three layers that alter the face of the Earth. The felsic layer depends upon the force of the mafic layer and momentum for its movement. The ultramafic layer transfers energy to the mafic layer from the molecuumsphere. Generally, we will not consider the molecuumsphere as part of the movement. The magnetic energy shifts without affecting its rotation. The angular movement of the molecuumsphere remains constant. Even so, the molecuumsphere rotation will provide some input.

The general scenario is motion operating in a four layered plain. The bottom or fourth layer has the final prevailing vector of direction, moving the third layer magnetically in near unison but not rotationally. The asthenomoho provides a near frictionless rotational input outside the magnetic hold. The second layer is caught between the movement of the third layer and the movement of the surface layer. Each layer lags behind the movement of the layer underneath. After the molecuumsphere stops responding to the magnetic stimulus, the ultramafic layer continues to input its vector to the assembly of layers. The top two layers are the first to rejoin in vector considerations as there is more friction between them. The third layer joins shortly afterward having more friction with the second layer than with the bottom layer. Lastly, all three layers are accelerated by the vector of the bottom layer until they are all four back in unison.

Our next phenomenon describes the nature of the unified felsic landmass. Previous descriptions by other documents indicate brittleness of the felsic material. Contrarily to previous concepts, it appears to be surprisingly ductile (able to stretch out into thin wire-like strands). Mafic rock behaves more brittle than the felsic layer via density of material, even so, the mafic material is not as brittle as we might suppose. Imagine a non-elastic (a quality of a substance to stretch without returning to its original shape) stretchable rubber band with cracks at various locations and of different depths before being stretched. Unlike a true rubber band, it does not spring back into shape. If the crack completely severs the band, no stretching occurs. The band will stretch with cracks in it. The extent of the stretch per crack depends upon the severity of the crack

Chapter 18: Peleg's Earthquake Overview

and the magnitude of the force pulling the band apart. For instance, a crack of the same depth pulled by a slow force will stretch farther than one that pulls fast. This is because the molecular structure of the material has time to readjust before the limits of its adaptability are reached. Because of this nature, it has the ability to form string-like formation, showing it to be very ductile. Examples witnessed today are the Sandwich Island formation, Aleutian Islands, and the Kuril Islands to name just a few.

Another phenomenon that occurs is opposite in nature to the ductility phenomenon. This is the ability to wrinkle via pressure. The pressure normally exerts laterally to the substance. As this occurs, the altitude increases, and the surface area decreases. The inverse is true with ductile phenomenon: the landmass gains surface area but loses altitude. The gain in surface area hinges upon its ability to stay above water. Not only can the cap wrinkle, but it can also slide over land. However, during this process it also wrinkles both the felsic and mafic material.

Lastly, the felsic cap's Moho has uneven movement factors. There are three major inhibitors of movement. The Havilah spiral ridges have the greatest strength; the original center under Eden is second, and lastly the Hudson Comet Crater. The least inhibited is the outer edge of the Kadummagen shield. Of the three major inhibitors, the most successful are the Havilah ridges, as they exist at and near the rotational center of the mafic plain. By these, the main landmass moved more with the mafic plain than the pieces that break away. In fact, the breaking away occurs as the difference in friction between the various felsic landmass pieces and the mafic plain become extreme.

The movement of these felsic fragments is also controlled by another factor. This is the resistance of the surrounding ocean and settling back unto the mafic plain. Smaller pieces are moved farther from the landmass given the same conditions of mafic stimuli as they easily settle back onto the mafic plain which is still moving. The Hawaiian Ridge is a prime example which formed the original coastline of the Australian felsic fragmentation continental sheet.

Rotational Phenomena Set

Movement of a spherical surface within the confines of a sphere always results in some rotation. The rotation phenomena concerning the Earth during this period of history have some aspects seldom considered. Usually, we consider only a single rotational axis. During the "dividing of the Earth," we have four major axes to consider. The first obvious axis is the original rotation of the Earth. The second is the rotation caused by the magnetic forces. The third is the rotation of the felsic cap when it becomes its own entity. Then we have the rotation of magnetic response as the two planets pass each other.

The first phenomenon that we must consider is the nature of two rotational axes existing upon the same sphere. Our first consideration is the nature of the two axial

Part 6: Creation of the Macrocosm

movements. We may assume that the two axis work independent of each other in a stationary style. However, this is not the case. We have a major rotational axis and a minor rotational axis. To visualize this phenomenon, we will start with a motionless sphere. Our next step is to establish a grid over the sphere with longitudes and latitudes, placing the equatorial latitude horizontal, which does not move, with the sphere. Instead of starting the rotation upon the polar axis of the grid, we will start our rotation sequence at the equator. If we could distinguish features upon the sphere, we would see a vertical rotation of the sphere. Now, we are going to shift the rotational axis around the equator. As this occurs, the sphere never stops rotating around its axis. As noted earlier in flipping an axis $180°$, the rotation in respect to the grid reversed itself without stopping the original spin. Actually, the flip in rotational direction begins at $90°$. At this angle the rotation is perpendicular to the original movement. When we move another degree farther, we start to see an inversion of the original rotation. This is the nature of the magnetic flip. By this nature, we understand that the process did not take more than a day; otherwise the surface would be totally destroyed.

Our second phenomenon in this scenario concerns the off centered rotation of the felsic cap. Instead of rotating in the center, the center of rotation is about $30°$ off the center within the region known as Havilah. This means that the western region of the felsic cap moves at a greater linear speed even though it does move at the same angular velocity. Land east of Havilah experiences a different phenomenon as it moves against the rotation of the magnetic flip. Initially, this region experiences the greatest amount of catastrophic reformulation.

Jupiter Arrives

As Jupiter continues to attempt to pass by our sun, it becomes trapped by the gravitational pull of the sun. Its initial trajectory, even without the gravitational pull, would have crossed the asteroid belt and even to the Martian orbit. With gravitational pull of the sun, the trajectory caused Jupiter to pass near the Earth's orbit. However, it did not come into close contact as Venus did to the Earth. If it did, the Earth would be orbiting Jupiter instead of the sun.

Another interesting observation is that Jupiter was moving slower than the Earth during this encounter. Actually, the Earth had to catch up to Jupiter in order for the phenomenon to occur. This the opposite of the most encounters with comets, even Venus moved faster than the Earth. Again, this takes planning on the part of Elohim to keep the Earth from being totally destroyed.

Jupiter has an extremely large magnetic field, which affects the Earth's atmosphere long before the ground quaked. As noted earlier, water molecules are magnetic; this is because water molecules are bent in structure. The two hydrogen atoms do not collect on opposite sides of the oxygen atom. They collect at an angle of about 105 degrees

Chapter 18: Peleg's Earthquake Overview

apart. This gives the water molecules the ability to respond magnetically. The magnetic response is that the water molecules draw toward the object along with the gravitational pull. The result is an excessive buildup of water vapor above the ozone. This excess causes the ozone to buckle under allowing the water vapor to return to the Troposphere. In turn, clouds of rain cover the planet again. It began to rain like in the flood days. However, there was less of it. Shortly afterwards, the landmasses began to move. As we shall observe, the entire breakup of Kadummagen occurred in one day. Afterward, other phenomenon occurred initiating the ice ages.

Preceding Jupiter's interaction with the Earth, it traversed through the solar asteroid belt and the Martian orbit. When Jupiter crossed this asteroid belt, it disrupted the orbits of many asteroids surrounding the planets immediate path. Numerous asteroids crashed into the planet's surface and its moons' surface. Many more asteroids formed a trailing stream following the planet. After cutting through the asteroid belt, Jupiter passes close to Mars. As it crossed the Martian orbital path, it temporarily disturbs the Martian orbit. Mars runs into this stream of asteroids following Jupiter. As scientists have noted, half of the Martian Surface is bombarded with meteor craters. On the other hand, volcanic activity prevails in covering the surface on the other side of the planet. Many meteorites would have partially orbited Mars before impact. These would be covered by the later volcanic activity.

Visualize the width of the colliding asteroid stream being about half (at most) the diameter of Jupiter. The diameter of Jupiter at the equator is about 88,800 miles. The orbital speed of Mars is nearly 930 miles per minute. An asteroid stream at 44,400-mile thick would take Mars a little under 48 minutes to pass through. Since Mars rotates a little slower than the Earth, its rotation would be only about 12 degrees (1/30 of a circle) during that time. This calculation is the extreme large-end scenario. The thickness could be much less depending upon the distance from Jupiter that Mars crossed the trailing stream and the thickness of the stream.

Consider the observation that one hemisphere of Mars is indented with a multitude of craters while the other hemisphere is relatively clean of such craters. Some contend that this crater phenomenon occurred because that Martian altered orbital aphelion (the farthest point in the orbit from the sun) runs into the asteroid belt. The primary problem with this reasoning is the following: if Mars moved into the asteroid belt, its orbit would hold it within the asteroid belt for quite a lapse of time. Within that period, Mars would have to rotate at least once due to the nature of an orbit before the orbit's departure from the belt. Even if the orbit could only pull asteroids for only one degree of its orbit, Mars would rotate nearly two times. This rotation would cause craters to form all over the planet relatively evenly.

When we examine the trail of asteroids following Jupiter that crashed into Mars, we find that among these asteroids was one sizable planetoid, which crashed into the Martian surface. From geological observations, we can conclude that at the place of the crash,

Part 6: Creation of the Macrocosm

there existed an ocean-like body of water. After the crash, much of the water evaporated as the planetoid broke through the crust into the inner magma ocean. The water provided the necessary quick cooling process holding the interior plasmatic pressure intact. Even so, the destruction upon the surface was severe. Moreover, all of this effort was in vain.

Before this occurrence, Mars had an atmosphere containing large quantities of oxygen, carbon dioxide, and water vapor. One important item is that the quantity of carbon dioxide (CO_2) within its atmosphere was much greater than that of Earth. Knowing that CO_2 causes the Greenhouse Effect, on Mars, this effect greatly increases with its larger quantities. If it were not for this effect, Mars would have been a planet in a perpetual severe ice age. Instead, its atmospheric temperature was comparable to Earth's atmosphere. Secondly, its atmosphere was also much greater, event greater than that found on Earth today (probably near twice). Lastly, humans from Earth could live on Mars without altering atmospheric conditions for the living environment with enclosed domes. Whether humans did or not live on Mars cannot be scientifically proved because of the extreme catastrophe that occurred during this period of time. Any such record would have to be found on Earth.

The crash of this planetoid cracked the opposite Martian planetary wall via mass of crashing asteroid. These cracks allowed the magma from the inner ocean to seep toward the surface via inner plasmatic pressure. The heat weakens the upper crust though its internal fractures releasing the pressure from the core. It does not take long for this magma to seep to the surface; otherwise, it would have cooled and sealed the cracks. As we observe upon the Martian surface, it found four locations to spew out magma. Unfortunately, the inner magma ocean was relatively thin and ran out of material holding back the plasmatic pressure. Plasma then spews out from the volcanoes into the atmosphere. The nature of this plasma was a mixture of heavy noble gasses with a large quantity of iron. As the iron plasma cooled, it requires electrons to fill its atomic structure. The closest electrons primarily were attached to the atmospheric molecules, which was primarily oxygen. From this, hyper-oxidation forms of rust form (iron with many oxygen atoms attached). In essence, the iron plasma captures the atmosphere of Mars. This interaction formed particles that rained to the surface and painted the surface red as seen today. Another observation of this "rain" is that its behavior resembles that of a multitude of projectiles shot from the ground into the air rather than formulating in stillness above the atmosphere and dropping to the ground. The image is that the plasmatic iron atoms shoot out from the volcano. As it moves in its trajectory, the iron atom collects oxygen atoms. This makes the particle heavier altering its course downward to the surface. Upon crashing upon the surface, the collection of particles splatters its "paint" upon the surface.

This is not the end of the planetary interaction with Jupiter. To be close enough to the asteroid stream (before their dispersion via their individual trajectories) generated by

Chapter 18: Peleg's Earthquake Overview

Jupiter for such destruction, Jupiter would also affect the orbit of Mars. Jupiter pulls Mars toward the Earth's orbit; the reconfigured orbit seen today is closer to the sun than the original Martian orbit by accelerating the planet. In Essence, the Martian orbit stabilized before Mars could return to its original orbit. If Mars had reached its original orbit, it would have continued moving outward away from the sun, never to be seen again due to its velocity.

Scientist have noted that Mars has had a period of time when its orbit was greatly disturbed to the point of passing close to the Earth at regular intervals. This coincides with Jupiter pulling Mars out of its orbit. As stated earlier, its orbit stabilized to its present day's course. It moves closer to the sun by about seven million miles as its velocity increased by the pull of Jupiter. We can derive this from examining the nature of orbits. For example, the orbital relationships between Uranus, Neptune and Pluto are nearly exact one billion miles. As stated earlier, Venus was not part of the original equation of the inner planets. Using the same near exactness among Mercury, Earth and Mars, we get 36, 92 and 148 million miles. However, today Mars is only about 141 million miles.

The catastrophes that the Martian perihelion (the closest point in its orbit toward the sun) caused on Earth are minor when compared to Venus and Jupiter. The Martian catastrophes were not caused by magnetic disturbances but from gravitational disturbances. Consider for a moment that Mars had a magnetic field (which it does not have due to the eruption of the plasmatic core). The magnetic flipping interaction between Mars and Earth could only occur once. In order for the Earth's magnetic poles to flip, the magnetic poles had to start in the same alignment. The purpose for the magnetic flip would be to reach the optimum magnetic positions. Once achieved, there is no more reason for another magnetic flip to occur. From this standpoint, we observe that this kind of disturbance could only happen once.

The gaps, in the asteroid belt from both entering inward to a closer orbit and leaving by Jupiter, eventually refills, as the orbital speeds of the particles vary. This explains the disbursement of asteroids out to the orbit of Jupiter and inward from the orbit of Mars. This is because outside the range of those asteroids trapped within the gravitational field of Jupiter are those affected by the gravitational field but not strongly enough to become trapped.

The question then arises: Why would Elohim destroy the Martian atmosphere? HE could have chosen a different path for Jupiter and accomplish the same task. Perhaps, pre-flood man developed space travel. We know little about the technological advances before the Flood of Noah. The Torah is comparatively silent on this issue, but archaeological data indicates that it had technological advancements surpassing and often dissimilar to the capabilities of today. Scientists, promoting evolution, tends to paint them as primitive beings. Imagine, people escaping the flood by going to Mars, and being forced back to Earth to meet annihilation by Edom (descendants of Esau)

Part 6: Creation of the Macrocosm

and Israel later on in history, using technology alien and even impossible to us. Another answer could be to keep us from colonizing Mars bringing sin, yet, to another planet.

Meanwhile back on Earth, observations of Jupiter were never beyond that of being a bright star that was growing in brightness. Note: there is no documentation substantiating the approach of Jupiter (or any other planet) toward the Earth. One reason is that languages were freshly confused. Another reason is the cloud coverage was so extensive during its interaction. As such, there were no writing conventions other than ancient Hebrew (before Hebrew was known as Hebrew). The Hebrews saw the entire event as an act of Elohim, for it is. Others thought that Elohim went back on HIS Word as water washed over areas of the fragmented landmass. Any external account from other sources would probably be infused with pagan deities obliterating the account.

Jupiter does not escape without encountering changes. After passing the Earth, it continues its modified trajectory toward the sun. Its path is not directly aimed at the sun; otherwise, Jupiter would crash into the sun. However, it did come within the vicinity of the orbit of Venus. In doing this, the magnetic energy of Jupiter interacts with the sun and flips. Now, the north pole of Jupiter faces the same direction of Earth. Again, the magnetic alignment with the rotational axis is thrown off. As observed, the rotational axis is $3°$ the solar rotation and magnetic axis is about $7°$. The result is that Jupiter's axis flipped about $176°$ instead of the $180°$ required to maintain its magnetic axial alignment with the rotational axis of Jupiter.

Time Measurements During the Interactions

To determine the duration of time that occurred during Elohim's "dividing of the Earth," we must find measurements that are unaltered by friction and secondary inertia. In other words, we cannot purely measure the distance that the felsic and mafic material moved to derive time. We have only two factors that we can count upon for this data. They are the rotation of the Earth by the molecuumsphere and the movement of the magnetic pole, itself. However, these can only be reasonably estimated as we still need to rely upon surface manifestations for the data.

Despite the complex movement of the surface layers of the Earth, there is an element in the nature of the magnetic flip that simplifies the problem of finding the starting point. That is the magnetic flip has a rotational axis. As such, there is no vertical movement at these locations. Because of momentum and friction the ultramafic layer keeps the rotation of the Earth despite its rotation. Recall that within the molecuumsphere rotation was impervious to the shift as electrons shifted within the mass. The ultramafic magnetic hold is as polarized points exerting force upon its layer. Along with the slow acceleration rate, the rotational axis of the Earth acquires a grip on the surface movement. While the mafic layer lags behind in the flip, it shares the same initial

Chapter 18: Peleg's Earthquake Overview

angular direction of the ultramafic layer in respect to the axial rotation of the Earth. We should also note that in the Torah, there is no mention of a "long" or "short" day associated with this event. By this, the flip expressed upon the surface moves with the normal rotational axis.

Our first point of interest is the starting position of the flip. Obviously, it started at the rotation axis and moved "southward," orientated from Havilah, in our orientation from the southern rotational axis northward. Naturally, the opposite pole moved in the opposite direction. For our point of reference, we will use the Havilah rotation axis and count it as moving northward. Moreover, the magnetic pole can only move north. It moved to a position about $23.5°$ from the northern rotational axis, plus or minus $3°$ depending on the rotational axis of Jupiter is to Earth. Since the southern magnetic pole existed outside the Antarctic Circle in 1831 AD, we can assert that we need to add instead of subtracting from the tilt of the Earth. For simplicity, we will assign the angle of $24°$. The actual movement of the magnetic axis is half of the measured arc from the rotational axis giving about $12°$.

We still need to determine the position of the felsic cap to this angle. According to our constructed map, the angle is $11°$ east from due "north" (see page 292). We allowed one degree for the transfer of energy to the surface. It moves in line with the division between Antarctica and Australia. We will also need to look at the coastal region in which this division terminates. This brings us to Fijiland and its associated pieces. In today's geological placements, we find that the upper intersection of the coastline and the division between Australia and Antarctica a little below the Samoa Islands (about $172°$ West and $14°$ South). The "dividing edge" is found south of this location known as the Tonga Trench and the western coastline of New Zealand. Using this trench, we project northward and find the Hawaiian Ridge existing at a near $90°$ angle to the trench. Using the island of Hawaii, the largest piece of the original coastline residue, in conjunction with the Tonga Trench for angle analysis, we find little variation in angle.

Now we need to determine the original location of Hawaii to the felsic cap. To accomplish this task, we project a line back to the Tonga Trench from Hawaii. This line is not parallel to any longitude line, so we need to find a latitudinal cross-reference. For this cause, we need to know how far away from the rotational axis, Hawaii was, as part of the felsic cap. Since the coastline of the felsic cap ranges from $47°$ below to $6.5°$ above the equator and Hawaii existed near halfway between the two extremes, we get about $20°$ below the equator. Depending on the way we draw our lines, we will get a measurement of about $170°W$ to $172°W$ (west). This is our first measurement.

The second measurement is at the end of the magnetic flip. For this information, we need to examine the Arctic Ocean Floor. Looking at this region we do not only see the end of the flip but the compression of the poles as well. For our purposes, we will choose the Mid-Polar Ridge. Upon the surface, it aligns with the eastern cost of Greenland and crosses the northern coast of Russia at the peninsula east of Tiski,

Part 6: Creation of the Macrocosm

Russia. Our second formation to examine shows the secondary western movement of Greenland. The Magnetic Compression Great Circle (secondary vector) matches the arc formulated between the locations Osmo, Norway and Inuvik, Canada. By this, we know that the magnetic pole was in Norway before the compression commenced into action. These two lines form an angle of about $65°$. This is $25°$ below the expected $90°$. This is because the Mid-Polar Ridge formed from the position of the rotational axis and the magnetic compression angle from the position of the magnetic pole, which is near the expected location from the rotational axis.

For finding the longitude of the magnetic pole, we need to work backwards from the magnetic compression. The initial expression of the compression is that of moving within the framework of the offset form the rotational axis. In this, they moved closer together via moving in a latitudinal direction rather than longitudinal. Example, two objects existing $180°$ apart at a $23°$ angle from the rotational axis will exist at the 23rd parallel north and south of the equator at opposite longitudes. By moving one object to the same longitude without deviating in latitude, we have moved the objects closer together. They are actually only $134°$ apart, which is $46°$ less than $180°$. Consider the magnetic poles at 1831 AD were $151°$ apart or $29°$ less than $180°$. This implies that there was no need for longitudinal movement of the axis. The northern magnetic axis worked against the rotation of the Earth and the southern magnetic axis compression movement moves in the same direction of the rotation of the Earth. If the compression were to continue beyond $46°$, another tear pattern would occur as both pole leave their respective latitudinal parallels toward each other. From examining the Earth's ocean floors, we find that this extension did not occur. In this, we establish a limit for the compression of about $46°$. This is the most constrained limit, as the actual angle can be about six to seven degrees more because of the tilt of Jupiter giving up to about $54°$.

Using our $151°$ position, we can reestablish the optimum position. When we add the $29°$ back into the distance to the northern position in the direction of the compression, we find the magnetic pole relocates to the east coast of Greenland. The reason we are able to add the full angle and not the half-value is that the northern pole tore free while the southern pole remains embedded. This is as it should be; however, it does not reach to Norway to cover the entire compression. The reason it does not cover the entire compression is that there is also decompression that occurs only in the northern hemisphere. By this, our reading at 1831 AD is incomplete.

When we measure back to Norway from the northern magnetic axis at 1831 AD, we find that the angle is about $35°$, well within the range of the established limit. However, the present-day coastline of Norway is also stretched beyond its original position. We can approximate the answer by adding about three degrees or about 207 miles. This accounts for the movement forming the Baltic Sea. By this the total compression is about $38°$. This gives us the longitude of about $20°$ east. Recall that the differences in

Chapter 18: Peleg's Earthquake Overview

degrees longitude vary from 69.1 miles at the equator, a Great Circle, to zero at 90° from the equator. Our measurements are orientated primarily to a Great Circle. Since Magnetic North moved from the Southern Rotational Axis to a location 23° from the Northern Rotational Axis, we get an angle of 157°. If we multiply 157 by 69.1, we arrive with the answer of 10,848.7 miles.

Magnetic Flip Calculations. Using the data from Hawaii and Norway, we have the data needed to establish a time factor of the magnetic flip. The measurements are 172° west and 20° east respectively. We convert the eastern angle to a negative 20° west. Then we subtract negative 20° from 172°, which is adding 20°, arriving with 192° movement of the Earth during the magnetic flip. We divide this answer by 15° per hour giving 12.8 hours or 12 hours and 48 minutes for the length of time for the magnetic flip. If we were to measure time by looking at only felsic formations, the time factor would appear much shorter via friction in movement. Moreover, we will acquire different values in time depending upon which felsic formation we used. This factor of time derives from the movement of the ultramafic layer.

Now we can establish the acceleration of the movement. Since the Earth moved 192° while the magnetic flip moved 157°, the average movement of the flip was somewhat slower than the rotation of the Earth. The rotational velocity at the equator is about 1,037 miles per hour. To find the average velocity of the magnetic flip, we divide the rotation velocity of the Earth at the equator by the ratio between the two movements. The result is 1,037 (157/ 192) = 847.96 or 848 miles per hour. The peak velocity is twice the value of the average velocity as the magnetic flip started at zero and ended at zero miles per hour. Therefore, the peak velocity is 1,696 mph.

Our next objective is to find the acceleration of the flip. Since the magnetic flip started at zero and end at zero, the peak velocity occurs at some midpoint between start to finish. For our purposes, we will choose the standard midpoint of distance. Therefore, the peak velocity occurs at 6.4 hours into the motion being half the time. Using the formula v=at (velocity equals acceleration multiplied by time). We need to manipulate the equation to solve for acceleration. The equation then becomes the following: acceleration equals velocity divided by time. Moreover, we need to convert miles to feet and hours to seconds. The equation then becomes the peak velocity of 1,696 mph multiplied by 5,280 feet per mile divided by 3,600 seconds per hour giving feet per second and this is divided by the multiplication of 5.6 hours by 3,600 seconds per hour. This gives the answer of 0.10796 feet per second squared. If we were driving a car at this acceleration, to reach the velocity of 60 mph or 88 feet per second, it would take us 13 minutes and 35 seconds. This is found by dividing 0.10796 into 88, as time equals velocity divided by acceleration.

Magnetic Compression Calculations. The Measurement from the western coastline of Norway to the northern coastline of North America under Victoria Island crossing the Magnetic North at 1831 A.D. is about 40.2°. We will use the established

365

Part 6: Creation of the Macrocosm

acceleration of 0.108 fps² to determine the acceleration of the compression. Given each great circle degree represents 69.1 miles; the total mileage is 2,777.82 miles. Since peak velocity is reached at the midpoint, we divide the distance by two giving 1,388.91 miles. Finally, the magnetic energy of the two poles is pulling in independent directions instead of being in unison; we divide the acceleration by one half. The resulting acceleration is 0.054 fps² (feet per second squared).

We could look at the average latitude, but both poles would annihilate the effect, and there is no initial displacement. Therefore, both v * t and v_o are zero. The formula defining distance then becomes acceleration multiplied by time squared, we can use this equation to solve for time. This gives time equals the square root of the product: two times the distance converted into feet divided by the acceleration. Our calculation becomes the following: $(2 * 1,388.91 * 5,280 / 0.054)^{½}$. This gives us the time of 16,480.6 seconds. We then divide this answer by 3,600 seconds giving 4.578 hours or about 4 hours and 35 minutes. Remember, this is to the peak velocity. We need to double this time for the entire event, which is 9 hours and 10 minutes.

Since we have acquired the time to reach the peak velocity, we now can determine the speed of that velocity. Since velocity equals acceleration multiplied by time, we plug in our values with the adjustment between miles and feet plus that between seconds and hours and arrive with the answer. Our acceleration is 0.054 fps²; our time is 16,480.6 seconds. Multiplying these together, we get 889.95 or 890 feet per second. To get miles per hour, we need to multiply by 3,600 seconds and divide by 5,280 feet. This gives us an answer of about 606.8 miles per hour. This also provides us with an average velocity of 303.4, when we divide by two.

Decompression Calculations. Since the magnetic forces are not aided by an external force, the movement will be comparable to that of the compression. However, the compression is beyond the "norm" of tolerance of the magnetic field generating a greater magnitude of movement. The time and acceleration of the decompression, at this point, can only be estimated. For our purposes we will use the rotational axis of the Earth as a guide to the force of magnetic movement asserted by the poles. The primary reason the force stems from within the mechanics of the Earth. The response is only to a correction of imbalance by magnetic energy existing within its magnetic field. This correction moves with the rotation of the Earth making rotation translucent to the equation. Since the decompression force has no external assistance, we will cut the acceleration of independent movement in half giving 0.054 feet per second. To move 6.6 degrees, the distance of the movement becomes 456.06 miles. We take half of this distance because we are accelerating from zero and end with zero miles per hour. We

Chapter 18: Peleg's Earthquake Overview

will use the standard formula: $d = (1/2)at^2$, solving for t: $t = (2d/a)^{0.5}$. This, in essence, is the square root of the product of 256.06 miles multiplied by 5,280 feet per mile divided by 0.054 feet per second squared. Note: 2(1/2)456.06=456.06. The result is almost exactly two hours. We need to double this answer for the total time of the interval giving 4 hours 00 minutes.

Other Information. Up to now, our calculations have been concerned with the original position of the planet and the movement of the magnetic poles. The ultramafic layer moves nearly even with the magnetic responses via magnetic alignment with the molecuumsphere. The mafic acceleration is expected to be slower than the acceleration of the ultramafic layer. The felsic is the slowest because of inertia and virtually no magnetic influence. However, friction of the molecuumsphere is considerably less than that of the other layers, generating a slower acceleration. The driving force is the ultramafic layer. Calculating the velocity of the ancient mafic poles is much harder as there are many different factors involved even up to today's plate tectonic movement. These poles attempted to stay in unison with the mafic poles, but the inertia of the mafic plain coupled with the friction between the felsic cap and the mafic plain, which pulled the poles away. The result is that we have two magnetic "hot spots" outside the normal magnetic structure of the planet. As noted earlier, one is located near Bermuda and the other near the Mariana Trench. The one near Bermuda was formed originally about 30° outside the opposite of Havilah location, or the **Mafic Havilah**. Contortions of the mafic layer have moved these "poles" from their original positions within the mafic layer, but they do remain somewhat close to their respective Havilah formations.

"Dividing of the Earth" Time Table

Event	Distance (miles)	Peak Velocity	Acceleration (ft/s²)	Time
Magnetic Flip	10,849	1,696 mph	0.10796	13 hr 35 min
Compression	2,777	606.8 mph	0.05398	9 hr 10 min
Decompression*	456	347.7 mph	0.05398	4 hr 00 min
Total Time				26 hr 45 min

* Estimated Time

Part 6: Creation of the Macrocosm

Looking at the "hot spot" near the Mariana Trench we find that the pole moved about 15° less than the felsic counterpart. It also moved eastward more than the felsic cap by about 60°. This is consistent with its continuous input of the eastern movement by the molecuumsphere as it is moving eastward with an angular velocity of 15° per hour. Bermuda's hotspot is more northern in location, this is to be expected as the mafic plain was moving northward through Norway. Magnetic North appears to move through the felsic cap by only 25° before the compression began because of cap's lagging nature.

From the table above, we see that the entire process of dividing the Earth during Peleg's time took less than a day. There are modifying factors that continue to alter the Earth's appearance even up to today. As noted by scientist, if these factors were responsible for the total movement of the landmasses, it would take millions of years. The movement of individual fragments will vary in velocity and in time of movement.

The velocities of the felsic material can be calculated from Havilah, which is located in Tibet. Note that it is generally 30°+ north of the equator for a total movement about 120°. Rounding the magnetic flip measurement of 157° to 160°, we get a ratio of 3:4. In this, we can say that the peak velocity of this region was about 3/4 of 1,938 mph. This rounds off to about 1,450 mph. If we look at the speed of Alaska, it moved somewhat faster. On the other side, we have Australia which hardly moved almost zero, comparatively speaking.

Magnetic Poles of Today

One last point on the magnetic poles, it appears that within the last 100 years that the magnetic poles "broke free" and are attempting to return to their 180° positions from each other and to realign with the rotational axis. The reason that they are able to "break free" is found in the nature of the ultramafic magnets. The Polar Regions have been frozen so long that the plasmatic nature of the ultramafic magnets has greatly reduced via less plasmatic enhancement. Today, the weak ultramafic magnet moves with the magnetic plasma like an electric magnet changing with the movement of the plasmatic magnet. If there were to be another magnetic encounter, the crust would not move, only the plasmatic magnetic pole would flip with the molecuumsphere moving along only like a semi-plasmatic electric magnet. During Peleg's time, the ultramafic magnet acted more like a permanent magnet as the plasmatic magnet underneath was much stronger. It took longer for the magnetic neutrality of the ultramafic layer to return to a given location.

The process behind this is that the electrons are held by the elements have not collected all the electrons needed to become a true non-plasmatic element. The difference of today's magnetic hold in comparison to the time of Peleg is that the elements lack fewer electrons to fill their shell. The reason for the "leaps" of the poles in its oval-like movement during the year is that the electrons are forced to move from one set of

Chapter 18: Peleg's Earthquake Overview

atoms to another set via the demand of the magnetic structure. The image is like a balance scale, an electron is moved from one side to the opposite side, one by one. After reaching the threshold of equilibrium, another electron moves "tipping" the scale and the other electrons run downhill causing the entire assembly to shift. Each time this occurs, the molecuumsphere seeks equilibrium altering the locations of the poles. Notice that today; both poles sit about at the 80th parallel in their respective hemispheres instead of being around 70. The ending positions of these movements will be when they have returned to be in unison with the rotational axis.

Magnetic equilibrium within the molecuumsphere is similar to the nature of gravity. Each atomically free electron and under-filled nuclei contributes individually to the collective effect. However, they also interact with each other within their vicinity of the structure. The electron distribution was originally like a color being 100 percent strength at one pole fading as it approaches the opposite pole to its absolute absence. When the magnetic poles were "bent," the magnetic pressure on one side became greater than that on the side of the greater arc. The objective of this whole process is to restore the original pressure "equilibrium" between particles.

Chapter 18 Quiz

1. Peleg's Earthquake happened when?

    A. After the Tower of Babel
    B. Before the Flood of Noah
    C. During the Creation Period
    D. Long before Adam was Born

2. What Natural Phenomenon caused the Earth to Divide?

    A. Solar Magnetic Activity
    B. Passing of Jupiter
    C. Meteor Crashes into the Earth
    D. Interaction between the Sun and another Star

3. How long did the Magnetic Flip last?

    A. 13 ½ Hours
    B. 9 Days
    C. 1 Month
    D. 50 Minutes

4. At the Flip's Acceleration, how long would it take a Car to reach 88 mph?

    A. About 1.735 Seconds
    B. About 7.9 Hours
    C. About 13.5 Minutes
    D. Little Under a Month

5. The Asthenomoho has Friction that is:

    A. Almost Nonexistent
    B. Greater than that of the Moho
    C. Less than Zero
    D. Unusually Strong

6. (T/F) Peleg's Earthquake was caused, in part, by the Moon.

7. (T/F) Many layers of crust moved independently during the Earth's flip.

8. (T/F) The rotation of Kadummagen aided its demise by the flip.

9. (T/F) Jupiter moved by the earth faster than the orbital speed of Earth..

10. (T/F) Magnetic Decompression is the poles attempt to separate to its norm.

Chapter 19
𝕶adummagen's 𝔇emise

As noted earlier, Kadummagen has already experienced alterations by the flood. The sedimentary material acquired by the flood divided Kadummagen the two shields of Laurasia and Gondwana. These shields, in respect to the breakup of Kadummagen, are primarily irrelevant. They record only the activity of the flood. The felsic cap defining Kadummagen breaks into felsic fragment continental sheets called **felfraxim** (fell-frakseem') or **felfrax** (fell-fraks', singular) from **FEL**sic **FRA**gment and **X** is to combined **C**ontinental **S**heet. A group, or complex, of connected continental sheets form a **felfraplex** (fell-fra-pleks). A fragment of a particular felfrax is a **subfelfrax**. Strips of felsic land masses containing a multitude of island formulations forms **felfraxim belts**. Hierarchical layout of Kadummagen's breakup is as follows:

    A.   The Lone Felfrax
          1. Australian Felfrax
              a. Fijiland Subfelfrax
              b. New Zealand Felfraxim Belt
              c. New Guinea Felfraxim Belt
              d. Java Felfraxim Belt
    B.   The Western Felfraplex
          1. Antarctican Felfrax
          2. South American Felfrax
          3. North American Felfrax
              a. Caribbean Felfraxim Belt
              b. Greenland Felfraxim Belt
              c. Bermudaland Subfelfrax
    C.   The Eastern Felfraplex
          1. Eurasia Felfrax
              a. Indonesia-Japan Felfraxim Belt
          2. African Felfrax
              a. Madagascar Felfraxim Belt
    D.   Icelandic Formulation

Felfraxim are not plates as known in plate tectonics. Normally, they are partially responsible for some of the plate's existence. Plates exist in the mafic layer and should not be confused with the felsic layer's felfraxim. Secondly, some mafic plates also

Part 6: Post Creation Activity

formulate by other phenomena as will be seen. Some felfraxim exist on more than one tectonic plate.

In order to describe the sequence of events, we will use a pseudo division of a timeline numbered from zero to ten. The divisions start at zero and at ten the catastrophic processes ends with the "residue" phenomena that we observe in today's formulations. The reason for this timeline is that there are several events occurring simultaneously at different locations. The timeline markers are not uniform intervals of time but are event orientated. Below is an overview of the entire breakup of the felsic cap. Notice that Africa is not mentioned. The reason is that Africa is the only felfrax that stayed virtually intact. All the other felfrax pieces evacuated away from it. Asia is also not mentioned because all other landmasses evacuate from it. Asia has its modifications when we consider Havilah.

Breakup Sequence

0	Mafic Plain begins to Move
1	Australia Breaks Away
2	Antarctica Captured by Rotation Axis
3	South America Forms
4	Alaska Crashes into Earth's Rotational Axis
5	Greenland Separates from Russia
6	Compression Phenomenon Begins
7	Europe Forms
8	Americas Remove
9	Decompression Begins
10	Miscellaneous Ending Events

Mafic Plain Moves

As indicated earlier, the Hawaiian Ridge is the first indicator of the movement of the mafic plain underneath the felsic cap. The island of Hawaii actually is the corner of the Australian Felfrax facing Antarctica and the Ocean. There are also islands east of this longitudinal parallel that separated later as Antarctica begins its journey away from the main mass. However, this occurs later bringing their latitudes closer to the southern rotational axis.

Another interesting observation at this point, it appears that the felsic cap became independent of this movement almost immediately. This appearance reinforces the idea of a **Felsic Moho**, as the inertia of the cap with the smaller inertia of the surrounding water cause the cap to remain in location for an interval of time. Even so, the Hawaiian Ridge moved away from the cap.

Chapter 19: Kadummagen's Demise

Lastly, the rotation of the magnetic flip also has an axis. This is the dividing line between the mafic plain pulling away from the felsic cap and where the mafic plain pulls into the felsic cap. If we look at our constructed map, we will see that the multitude of islands formulated by the felsic cap is from the side in which the mafic plain moves away from the cap. While the opposite side of the felsic cap, the coastal region holds together. These regions of the felsic cap are only about 120° from end to end. The island forming movement of the mafic plain is aided by the initial rotation of the felsic cap.

Australia Breaks Away

As noted, the island of Hawaii begins a ridge that reaches to Kamchatka via Emperor Sea Mountains, and marks the leading edge of the Australian Felfrax. This ridge is the outline remains left behind similar to a mud-ridge outline left from a removed door mat. In this, it stays with the mafic plain as it moves. This makes it an excellent record of the initial location of the felsic cap. Fijiland is the subfelfrax that fills this corner of the Australian Felfrax. Fijiland was the first major fragment to depart from the cap. Its low altitude (under water) is indicative of its violent separation. In essence, it sunk underneath dividing its mass into fragments. The cause for the violent movement of the subfelfrax is that the inertia of being the leading edge, which caused the ultramafic layer to buckle downward slightly as the mafic plain collected and sank underneath. The largest fragment of Fijiland forms the Fiji Plateau in which their islands sit.

New Zealand separated with Fijiland from the main felfrax. Unlike Fijiland, New Zealand was somewhat protected, causing the mafic plain to build up material in front of its movement. When we examine the ocean floor above New Zealand, we find that there is a near linear trench. This trench continues to a point east of the Fiji Islands just south of the Samoa Islands as expected. This trench, the Tonga Trench, with the eastern coastline of New Zealand is the result of the Australian/ Antarctican fracture.

During this period, Australia was partially pulling apart as Fijiland clung to the present day northwestern coastline. Hence, we have the land feature known as Cape York Peninsula. Meanwhile, New Guinea was also pulling backwards upon Australia forming the region known as Arnhem Land at the northern coast west Australia's notable northern gulf the Gulf of Carpentaria.

The next fragmentation appears in the separation of the New Guinea Felfraxim Belt. The separation of the Java Felfraxim Belt follows this event. These belts separated as the mafic plain drug against the felsic cap. The cap's rotation aids this force. The Java Felfraxim Belt exists as the direction of force shifted enough to generate another fracture upon the Australian Felfrax, separating Taiwan down to Papua New Guinea.

By this time, both Australia and Antarctica were moving away from the main landmass. However, the greatest impact of force moved Australia. This force is moving the mafic

Part 6: Post Creation Activity

plain into Australia as the felsic cap rotates against the mafic plain. Because of this, Australia moved away from the main landmass at a faster rate. Sri Lanka (Ceylon) broke away from the Australian Felfrax before there was much movement and stayed with the coastline of India. The last place of the Australian Felfrax to break away is, in today's terms, the southwestern corner of Australia. This is accompanied with the islands of Sumatra, Java, Borneo, Celebes, the Philippines and all of the Indonesian Islands breaking away.

The image is that of an independent rectangular object resisting the continued momentum of rotation by the main landmass. The reason for the resistance is the friction underneath overpowers the smaller land pieces faster than larger pieces. This also gives the apparent outward thrust more strength as the felfrax moves more at unison with the mafic plain underneath. By this, Sumatra forms as pressure by Australia pulls western coastal land of Southeast Asia away from the main landmass. This result is the formation of the Malaysian Peninsula. Part of the stretching ability of this land feature was the availability of landmass supplied to it as Australia pushed against the future peninsula. By examining the nature of the peninsula, we observe a sudden bend of the arm. This indicates the sudden vector change of pressure as Australia became free of the main landmass and acted independently. This also is indicative of the crashing of Alaska into the rotational axis we shall expound later.

The fracture separation between the main landmass and these two felfraxim made a deep canyon reaching down to the mafic plain. The rotation of the felsic cap pushes against the movement of the felsic cap at the Australian Felfrax. This movement opens the fissure between Australia and Antarctica widening the canyon-like formation. Naturally the ocean waters poured into the canyon. This process was very violent in itself. In fact, water rushed over segments of both felfraxim leaving more deposits of sedimentary material as witnessed by science. The extreme occurrence being within Australia as it is moving the most contrary to the ocean's current.

The path of the rotational axis over the felsic cap plays an important role. As the rotational axis departs from Havilah, it traverses past the Eurasian Felfrax above the region known today as Arabian Peninsula and passed upward toward Greenland.

Antarctica Captured

By the time southern rotational axis was ready to pass over Australia, Australia was already displaced. The next contact the rotational axis has with the felsic cap is when Antarctica moved into it. By the time Antarctica was about half across the rotational axis, the felfrax was captured by the conflict of vectors by its crossing. By this, Antarctica pulls away from the main landmass creating the eastern coastline of Africa. Note the fracturing at the southwestern coastline of Africa forming the Madagascar Felfraxim Belt. This belt of landmass derives its existence from the departure of

Chapter 19: Kadummagen's Demise

Antarctica from Africa and the capture of the Antarctican Felfrax by the southern rotational axis of the Earth.

Let us return to the location in time when Antarctica separates from eastern Africa. Initially, Antarctica drags behind the southern tip of Africa as the mafic plain accelerates underneath. During the initial separating process, the Antarctican "tail" remains attached to South America. The southern rotational axis of the earth crashes into Antarctica shortly afterwards. The location of the impact is known today as Edith Ronne Land. In examining Edith Ronne Land, we find that it is an island connected to the mainland of Antarctica by a fairly long and thin isthmus surrounded by an ice shelf on both sides. The island is the impact zone of the rotational axis. The isthmus forms as part of the path of the rotational axis passing through to its present location. The reason that the isthmus forms is that the energy of the rotational axis pushes Antarctica away from the impact island. Mass draws from the island to formulate the isthmus, until the rotational axis enters the felfrax. From this location to the Antarctic Arm (Palmer Peninsula) is small showing that a larger portion of Antarctica Felfrax existed upon the opposite side of the impact area. Antarctica becomes trapped by the rotational axis as it flings around the southern tip of Africa. The impact phenomenon gives two results: one, it causes the land to fold into itself around the rotational insertion; two, it causes the coastline to tear as it folds around the axis. The largest tear forms the frozen inlet forming the Ross Sea and the Ross Ice Shelf. Another smaller split occurs at the edge of felfrax, it is known today as Pryde Bay. Despite the force involved, Antarctica stays connected to South America during this period.

South America Forms

Eventually, being trapped by the rotational axis, Antarctica twists in response to the rotational acceleration upon the felfrax. Note: this is opposite of the impact influence. The impact velocity was greater than the rotational vectors generated at the polar region as these are angular velocities multiplied by a near zero trigonometric function (the cosine in respect to rotational latitude). The connecting arm of Antarctica pulls down on South America extending the fracture around Africa.

Starting this process, Argentina separates from the western coastline of Africa. As the process continues northward, the width of the landmass removing from Africa increases. Large inlets form as the South American landmass tends to break westward as it is literally being peeled away backwards from Africa. This can be seen as the straightness of Chile from its curvature of the felsic cap. The largest of these forms the inlet that separates Uruguay from Argentina. The process of peeling away South America from Africa actually stops when it reaches the northeastern corner of Brazil. By this time, another process comes into play that causes the corner to exist. There are also other phenomena that alter the formation of South America. These will be expounded shortly.

Part 6: Post Creation Activity

Alaska Crashes

As the mafic plain moves northward (today's orientation), it moves the felsic cap into the northern rotational axis. This interrupts the rotational motion of the felsic cap, primarily by the Americas. The vectors of force invert as some of the landmass crosses over this axis. Example: in spinning a disk, the bottom most part of the edge is moving in the opposite direction from the upper most edge. The same is true for all sets of opposite points orientated from the rotational center. This can be observed by spinning a sphere between your fingers and turning the sphere upside down. The result is the sphere rotates in the opposite direction without stopping and re-accelerating.

The felsic cap continues to rotate as it moves northward. The minimum movement northward possible to occur, before the magnetic compression of the axis, is about $83.5°$. This is because the rotation focus shifted to the center of the disc $30°$ away from Havilah at the "southern" rotational axis, today's view. The coastline is about $66.5°$ from the center. This puts the northernmost coastline at the equator leaving only $83.5°$ of movement to the opposite rotational axis. As noted today, the Sinai Peninsula is the center of the felsic cap, existing at $30°N$. This means the felsic cap pushed about $6.5°$ beyond the rotational axis. It is like the difference between spinning a tire an inch away from the pavement to touching the pavement. It crosses over just enough to experience the impact. After crashing into the opposite rotational pole, the movement of the cap fragments into many different vectors of motion. However, much of the main mass continues to rotate reaching approximately $120°$ of rotation. Some of the movement does not stem from the original rotation vectors of Kadummagen, but from the crashing into the opposite rotational axis. The actual Kadummagen rotation is about $75°$. Some of the fragments of Kadummagen moved less, and others more. This depends upon location and friction.

Visualization of the breakage by Kadummagen at the rotational axis exists by examining the process of breaking a cookie. Take a cookie (disc shaped), and grab both sides of the cookie. Holding each side with one hand and placing the thumbs of your hands underneath the cookie exert pressure upward with one's thumbs while pressing downward upon the edge of the cookie with the fingers. Examine the resulting breakage in slow motion. Tension upon the upper surface is beyond its structural capacity creating a crack. As the crack deepens, the disjoined surface spreads further apart. Note the bottom surface still attaches. Eventually, the crack reaches the bottom surface.

Examine the cookie from a cross section view perpendicular to the crack. The cookie forms an angle less than $180°$. Pull the edges downward until the two pieces form a right angle ($90°$). Closer examination of the cookie, we observe another much smaller right angle. This right angle represents the canyon like formation generated by the crack. Note that the pressure point underneath the cookie is the focal point of the split.

Chapter 19: Kadummagen's Demise

Now, let us translate the same phenomenon in a fashion that it occurs with the felsic cap. Using another cookie, break the cookie by exerting pressure upon its edge instead of underneath. The cookie opens at the opposite end of the pressure point. Instead of a canyon, the canyon engulfs the entire cookie laterally. The same occurs when the felsic cap encounters the rotational axis.

Unlike the cookie pieces, Kadummagen divides into two unequal felfraplex forms. The smaller western felfraplex contains the Americas, Bermudaland, and Antarctica via its attached arm to South America. The larger eastern felfraplex contains Africa and Eurasia. Australia is external to both felfraxim. Recall Australia separated first. Influences upon the Australian felfrax are only by the resulting change in direction of the mafic plain underneath. The two felfraplex forms acquire two opposing vectors. As seen in studying the ocean floor of the Atlantic Ocean, the two felfraplexim moved away from the point of separation virtually equally. However, in examining the vectors involved, we observe that the larger felfraplex actual motion was nearly still as the mafic plain continues its western movement. The smaller felfraplex movement accelerated beyond the mafic direction because the two vectors add together.

When Kadummagen traverses the northern rotational axis, there is a jolt. Direction of the South American fracturing shifted nearly perpendicular to its previous course. The fracture moved directly westward toward the original coastline. South America remains connected to North America as Antarctica stretches releasing the stress of the vector conflict generated by the rotational axial crossing. Vector conflicts exist for the following reasons: Kadummagen's rotation decelerated dramatically, and its western velocity diversified. Diversification of the westerly velocity occurs as the cap divides into two unequal pieces, the northern and southern felfrax forming the Americas.

Australia experienced a different angle of torque. This is seen in the angle of Sumatra and the Malaysian Peninsula. Before this time Australia was sliding outward away from the main mass as the mafic plain pulled it. The image is that of rotating a rectangular object on a flat surface. Rotation shifts from one corner to the next without any real transitional rotation by Australia.

During this period, Antarctica still was attached to the southern tip of South America. Just as Australia experienced a difference in torque applied to it, South America also experienced the jolt. By this time the separation asserted upon South America reached the northwestern corner of Brazil. The jolt caused the fissure to change directions causing Brazil to break away from the western coastline of Cameroon and westward.

By this time, the magnetic flip is approaching its completion. The dividing fracture arched around western Africa separating Bermudaland forming primarily the coastline of Mauritania. This particular arching of the dividing fracture was the existence of a transitional time between the old directional vectors and the optimizing of the new vectors. Before this completes, the dividing fracture encounters the old fracture rings

Part 6: Post Creation Activity

formed by the Hudson Comet. In this, these fractures activate. The primary application of force was to the fracture separating Europe from Africa. There is a suction-like effect to the southern coastline of Europe stretching the landmass.

Another interesting observation is the diminishing of the splitting force as its fracture moves northward. This is directly proportional to the distance from the rotational axis that is the focal point of the split. The result upon North America and northern Europe is that the felfraxim separated slowly allowing the felsic material to stretch. The Gulf of Alaska forms by the initial crossing over the rotational axis by Kadummagen. North America and Asia bends southward during this time while Alaska remains locked in place by the rotational axis northern movement.

Siberia and Greenland Separates

The rotational axis is offset from the direction of the flip. When Alaska hits the rotational axis, it did not stop the move of the magnetic forces underneath towards its goal of reach its optimum at some location near the 23rd parallel latitude. Therefore, the angles of the fissures generated from the impact are offset from the angle of the magnetic flip. They run at an angle $23°$ to the magnetic flip. If we examine line-b on the map above, we will find some interesting connections. One, a Korean Fissure forms as the leading edge of the mafic direction. Two, the southeastern coastline of Greenland aligns with the line. Three, the coastline of eastern North America is aligned with this line. Finally, the southern coastal region of Central America crosses this line.

Our first objective is to determine if Greenland and the associated landmasses moved away from Russia or did Russia move away. Actually, Russia moved away from Greenland. We find this out by looking at the ocean floor of the southern coastline of Russia and our reconstructed map of the felsic cap. We will start our examination using the reconstructed map. The mafic plain initially moves away from the felsic cap center at the location under Australia and Antarctica, as noted earlier. The felsic cap moves clockwise pulling Russia down toward the position of Australia, but the cap and the magnetic pole are also moving northward. In reference to the felsic cap, it appears that the outward movement of the mafic plain is rotating counterclockwise decreasing in angular velocity as it travels. In this, the felsic cap pushes downward upon the Russian landmass. As noted earlier, Korea is the initial hint that there is southern movement in the direction of Asia.

The following events happen nearly simultaneously. The severity of the phenomenon depends upon the angle of the coastal region to the movement of the mafic floor. For this reason, the Korean reaction to the phenomenon is less than that of Kamchatka, and the reaction of Kamchatka is less than that of Alaska. Korea is pulled away from China as there are differences in vectors of China and Korea and the northern landmass. The Chinese vectors are uninhibited by the friction between Russia and Alaska. The island

Chapter 19: Kadummagen's Demise

of Taiwan separates from both Japan and the Philippines. Another chain of islands formulates by the ductility known as the Ryukyu Islands near China.

Japan is pulled away from China and Korea via movement of the mafic floor. The Japanese islands and Sakhalin Island separate from Manchuria and Korea. The southern half of the Sakhalin Island stretches via Japan's downward movement. Korea separates from China via Japan's separation. The Yellow Sea unfolds into existence by the continued rotation by the eastern felfraplex. Korea and Japan connect to mainland China just above Shanghai. The locations indicated by the city of Yancheng, China and Pusan, Korea were once next to each other. After the separation of Japan from the mainland, the main island of Japan, Shikoku, squeezes into the northern island of Hokkaido. The reason for this change is the downward motion by Japan from the felsic rotation.

Kamchatka is, in turn, pulled from Russia and Japan. This leaves another trail of islands called the Kuril Islands connecting the southern tip of the Peninsula of Kamchatka to the northernmost island of Japan. This, in turn allows Siberia to slides downward, it also pulls upon the western coastline of Alaska. Alaska is pulled away from Russia by the continuous dragging of the caps coastline, described shortly. As Russia leaves, the Aleutian Islands form leaving a trail to the position in which Alaska was connected to Russia. This location is a little above the midpoint of the Peninsula of Kamchatka.

All of the previously mentioned coastal changes occur nearly simultaneously with our next set of movements. Between crashing into the polar axis and the compression of the magnetic axis, they comprise all the major changes that occur in our present-day northern hemisphere.

The next objective is to show why the rest of the landmass did not follow with Russia in its southern journey. The answer is that it actually did, but had to travel northward in order to do so. The real stoppers are the rotational axis at the Gulf of Alaska and the rotational axis of the felsic cap. Russia slides down the western coastline of Alaska, during the same time Havilah slides downward over northern India forming the Himalayas and the Tibetan Plateau. This movement is complete after the compression of the poles.

Moving further westward, we find the Scandinavian Peninsula pulled upward to the northern half of the western coastline of Greenland. This particular phenomenon has an added feature the fracturing of the felsic cap by the Hudson Comet. There is a slight increase of strength in the felsic cap within the impact ring section found under Finland. It was a trough in the twisting stress of the impact. The ring generates the separation between Greenland and the North American Islands to the land of Siberia. However, near the stress-trough of the twist are transitional zones. Beyond the Scandinavian Peninsula we find stretching between Denmark and England, and between England and Newfoundland. The land labeled as Zerosia is the largest buffer zone being stretched

Part 6: Post Creation Activity

forming islands and sub-sea-level formations between the Murmansk Peninsula (eastern Scandinavian Peninsula) and Greenland's old position.

Once again, it is not Greenland that moves. It is Siberia and the Scandinavian Peninsula moving together, somewhat, at least. Even though the distance between Denmark and England grows, England is not moving southward. It is moving northward as it is also being pulled in the mafic plain underneath. Moreover, the entire coastline of North America is being pulled northward. This brings us to the southern coastal region of Central America. It also pulled by the northern mass and fractures as it moves northward by the impact of the rotational axis by the felsic cap.

While all these events were occurring, the felsic cap continues to rotate about another $20°$. This drags against the coastline of western Canada over the rotational axis of the earth. It "wrinkled" the coastline starting at the island of Vancouver before it plunged into Alaska to reach its present position. Later the ice ages will accent their features. The Gulf of Alaska moves downward near the 60th parallel north in today's global grid.

Compression of the Magnetic Poles

For the first time within this catastrophic period, the mafic plain is forced to move in dissimilar directions. Before this time, the angular velocity of rotation dominated its movement. Eventually, the conflict of movement within the mafic plain will cause separations, subterranean mountains, and trenches via subduction (plate sliding under another plate, like India). From this phenomenon, the tectonic plates formulate. Notice that they do not always lineup with the felsic separations. The primary compression area of this phenomenon occurs in the Pacific Ocean forming the deep trenches. However, the weakened region between Antarctica and South America does not escape this phenomenon. The southern magnetic axis pushed ahead via magnetic energy forcing the southern surface to rotate slightly faster temporary by about $12°$.

The first event of this magnetic compression occurs with Greenland. Before Greenland severs from Norway, it pulls on Europe. The pull extends throughout all of Europe, Asia Minor, and into the Saudi Arabian Peninsula. This pull was both northerly and westerly. Turkey pulls away from Israel, and Israel rotates upright. Greece pulls away from Turkey and Egypt. Greece also rotates as it is pulled upward. Italy separates from Greece and Libya taking with it the major Mediterranean islands that broke away from Italy's southern Coastline as Italy pulled away from Eastern Europe. The Bay of Biscay forms as Spain rotates away from France's western coastline as France is pulled northward and somewhat westward. England is pulled away northward from France. Also, Ireland moved northward slightly as well. Recall the rubber band with cracks.

During the same time, The Black Sea Forms as Ukraine breaks away from Turkey. The Crimea Peninsula breaks away from the western coastline of the Black Sea. Likewise, the Caspian Sea Forms as Europe moves northwestward. Azerbaijan's capital, Baku,

Chapter 19: Kadummagen's Demise

fits into the southeast corner of the sea in northern Iran. The Aral Sea forms. Imagine the border between Kazakhstan and Turkistan, as it intersects with the coastline of the Aral Sea in two locations, draw back together. This gives the approximate image of the original land formation.

After this point, the Baltic Sea formulates as Norway moves northward still captured by Greenland. The large lakes of Finland (now Russia) formed. The White Sea forms as the Kola Peninsula pulls outward from the mainland. Estonia separates from Finland and western Latvia. The western coastline of Finland separates from Sweden. Sweden separates from the Baltic Countries and today's Poland, northeastern coastline of Germany, and from Denmark. Norway also separates from Denmark. These separations form via the Hudson Comet's impact twisting fracturing phenomenon internal to the fracture rings. These rings form the top of the would-be crater's edge.

England becomes further removed from Norway and Greenland. From this point, the land has tension upon the fracture between Greenland and Norway increases dramatically. The movement of the magnetic poles has overcome inertia. The land had stretched to a certain capacity and stopped. The next force asserted was greater than the fracture could withstand; thereby, it snapped as can be viewed by examining the continental shelves.

The mafic plain breaks along the Norway Fracture. Greenland moves with the American Felfraplex westward. South America collides into the "Mafic Havilah" of the ocean floor. Recall that Havilah is a spiral-like formation of the felsic cap. The Mafic Havilah exists at the opposite pole. The formation moved with the entire mafic plain. Result of this collision is the Peruvian Bulge. This movement westward by the felfraplex included Antarctica even after separation from the landmass. By this, we observe some initial momentum generated by the separation of the felfraplex from the main cap.

Another interesting observation is that the Antarctican Arm did not abruptly break. The isthmus between Antarctica and South America moved approximately a $40°$ arc before it stretched beyond its ductility. However, the distance between the two felfraxim in itself was not enough to cause the phenomenon. Later, other phenomenon resistance the western movement of the isthmus stretched the isthmus "backwards." This backward stretch is in appearance only. The prevention of the isthmus advancement by the mafic plain occurs during the magnetic compression. South America and Antarctica advance about 25 degrees further during this process. South America continues to move a few degrees further than Antarctica via momentum of its mass. The South Sandwich Islands are the remains of the continued connection maintained between the two continental felfraplex. A secondary note is that the mafic plain weakens within this region between Antarctica and South America felfraxim because of the phenomenon's resulting surface distortion of the mafic plain resulting from the southern push by the Americas.

Part 6: Post Creation Activity

Americas Remove

When the compression started, the movement of the felsic cap experienced another jolt. The jolt was even stronger than that of crashing into the northern rotational axis. The Americas illustrate this in the large movement westward. This is because we now have two pressure points at or near the edge of the felsic cap. The space between these two points is entirely at the mercy of the two sources of force. However, one of these forces is not stationary. This is, of course, the magnetic pole moving to a more compressed form. In essence, its force extends down into the Americas because the Americas are rotating into the stoppage. This provides the energy to pull the Americas away from the main landmass as the magnetic pull moves Greenland away from Norway.

We need to retrace to the point before Greenland severs its connection with Norway. The Bermudaland Subfelfrax has not disconnected from North America or from the Caribbean Subfelfrax. As the felfrax of North America moved westward, Bermudaland pulled upon the eastern coastline of America from Georgia to New York. Results of this extra drag upon the main felfrax are the pulling out of Florida from the Gulf of Mexico, and the formation of the Appalachian Mountains. Note: The Florida Peninsula is much bigger than that which is above water.

Along the southwestern edge of the Bermudaland subfelfrax existed the future Caribbean Sea islands (Cuba, Haiti, West Indies, Greater and Lesser Antilles, and even the Bahamas). These remained connected pulling them away from South America. Greenland breaks away. Bermudaland begins to break away, continuing the pull upon Florida. It breaks away from the coastline of America and of the Islands outlining the Caribbean Sea. North America continues to move beyond South America's limit.

The last location of the North American Continent to separate from Bermudaland is the coastal region from Georgia to Virginia. After the separation was complete, the remaining felfraplex continued its western movement. Bermudaland later disintegrated without known cause other than some kind of Divine Judgment. Unlike Fijiland that sank at the base forming a lower underwater plateau, this subfelfrax is totally annihilated. Bermuda and other islands in this area are not part of the original subfelfrax landmass. They are the volcanic remains of its destruction. The felsic material of Bermudaland is today nothing more than what appears to be underwater eroded sediment.

Central America stretches as South America collides with the Mafic Havilah. The Yucatan Peninsula splits from the northern Honduras coastline via Bermudaland pulling upon Florida. Panama's isthmus shows again the ductility of felsic material. It was first pulled straight, and then buckled after the continued compression of the magnetic poles. Actually, the land from Mexico to Panama buffered the conflict of velocities within the American Felfraplex. Baja California separates as Mexico stretches to buffer

Chapter 19: Kadummagen's Demise

the conflict between the Americas as North America continued western motion. Bulging of the western coastline of United States is indicative of this conflict.

Recall that the movement of the separation between Greenland and Norway was about 38°. With the additional influence by rotation, South America moves about 45° from Africa. Another interesting note is that the division between the two landmasses is nearly even, expressed as the Mid-Atlantic Ridge. This shows that the "stopping phenomenon" originates with the felsic friction against the mafic floor. The mafic floor absorbed half the force slowing the movement eastward by felsic cap via force conflict.

Iceland forms as the magnetic compression forces the westward movement of Greenland drags it against the fracture generated by the separation of North America from the main mass via the Alaskan collision with the rotational axis. The image is like the letter "T." Near the letter's intersection is where the mantle crumbles releasing magma to flow to the surface from the asthenomoho.

Africa

The Felfrax of Africa escaped virtually all of the turbulence during the catastrophic events. This is one reason there are so many ruins of ancient cities left desolate in the jungles. However, it also rotated with the cap, and was one of the last pieces to stop rotating. The Arabian Peninsula also rotated with Africa before stopping, forming the Persian Gulf. The Red Sea formed from a radial fracture in respect to the center of the felsic cap, namely the Sinai Peninsula, as Africa continued to rotate. Africa pushed upward into Spain for a short time before stopping. This movement upward by western Africa generated the Atlas Mountains of Morocco and Algeria. The phenomenon also contributed to the fissure underneath dividing the Horn of Africa from the rest of Africa. This fracture extends northward under and beyond the Dead Sea east of Israel, and eastern African Rifts forming valleys west of and including Lake Victoria. The greatest loss to the African Felfrax was Madagascar as Antarctica pulled it away.

During the upward pull of Europe, as noted earlier, Africa rotated upward pivoting on the Sinai Peninsula. Eventually, Africa breaks away from the Europe as it was part of the fracture ring generated by the impact of the planetoid, Hudson Comet, during the flood of Noah. This fracture did not only form the northern shore of Africa, but it continued into the North American Felfrax at the location between Canada and the United States. Lastly, the river passing through Africa and South America was destroyed.

Decompression Influence

The basic concept behind the decompression of the poles is that they were deformed beyond their magnetic strength. After Jupiter's influence became removed enough for

Part 6: Post Creation Activity

the earth to "mend" itself, the magnetic poles attempted to return to a straight angle altered by its new orientation about 23° off from the rotational axis.

The primary effect is found in the present-day Northern Hemisphere. Recall all the activity of the compression was compensated by movement primarily in this particular hemisphere. The same applies to the decompression as the Southern Magnetic Axis movement with the rotation of the earth maintaining strong integrity of the mafic plain underneath.

The movement pulled initially within northern Canada east of the Great Bear Lake. It moved the mafic plate which aligns with the eastern coastline of Greenland. This movement pushes upon the newly formed mafic-like floor between Greenland and Norway. The plate does move northward about five degrees as mentioned earlier. In doing so, it pulls upward on the mainland of North America. The initial result is that it stretches the landmass east of the newly formed Rocky Mountain Range. This occurs as the friction under the mountain range is greater than the eastern plain via mass and contortion. The fractured regions created by the Hudson Comet formed the Great Lakes of Canada and those between Canada and the United States.

As the motion continues, Greenland pulls northward leaving the western island of Ellesmere behind. The image is that this island slid downward the westernmost coast of Greenland. At the same time, Baffin Island is left behind. Eventually, the fractured pieces of landmass north of the North American mainland break away as well forming small islands north of America and west of Greenland. The Hudson Comet radial fractures determined the size and shape of these islands. The drag of the felfrax against the mafic plain generated the final separation.

Meanwhile, it is not all quiet in Europe. Mountains form in Norway as it is being pushed southward. Murmansk Peninsula unfolds from the Russian Mainland. The Alps form as Italy gives the greatest resistance in Central Europe because of its long mass piece. The Caucasus Mountains form between Russia and Turkey. Spain pushes into Africa. Perhaps just as dramatic, the Ural Mountains with the northern islands of Russia, Novaya and Zemlya extending northward form Europe. These come into existence as Europe pushes into Asia via the new ocean floor formed by the separation of Greenland from Norway.

Within the southern hemisphere, the decompression phenomenon had little to no effect. The primary reason is that the southern magnetic axis moved in the direction of the rotational axis during the compression of the magnetic axis. The only real effect of the compression is the formation of the Sandwich Islands by the isthmus joining South America to Antarctica. However, when we look at the ocean floor between South America and Antarctica, we find that the compression expressed itself primarily in the easterly direction as the isthmus buckles under the stress. Because it did buckle, all the energy that would be used for decompression was already expended. The trenches also

Chapter 19: Kadummagen's Demise

were unaffected by the decompression as they too are places where the energy buckles via overlapping layers becoming places of subduction.

Other Influences and Factors

Before finishing the catastrophic scenario, there is a final factor involved in the formation of the present day landmasses. While this force is not part of the catastrophic influence, it provides the final touchup work. The reason centrifugal force provides these alterations are the plate nature of the surface and the ductility of the felsic material. Even the slow movement of the tectonic plates will eventually alter the appearance of the felsic fragments.

Consider the thickness of the felfraxim in comparison to their lateral dimensions. Let us examine an island like Madagascar. It has a length of approximately 1,000 miles and width just less than 230 miles. The average thickness of the felsic layer is about three miles. Scaling this information into a ratio of one inch per 1,000 miles, the object is one inch long about 1/4 of an inch wide and under 3/1,000 of an inch thick. We could stack 300 of these objects generating a height just under an inch. Paper is thicker than this object. Notice also that paper crinkles easily.

However, paper is not ductile. Consider the number of atoms required to compose the thickness of paper to the number of atoms required composing the average thickness of the felsic layer. The ratio is something beyond 50 million to one. The felsic layer has more than 50 million times the number of atoms to link than paper per pillar of atoms. Image trying to bend a block of wood that is an inch cube; it is nearly impossible. Now, imagine a one-inch square beam of wood twenty feet long- gravity will bend it without any effort by the means of gravity. The reason for the difference is the atoms are not truly ridged only nearly ridged. It is for this reason that the felsic layer easily bends to the enormous magnitude of the forces involved.

Centrifugal force pulls the continents toward the equator. Asia and North America sag away from the polar region. Antarctica gradually shapes more evenly around the southern rotational axis. The force pushes Australia northward to the equator. The coastline of Iran continues to shove into the Saudi Arabian Peninsula and increases the indention even more.

Centrifugal influence upon the various felfraxim is the greatest at a 45° angle North or South. The answer is trigonometric in nature as we examine the spherical form. Directly upon the rotational axis, centrifugal force is zero. Increasing the distance from the rotational axis increases the force. Upon a flat plane, this phenomenon continues to manifest itself without any inhibitions. However, the earth is a spherical object. Imagine existing at a point upon the equator. Centrifugal force is pulling straight out from the surface. This means that there is no lateral force applied; even though, the force finds its optimum at the equator. From this information, we arrive with an

Part 6: Post Creation Activity

equation. The factors involved in the equation are the following: The cosine of the angle of location upon the surface multiplied by the distance from the internal rotational axis. Multiply this resulting quantity by a constant representing the force generated by the angular velocity provided by the earth which also varies by the same surface location by its sine.

Another significant alteration in land formations is the amount of landmass available to the surface. There is a reduction of about 16.8 percent. This calculation may reach up to 20 percent. Our calculation bases itself on the total landmass of the surface and their continental shelves composing 35 percent of the earth's surface. Along with this, we are using the 23rd parallel as the measurement of the felsic cap. Some of the continental shelves are stretched formations giving a larger regional area than originally existed. By this, we can assume a larger amount of land disappears. If we measure from the Sinai Peninsula to the tip of Africa, we find that the measurement requires that the felsic cap to be near our estimation. The reason for the large loss of land is not only found in the continental shelves, but in the mountain building that occurred during the phenomena of "dividing the earth." Perhaps the largest of these is the formation of the Tibetan Plateau, in which the resulting surface area is only about a third of the original. Along with this, we have the new mountain ranges that formed. There is also the destruction of Bermudaland.

As noted earlier, if we look at the Hawaiian Ridge, we find another ridge called the Emperor Sea Mountains connecting the Hawaiian Ridge to Kamchatka at the same point that the Aleutian Islands point. Another feature illustrating the same phenomenon is the examination of the northern trenches. Starting at the Mariana Trench and moving upward, we find that at the end of the Japanese Trench that the trenches begin to run parallel to the island chains. Actually, the Mariana Trench once was next to the Ryukyu Island Chain connecting Japan to Taiwan east of the East Sea of China. These occurrences indicate the merging of the old coastline with the new as we move northward. This illustrates another result that Asia rotated back into the mafic plain which was removed. This means that the side of the mafic plain which existed opposite of the felsic cap in the Southern Hemisphere now dominates the Western Hemisphere. If we look at the placement of Havilah (Tibet) on the earth today and measure to the opposite location on the earth, we will find an interesting land formation. It is the place where the mafic "Havilah" moves into South America. Recall that in the formation of the earth, that this side faced the nitrogen plasma. One of the first mining sites for nitrogen deposits is in this region off the coast of South America, southern Peru and northern Chile.

Chapter 19 Quiz

1. Felfrax is the formation of::
   A. Mountains
   B. Continent Surfaces
   C. Felsic Caps
   D. External Atmosphere

2. South America "peels away" from Africa because?
   A. Australia flipped away
   B. Antarctica pulled on it
   C. Alaska Crashed
   D. It was a Summer Day

3. Which of these Events occurred first?
   A. Alaska Crashes into the Rotational Axis
   B. Mafic Plain Moves
   C. South America Separates from Africa
   D. Greenland separates from Baffin

4. When did the Alps Form?
   A. Magnetic Flip
   B. Decompression
   C. Compression of Poles
   D. Cosmic Radiation

5. Africa forms because?
   A. All other Landmasses moved away
   B. Magnetic Drifting
   C. Antarctica Crashed into it
   D. Ice Ages defined it

6. (T/F) Australia rotated away from Asia.

7. (T/F) Antarctica started at the North Pole.

8. (T/F) Alaska separates from Russia because of the North Pole.

9. (T/F) Africa pulled away from Europe.

10. (T/F) The European Alps form during the Magnetic Decompression.

Chapter 20

Ice Ages

Besides the aftermath of slow shifting movement by the mafic plates, there are other phenomena that resulted from the encounter. The most noticeable aftermath effect of the catastrophe, other than the apparent change of global features, is the ice ages. After the magnetic manipulation of the magnetic poles ended, the earth continued their catastrophic transformations including other changes more subtle. Their initial manifestation to human perception was the alteration in calendars. Apparently, the occurrence of the ice ages eludes us as early history focus in the middle latitudes. There is a reason for this. But for now, we will focus on the ice ages.

Foundational Information about the Ice Ages

The purposes of the ice ages were twofold. The primary purpose is that by generating thinner atmosphere. The means to accomplish this task was the act of removing water vapor from the upper atmosphere. If the water vapor were to exist within the lower altitudes of the troposphere, it would be almost impossible to breathe as we cannot acquire oxygen through water molecules. The purpose for thinning the atmosphere was to alter the lifespan of human and animal life on earth deceases to a much shorter period. The lifespan of human beings will be the primary focus as we do not have much in the way of historical data concerning animals. We have records recorded in the Torah that will help us tremendously. This, we will expound upon later.

As far as science is able to determine, there were four major ice ages. The forth ice age is a point of contention as it is relatively longer and warmer. Another point to consider is that the cycles could have diminished with cycles that did not manifest itself as an ice age. The mathematics will back this point as we examine the ice age cycles. Each ice age period has a freezing period separated by a warming cycle. This data has been interpreted into cycles in terms multiplied by a magnitude of extremely large periods of years. Our basic disagreement is concerning the cycles established by scientists is the scaling factor, which most people commonly accept as fact. The beautiful attribute of a scaling factor is that it does not destroy the actual ratios involved. By this, we can look at these cycles as a set of factors. For simplicity, we will divide the time given by 100,000 years. This will give us factors that we can analyze easily. Afterward, we can examine history to derive the scale needed. The following table provides us an illustration of the sequences of ratios established by these factors:

Part 6: Post Creation Activity

Ice Age	Duration Factor	Stage
1st	6.4	Freezing
	6.0	Heating
2nd	15.6	Freezing
	19.0	Heating
3rd	5.5	Freezing
	6.0	Heating
4th	9.5	Freezing
Total	68.0 Units of Time	

Example: Suppose the scale was one year per unit, the number of years of the first freezing stage is 6.4 years. If it was only a half-year per unit of scale, the first freezing stage duration is 3.2 years with the total interval duration of all ice ages being 34 years. In order to understand the various lengths involved within these cyclic ages, we need to examine the different phenomena affecting them. These phenomena focus upon the alignment of the rotational axis in respect to the sun. There are three basic alignments. They are the gyro alignment, the solar-centric alignment, and the kinetic non-alignment.

The **Axial Gyro Alignment Phenomenon** is the present phenomenon controlling the planet's rotational axis. Within this process, the rotational axis exists uniformly within Cartesian coordinate system. Imagine the sun being a small sphere within a large cubical chamber having another smaller sphere rotating around it. From a face of the cube that is perpendicular to the orbit of the smaller sphere, we will construct another plane parallel to this face passing through the smaller sphere at its center. Within that smaller sphere, we can draw a line, to represent the rotational axis, which exists within that plane at any angle except that it cannot be purely vertical. This is because if it were truly vertical, we could not tell if it behaves in the Axial Gyro Alignment or any other alignment. We observe the smaller sphere revolves around the larger one. The plane passing through the sphere does not turn with the rotation. By this, the rotational axis is held in position regardless of its location within its orbit. Actually, the direction that the axis faces is in the direction by which the sun travels through the galaxy. This alignment has become the natural alignment observed in today's solar system. As such, any other axial arrangement will accelerate to reach this favored alignment.

The **Axial Solar-Centric Alignment Phenomenon** describes the alignment existing before the fall of Adam. Operating within this process, the rotational axis responded uniformly using the Polar Coordinate System. The rotational axis is preferably not perpendicular to the plane inscribing the orbit of the sphere because at this condition the influence is unnoticeable. Secondly, the axis is always inscribe within a plane defined by the imaginary line connecting the centers of the two spheres and is perpendicular to the plane of the orbit. In essence, one pole is always pointing toward

Chapter 20: Ice Ages

the solar center. This alignment can be visualized as an air balloon moving around the earth. The basket is always pointing toward the earth. This alignment has the unique effect of zero seasonal change. This lack of seasonal change promoted from this alignment is the phenomenon defining the ice ages.

The **Axial Kinetic Non-Alignment Phenomenon** is a variation of the previous phenomena. As the appellation suggests, the rotational axis has neither gyro nor the solar-centric stability. An external force influences the axial movement generating an axial drift. This drift can move either toward the gyro alignment or away from the alignment in a positive or negative direction. This phenomenon initiated the scenario.

These three describe the primary factors involved within the ice ages. We will also find an initial position, as well as, an initial angular velocity to the movement of the rotational axis. This is followed by its acceleration toward the gyro alignment. Then we shall see the passage through the solar-centric alignment. Then lastly, after reaching the gyro alignment, the momentum of the drift will create some post deceleration to achieve zero acceleration. This will fluctuate back and forth over the exact location of the gyro alignment losing energy each time it crosses generating a wobble that eventually deteriorates to zero.

Now we are ready to look at the intervals under the assumption that scientists correctly determine the derived ratios between freezing and warming intervals. The general format between the intervals is -6.4, +6.0, -15.6, +19.0, -5.5, +6.0, and –9.5 (remember these are ratios and not actual years). The negative values represent the time freezing and the positive values for warming. From a casual glance, the numbers seem to conflict with any organized progression. However, in examining the nature of the transition, we can see a progression form.

Let us assign the value of zero to the gyro pattern meaning that the drift by the rotational axis is stationary within the yearly orbit of the earth. Let us assign the value of minus one to the solar-centric pattern representing one revolution of the rotational axis thus keeping one pole toward the sun. Imagine a rotational value of minus two. A given rotational pole would face the sun once a year just as the gyro pattern; the difference is that the axis is not stationary. At the value of minus three, the number of winters and summers per year becomes two each. At the value of minus four, three winters and summers occur per year. Graphing the data derived from this information gives the following equation: $y=|x+1|$. The equation reads y equals the absolute value of x+1; thereby, if the result of the right side of the equation equals negative one, the answer becomes one. The y-axis represents measurements of angular revolutions in complete seasonal cycle terms and the x-axis represents the angular revolutions the axial drift performs per year. These negative values are approaching zero, where the final gyro pattern is the optimum goal. This reads: y equals the absolute (positive) value of x plus one. The graph's image typically has two line segments meeting at perpendicularly at a 45-degree angle at the coordinates of x=-1 and y=0.

Part 6: Post Creation Activity

Consider the nature of the revolutions from the standpoint of the solar-centric alignment pattern, in other words, the number of years that the axial drift needs to complete one revolution. This can easily be seen in the graph of $y=|1/(x+1)|$ representing yearly seasonal change on earth or $y=1/(x+1)$ representing the actual yearly movement. Both have the graphic similarities of $y=1/x$. The asymptotes (where the values approach infinity, negatively and positively) of the graph are $x=-1$ and $y=0$. As described earlier, it is a near horizontal line at large x-values near $y=0$ that bends to a near vertical line near, in this case $x=-1$. In the graph of the absolute values, both near vertical lines approach $x=-1$ at the same side of the x-axis approaching positive infinity. The reason is that at $x=-1$ we have complete solar centric alignment. At this velocity, we could go to the end of time without a single revolution of the axial drift to generate a seasonal change. In this graph, the x-axis represents the angular velocity measured in revolutions per year and the y-axis represents the number of years that actually transpire per revolution.

Let us examine a rotational drift moving in the opposite direction from the eternal seasonal position in relationship to the gyro pattern. The first whole number value gives the drifting rotational pattern a value of negative two. Since the transitional pattern between summer and winter are symmetrical, the negative direction is perceived within the same framework as the positive pattern. Therefore at $x=-2$ the y-value is actually -1. The result is a continuous line of the equation, $y=x+1$ giving a slope of one ($m=1$, a $45°$ angle). This line represents angular acceleration describing change in direction that the earth's rotational axis from a kinetically disturbed alignment toward the favored gyro pattern. If there were no change in the seasonal pattern, the line would be horizontal meaning whatever velocity the "axial drift" was moving, no change.

We know the instantaneous velocity of acceleration is found in its antiderivative. Taking the antiderivative of $y=x+1$, we get $y=x^2/2+x+c$, and c is a constant. This can be written as $(x^2+2x+c)/2$. We can assign the value of 1 to c because the difference between the rotational axial variation is one revolution $x=1$. The equation then is written as $(x^2+2x+1)/2$ or $(x+1)^2/2$. This is a **parabola** opening upward from a point one unit in the negative direction from the origin on the x-axis (coordinates (-1,0)). We have initially ascribed the acceleration directly to the one-to-one relationship between the gyro and solar-centric alignments. The division by 2 is a scaling constant dividing the equation in respect to the number of years represented that a single revolution axes drift. We will assign the letter d, the initiated force of the drift, for the constant giving the equation $y=(d/2)(x+1)^2$. Graphically speaking, this factor changes the slope of the "acceleration" line to d instead of one, our original showing the direct one-to-one relationship between seasonal change and axial rotation.

Now, we need to look at the d factor or the number of years. This is actual rate in which the drift moves across the x-axis. For an example, it can take 2 years, 25 years, or some other length of time to traverse the segment of the x-axis inscribed between a

Chapter 20: Ice Ages

value of x and x+1. As the length of the transverse time increases, the velocity of the drift decreases. As the number of years increase, the forces operating the phenomenon become smaller. If the forces were great, there would be no ice ages via smaller cycles.

An interesting attribute of a parabola is that it remains proportionately the same regardless of the constant multiplying its equation described by d. This means that the parabola shrinks or grows proportionately depending on the constant. If the constant is less than one but greater than zero, the parabola grows. Inversely, the parabola shrinks as the constant increases. To illustrate this principle, we will look at two points $x = 2$ and $x = 3$. When we square these points, we get y-values of 4 and 9 respectively. Now, let us double the dimensions of the parabola. Instead of $x = 2$, we get an x-value of 4 and a y-value of 8. According to our logic, we can double the size of the parabola by dividing the equation in half giving $y = x^2/2$. Note: at $x = 4$, $y = 16/2$ or 8. Using our second sample $x = 3$, doubling the dimensions we get 6 for the x-value and 18 (2*9) for the y-value. We then insert into the equation the value of 6 and get $6^2/2$, which equals 36/2, which gives us the expected value of 18. Therefore, there is no alteration of the shape of the parabola except for its size.

For the reason of simplifying our examination process, we will utilize the standard parabola $y=dx^2$ because the process resembles a parabola and is consistent with the nature of acceleration. Note: we can solve for the value of d in two ways by multiply x by the years transpired or divide the years transpired into x. It becomes a matter of scaling. The way we are approaching the equation, we will be dividing into x the number of years that the process transpire. Afterward, we can multiply the scale to a more accurate size.

Let the x-axis represent years expired by the drift's journey towards the rest. Let the y-axis represent the axial drift's angular velocity in terms of the solar-centric perspective. We chose this perspective as the seasonal changes are brought to naught when the axial drift moves in a solar-centric pattern. From the parabola describing the velocity, we can also describe the seasonal change in years by taking the multiple inverse of y (meaning 1/y). This graph looks like a cross-section of an upside-down funnel while graphing within finite space. The graph has an asymptote at $x=0$ generating a y-value of infinity ($y=1/0$) indicating the eternal season; as the line approaches the x-axis, it stretches out to infinity as a near parallel line to the x-axis in both directions. This means the drift causes the seasons to change in increasingly smaller cycles that, if possible, the change would occur several times in one day, hour and even within a fractional second as the line approaches the x-axis.

This formula further reflects in the raw data provided to us from geological studies of the ice ages. Note that the second ice age/warming is approximately three times greater than the first and second set. Also, note that the third set is nearly equal to the first set showing a nearly symmetrical pattern excluding the fourth set, which we will expound shortly.

Part 6: Post Creation Activity

Ice Ages Scenario

Our next observation concerns itself with the actual revolutions that occur in the drift by the rotational axis. At first glance, we might assume that each of the three ice ages represent a single drift revolution. This assumption is true for the first and third ice ages, but not for the second. The second ice age incorporates the eternal season phenomenon. Imagine for a moment, a single revolution by the rotational axial drift in which the drift achieves the eternal season. At this point, the drift is making one cycle per year maintaining the eternal season. The average between having a total seasonal cycle and no seasonal cycle is a half-seasonal cycle. Therefore, it takes two revolutions to complete the seasonal cycle of the second ice age. The total count of axial revolutions of drift within the first three ice ages becomes four.

We now need to translate the ratio between the second ice age and its adjacent ice ages into terms associated with the equation formulating the parabola. We have previously noted the ratio to be 1:3:1. The four revolutions then appear symmetrical about the parabola's center giving two revolutions per side. Our next objective is to find the ratio per side. This requires us to divide the three in half giving 1.5:1. We can convert this ratio into whole numbers by multiplying by 2 giving 3:2. Next, we need to find the actual distance that the farthest point away from the parabola's center that these ratios require. The distance reached from the parabola's center by the second ice age has the absolute value of 3 units. The adjacent ice ages reach 2 units farther giving 5 units (3+2). The ratio of these lengths is 3:5. We need to multiply this ratio by a constant d as required by the actual arc length defining a single drift revolution. This will alter the x-values by some factor multiplied by 3d and 5d within the graph.

Now, we need to examine the nature of the parabola as a geometrical object. Imagine two lines of equal lengths joined together forming a "V" shape. Open the angle of the shape, so that the base tip joins the parabola at the x-axis and resting the shape against the parabola. In this, both end points of the V-like shape are touching the parabola, as well as, the place where the two straight lines join. The result is that we cut the parabola at two location using two lines of equal length joined together. First, we will use lengths that are very large. If we were to compare the line lengths of the parabola that exist next to the opened V-shape, we find them to be near equal or 1:1 ratio. Similarly, if we were to make the opened V-shape an extremely small fraction of one unit, the result will be a near 1:1 ratio. However, if the V-shape lengths were to be an half of a unit, a greater variation occurs as the greatest change in parabola's curvature occurs at the base.

Using the basic parabola of $y=x^2$ from the point of symmetry, there is only one set of points that satisfy our ratios. When the uniform arc lengths approach zero the ratio approaches 2:1 between ice ages. This ratio becomes greater as the arc lengths increase. The actual ratio is not quite 3:1 as expressed by the ice age data. For simplicity, we will use the 3:1 ratio.

Chapter 20: Ice Ages

Note: A closer answer would be 71/128 and 47/128 totaling 118/128. While this is more exact, this value is still an approximation. Even if our 1:3 ratio is an exact evaluation, attempting exactness is a detail obfuscation of the data that can be observed today in our weather patterns. When the winter solstice arrives in December, it does not automatically become the coldest day of the year. Moreover, the coldest month is a month later in January. Even so, we can have colder days before the winter solstice in November. The same kind of phenomenon exists with the ice ages. However, winter prevails during the winter months, so does the ice age ratios.

Returning to our scenario, the x-values equal approximately 9/16 units and 15/16 away from the parabolic center. Note also that 15/16 - 9/16 = 6/16, 9/16 + 9/16 = 18/16 and (18/16) / (6/16) = 3; this is the expected 3:1 ratio. Another observation is there is a constant of 3/16 giving 5(3/16) = 15/16 and 3(3/16) = 9/16. The initial value of d is 3/16. The equal parabolic arcs between these values represent one revolution.

In the second ice age (18/16), we are adding two revolutions and treating them as one. The reason, as stated before, is the first revolution brought the seasons to naught and the second revolution brought the seasons back. In other words, the "eternal season" revolution swallows up one seasonal revolution. The first and third revolutions (6/16) represent one seasonal revolution (winter and summer).

The next factor involves the velocity, which is the rate that the earth moves through these ratios. Example: if it took the earth one year to move through the first ice age, then it will take the earth three years to pass through the second. The velocity then appears as a constant multiplying the ratio, or more precisely scaling the velocity/season ratio within the parabola. In other words, the parabola does not distort; it shrinks or grows uniformly maintaining the ratio curvature. This brings us to the existence of the fourth ice age.

The four uniform arcs occupy a portion of the parabola centered on $x=0$. The two endpoints marking this segment are equal in their y-value. Consider the reflective x-values of 7 and -7; when we square their y-values, they both give an answer of 49. If we were to connect these endpoints, we will get a horizontal line. Since 15/16 squared (225/256) is under 1, the horizontal line segment exists under $y=1$. As the rate of movement d decreases, the angle of the diagonal line cutting the parabola will increase remaining attached to the horizontal line at the shorter arm of the parabolic segment. This is because the end of the process has to achieve a 1:1 ratio in the end giving one year for one season. In this, the parabola shrinks in scale to the graphic grid bringing the intersecting horizontal line closer to the x-axis, and the diagonal line connects this intersecting line with the line $y=1$ as both intersect the parabola. The small parabolic arm segment exists toward the negative direction of the x-axis. We also need to move the parabola back to the left side of the y-axis such that the intersection of the parabola with $y=1$ occurs on the y-axis showing that the gyro "rest" has been achieved. Note: $1/y$ still equals the velocity in terms of years per revolution.

Part 6: Post Creation Activity

We now can examine fractional values. For example, imagine a near axial solar-centric movement that takes 10 or 20 years for the axis to make a complete cycle. Imagine a dot moving on the x-axis at an equal speed. If the pace of the dot were to be of a slower velocity, it would take it longer to pass through the four phases defined by the vertical lines as explained earlier.

The behavior of this phenomenon resembles somewhat a car decelerating. Consider a car decelerating from 60 mph to zero mph. It can accomplish this task easily within a couple of minutes. When the speedometer reads 30 mph on its way through the deceleration, it does not mean that the car has to travel 30 miles for one hour before continuing downward. Similarly, in deceleration, within the transitional speeds moving from 20 revolutions to 19 revolutions per year, 20 years do not need to transpire in order to accomplish this task.

Another observation to this phenomenon is the ramifications of taking many years for the axial drift to accomplish one revolution. Just as it explains the long periods of one pole or another pointing toward the sun, it also implies a long transition period between both extremes. It is for this reason that the domination in years of the warming cycle of the second than third ice ages. Actually it is felt beyond the two noticeable periods. Within the last intervals, the times were too short to form any major ice ace or warming cycle to occur.

Equipped with this information we are ready to reexamine the interval sequence. Notice that the first and third sets of numbers are nearly equal in span of time, and largest numbers occur in the second set of numbers. The conclusion is that the process started with the earth rotational axis was initially shifted one plus revolutions in the first cycle, during the second cycle the earth achieved the axial solar-centric orbit and then moved back out of it to achieve the gyro-orbiting pattern. Two numbers or intervals seem to completely contradict this pattern. That is the first and last intervals. If the pattern were not inflicted, these values should be smaller. There is a solution to these afflictions as well.

The first ratio value of 6.5- is greater than the expected 5.5- or the like because there is a counter momentum to overcome. Recall that the earth was moving faster than Jupiter. As the magnetic influence held the magnetic poles captive in the magnetic polar squeeze, this also pulls the earth's rotational axis backwards from the normal gyro pattern. After the earth's change in pattern establishes itself by the exterior planetary interference, the gyro pattern had to decelerate this motion to accelerate the rotational axis toward the "favored" pattern. This provides the larger than expected span of time for the first ice age.

The value of 9.5- operates on a different principle. Note: even though this ice age lasted longer. It was warmer than the others were, as indicated by its smaller ice sheets. This is a hint toward the reason. Imagine many mini ice ages existing close together

Chapter 20: Ice Ages

separated by many small warming ages. For argument sake, we will give them the values of -3.1, +2.6, -1.8, +1.2, -0.8. The last ice age exists because of two factors. These smaller values occur when the axis continues to draw near to the fixed gyro pattern because the symmetry starts at a location less than x=1. Two: The position of the earth during this occurrence favored the rotational axis of the northern hemispheres time facing away from the sun.

Incorporating yet another factor into the equation, the amount of moister available to each freezing cycle differs. Not only the water responsible for the flood froze within the polar caps but extra water also fed into the system. The water from the ocean retards the process of freezing the pre-floodwater vapor fed from the central vapor ring. Ocean water increases the vapor level of the atmosphere offsetting the drainage occurring by the freezing process. The largest amount of ocean water fed into the system was during the second ice age. However, all, if not more, of the water that was gathered from the ocean has returned. This can be determined due to the expansion of the planet and somewhat by the submerged continental shelves. Other factors diminish this particular reasoning concerning the submersion of the continental shelf.

Another item of interest is that the ice ages are hemispherical in nature. It is not that both hemispheres were not affected by the phenomenon simultaneously, for they were. The split is in the nature of the effect. For an example: while the Northern Hemisphere was experiencing the warming cycle, the Southern Hemisphere was experiencing its ice age. Within the Southern Hemisphere, the ice age affected primarily Antarctica which was already covered with an ice sheet. It is by these cycles that the polar ice caps completely engulfed the floodwaters from the atmosphere.

The information we have is that which is measurable. Notice that the warming cycle is omitted after the last ice age. This is because the freezing cycle never returned. However, this does not mean that there was some energy left to the axial drift. Moreover, another factor becomes involved in which the axial drift's momentum generates an additional axial wobble as the earth passes the gyro optimum before the rate of deceleration comes to a halt. However, the effect upon the earth and its inhabitants is unobservable to science as of yet.

Historical Accounts by Life

Lastly, by examining the Torah (B'reshyit (Genesis) 6:32 & 11:11-26,32 & 21:5 & 25:7) and some scientific correlation, we can determine the purpose and length of the entire ice age scenario. Before the flood people generally lived nearly 1000 years. Those born after the flood lived nearly 500 years. After Peleg's time, people lived 250 years and dwindled down to 120-year maximum seen today. The decrease is in result of atmospheric changes both in content and condition. Note: other scriptural data will alter the "timeline" results below, especially concerning Abraham and Terah.

Part 6: Post Creation Activity

Lifespan Chart

Patriarch	Having Children		Total Years	Life after Flood	
	Before	After		Start	End
Noah	500	450	950	0	350
Shem	100	500	600	0	500
Aphaxad	35	403	438	2	440
Salah	30	403	433	37	470
Eber	34	430	464	67	531
Peleg	30	209	239	101	340
Reu	32	207	239	131	370
Serug	30	200	230	163	393
Nahor	29	119	148	193	341
Terah	70	135	205	222	427
Abraham	100	75	175	292	467

Actually, this table can be extended down to Joseph: Isaac lived 180 years, Jacob (Israel) lived 147 years, and Joseph lived 110 years. Note that Moses and others still lived to about 120 years. However, the length of time that Moses and some of the others lived during his period of existence seems to be the exception rather than the rule.

An interesting sideline observation is that Noah, Shem and all the post-flood patriarchs were alive when Abraham was born. Furthermore, Shem (a pre-flood birth), Salah and Eber (both pre-ice-age births) were alive when Abraham died. Another and more pertinent observation is that Peleg and Reu both lived to the same age of 239 years. This item indicates that the thinning effect of the phenomenon was nearly immediate. However, we can only say that thinning of the atmosphere by the ice ages was primarily achieved within the first ice age since both lives (Peleg and Reu) were afflicted in the aging process within the same manner. Recall also that it rained only for 40 days and nights. When the earth was pulled farther away from the sun, this moister snowed via dictates of the axial alignment.

Some might argue that the shortening of the lifespans from 1,000 years to 100 years, although not many reach that age in the modern world, was caused by the change in solar radiation. Assuming that this is true, what would cause this alteration? This brings us back to the condition of our atmosphere. If our atmosphere caused the lifespans to shrink by 90 percent, then something occurred to generate this phenomenon. There are actually several different events that occurred. The first was the cursing of the ground with all of its ramifications and the limitation of the lifespan from being eternal to less than 1,000 years. The cursing of the ground did not cause the change in lifespan, sin

Chapter 20: Ice Ages

did. The change from eternal life to any limited lifespan has no natural cause. However, the shortening of an already demolished physical condition can be and is controlled by nature. The first event was the great deluge which by nature involves water. The end result in terms of lifespans was that the lifespans were cut in half. The second was the dividing of the earth, which does not require the involvement of water, cutting lifespans in half again. This is 25 percent. Then the ice ages affect the water in the atmosphere dwindling lifespans to 10 percent. The point is that it would take an atmosphere 10 times that of today to accomplish that task. Unreal as it might seem, this is near the water account of the frozen water on the earth. Note: the ice that melts into water does not count as vapor returning to the atmosphere as long as it raises sea level.

With all of this in mind, there seems to be no real data determining the actual length of the measurable ice age phenomenon. There are only three other factors to consider. One: it takes time for glaciers to carve. Two: the start of recorded history by those who migrated north after the ice ages (the second migration), the first being after the Tower of Babel fell. Three: the Torah accounts of the Seven-year plenty followed by a Seven-year famine during Joseph's time. The latter time is too short in span and too far past from the Peleg's catastrophe to be directly connected with any of the major ice ages. However, it could occur during the final and unrecorded freezing/warming cycles as it fades into the standard seasonal cycle. Therefore, the selected estimate is 400 years for the measurable time plus another 100 years of immeasurable time. Most of the 100 years involves the overshoot wobble caused by the axial drift's momentum. The seemingly one-decimal precision is only to maintain the ice age ratios observed by scientists with one as the base comparison figure.

The approximate timeframe of the smaller cycle measurements of the ice ages are from 2,400 BC to 2,000 BC. The actual length in time may even be much longer. Consider that history of the northern countries above the Mediterranean countries did not develop until much later. If this were to be accounted, the ice ages plus the uncounted time could have lasted up to 900 BC. This would give Germany time to develop into a civilization to defeat Rome, and the Vikings to develop even at a later time as the final ice sheet retreats. However, this would place the famine of Joseph's time about the beginning of the second warming. Long winters and summers could exist in cycles without glaciers.

A contrary argument to any larger figures is that it would take a much longer time for the gyro pattern to be reestablished. It would create more post ice ages after the three primary ones. The reason is that the axial drift would have to make several revolutions before reaching the gyro position, leave alone the counter momentum measures. Second, the idea that long periods of time being required to melt the ice sheets did not take into account a thicker atmosphere and the heat of a near eternal summer. The last ice sheet extension may have melted at a much slower rate, as the "eternal summer experience" was not a major factor. In this, the timing may still approach near 900 BC.

Part 6: Post Creation Activity

The first human migration met with mishap by the phenomenon. The people who migrated into Antarctica before the separation either died or found a way to leave. These were black people and may have migrated into Australia, New Zealand, or even Africa and South America and mingled with other blacks. In the north, we have the Lapps, Eskimos, Aleuts and similar peoples who survived the brutal ice age environment where shelter and food harshly overshadows any other economic and political development. Their isolation from the unaffected lands southward led to their survival. Others moved back southward away from the cold.

From examining the effects on lifespans, we observe that nearly half of the floodwater was lost before the flood. Venus partially ripped it from the earth passing by. Although, Venus did not acquire much of the atmosphere, the vacuum of space drew away much of the escaping atmosphere. As noted earlier, half again draws suddenly away by the initial freeze. During the ice ages, water vapor coming from the oceans interfered with the freezing of the floodwater via elongated exposures to solar radiation. The last of the water gradually dissipates from the atmosphere in the last ice age. Thereafter, less water fed into the atmosphere by the oceans to interfere with the process. History seems to favors the 1,500 year total, or even up to 2,000 years. We think, "How could the ice ages exist during the existence of civilizations, without any civilization recording it?" Despite our tendency to project modern information conveniences back to the past, we need to understand the profound effect of confusing languages had on technological advances. The following information gives the necessary time intervals for both scenarios:

Ice Age Duration Staged in Years

1st	37.6	112.8		Freezing
	35.3	105.9		Heating
	72.9	**218.7**		**Total Cycle**
2nd	91.8	275.4		Freezing
	111.7	335.1		Heating
	203.5	**610.5**		**Total Cycle**
3rd	32.4	97.2		Freezing
	35.3	105.9		Heating
	67.7	**203.1**		**Total Cycle**
4th	55.9	167.7		Freezing
	100.0	300.0		Mini-cycles & Wobble
	155.9	**467.7**		**Total Cycle**
	500.0	1,500.0		**Total Years of Intervals**

Chapter 20: Ice Ages

Changes in Time

There are two primary reasons for time's measurements to alter: the enlargement of the earth's orbit, and the expansion of the earth's diameter. A year before the breakup of Kadummagen was 360 days per year. Both the length of a day and year altered.

The alteration of the length of day began when the mafic plain divided into plates. The internal pressure of the earth was released by these fractures. The effect on time is that of lengthening the period of one day. The accomplishment of this phenomenon occurs by the separation and consequently rising of the crust. The pressure generated by the cooling of the crust (contraction) over the heated core caused the earth to expand. The percentage of expansion in comparison to the diameter of the earth is almost negligible. The increased time of revolution develops as the linear motion establish at a given radius increases.

Imagine a spherical balloon 10 inches in diameter rotating one revolution a minute. The surface at its equator is moving at a linear speed of 31.4 inches per minute. Let the pressure outside of the balloon drop to a convenient percentage. The balloon expands conveniently 10 percent. To maintain its angular velocity, the linear speed of the surface needs to accelerate. However, there is no force promoting such acceleration because the expansion is a force at a 90° angle to the rotational motion. For this reason, the linear speed remains the same. The distance that a point upon the surface completes one revolution has increased by 10 percent. The distance is about 34.5 inches at the equator instead of 31.4. This causes the angular velocity to degrease by 10 percent.

Using this information, we easily visualize a decrease of the ocean's water level as well. Returning to the previous illustration, imagine the whole surface of the balloon covered with a fixed volume of water. Let this extension be one inch above the surface of the sphere. After the surface expanded 10 percent at the equator, this decreases the thickness of the layer by water. It extends about 0.826 of an inch instead of 1inch from the spherical surface. Consider a volume 10 by 10 by 1 becoming 11 by 11 by 0.826+.

However, the earth's change is not as dramatic (even so, for the environment, this is extreme). It expands only 2 percent. The reason it expands is that the crust formed at a stage of being an extremely high temperature. After the crust hardened, it contracted as it cooled. When the crust cracks, the internal pressure released pushes the pieces outward. Scientists have determined that most metals contract to a measurement around 2 percent between its melting point and its measurement at absolute zero. While our temperatures are hundreds of degrees above absolute zero, they are also over about a couple thousand degrees away from the melting point by most of these metals. As slight as this expansion may seem, the water level does drop approximately 300 feet. Volcanic activity offsets this by a fraction. This fraction is nearly negated by the trenches and downward buckling of the ocean floor.

Part 6: Post Creation Activity

Using the expansion of about 2 percent, we see the surface moving outward maintaining its linear speed as ultimately it is setting on vapor/plasma leaving very little friction. In this, it takes 2 percent longer for a day to complete. This computes to approximately 29 minutes longer per day. This gives an alteration of 7 days per year. Instead of 360 days, it would become near 353 days per year. However, as we know the result should be 365 and a fourth of a day (365.2564 days or 365 days 6 hours 9 minutes and 9.54 seconds).

This brings us to the next cause of change in the length of the year. As mentioned earlier, Jupiter, further away from the sun, pulled the earth. The outward movement was about 3.5 percent farther. This naturally had a very significant effect upon the air quality of the earth. The earth moved about 3 million miles further away from the sun. This gives the colder climates and the polar ice caps. The slower moving in rotation gives a greater difference between night and day temperatures. After the ice ages rarified the water vapor in the air, night and day become even more differentiable in temperature. Another factor is that the air pressure becomes less dense letting more radiation from the sun through during the day increasing the ratio between it and night.

Another observation of time concerns the moon. The moon, before the Peleg's catastrophic earthquake, had an orbital period of 30 days. Taking account for the slowing of the rotational speed, the time would adjust to about 29.4 days. To have an orbit of about 27.322 days means that the moon had to move toward the earth substantially. It moves over 7.1 percent closer, giving a distance about 18,166 miles closer. Instead of being 238,855 miles as an average distance from the earth, it was 257,000 (derived from 257,124) miles from the earth. This increases the ocean's tides upon the earth. If these tides existed before the flood, the northern-most edge of the felsic cap of Kadummagen would have eroded away. In order for this kind of movement to occur, the moon had to be in front of the earth's movement away. Even though the moon was also being pulled away in front of the earth, the earth has greater attraction. While gravity decreases by the inverse square of distance, the ratio of distances between the moon and Jupiter and earth to Jupiter were nearly equal. As stated earlier, it was the ratio between the masses of the earth and moon that gave the earth a greater acceleration/ movement toward Jupiter during that interval.

Finally, if Jupiter was any closer to the earth than the experienced encounter, we would not be here discussing the issue. The earth would be thrown at an extreme vector either away or toward the sun depending of the length of time and position of its hold. It could even have been captured by Jupiter's gravitational force and become another moon to it. In either case, life would have been destroyed. By this, we know it was not just chance that the trajectories of these external objects only passes close enough to do the intended function described by the Torah. A variation of one degree, even calculated from the present day position of Jupiter, in any direction would make the catastrophe too extreme or inconsequential.

Chapter 20 Quiz

1. The Eternal Season occurs during Earth's:
   A. Gyro Alignment
   B. Solar-Centric Alignment
   C. Kinetic Non-Alignment
   D. Gravitational Alignment

2. The Ice Ages can be Mathematically described as a:
   A. Straight Line
   B. Circle
   C. Ellipse
   D. Parabola

3. What is the Ratio between the First and Second Ice Ages?
   A. 1:3
   B. 2:3
   C. 4:7
   D. 1:2

4. What caused the Ice Ages?
   A. The Passing of Jupiter
   B. Earth's distance from the Sun
   C. The Moon's gravity
   D. An Asteroid crashing into the Earth

5. How did the Ice Ages affect the Lifespans of Fallen Humans?
   A. Not at All
   B. Brings it to a Halt
   C. Cuts it in Half
   D. Dwindles it to 1/10

6. (T/F) The Solar-Centric alignment caused the shortest ice age.

7. (T/F) The longest ice age is about three times longer.

8. (T/F) All the ice ages can be expressed by a parabola.

9. (T/F) Lifespans increased after the ice ages destroyed germs.

10. (T/F) Months become longer because of Jupiter's influence,

Chapter 21

Resulting Atmosphere

The flood created much of the present day atmospheric conditions. The Ice Ages finished the transformation. Not only is the atmosphere thinner, the level of atmospheric activity increases. We have rain, snow, tropospheric wind patterns, and increased temperature variations. Atmospheric temperatures cool dramatically. The primary reason for the cooling is the increased distance from the sun by the earth. The Troposphere air masses divide. Instead of having two primary air masses, one per hemisphere, there are now four primary tropospheric air masses. We have the two polar air masses divided by two equatorial air masses. Because of this, the **jet stream** forms in the upper troposphere. Secondly, the air masses break into smaller air masses. These provide a medium in which cold and warm fronts form storms.

The initial concept to grasp is the ramifications of thinning the atmosphere. Imagine a square piece of plywood attached to a large firm mattress spring. Attach the opposite end of the spring to a platform. Compress the spring by exerting a five-pound weight. Let us say it moved the plywood downward one inch. Next, remove this spring and replace it with a weaker one generating the same altitude above the platform. Compress again with a five-pound weight. Measuring the distance in which the plywood moved toward the platform, it would be greater than the first measurement. The same is true with the atmospheric phenomena. It is for this reason that the jet steam did not reach into the troposphere as seen today.

Polar Pressure Phenomenon

The upper atmosphere at the Polar Regions pushes down upon the now thinner troposphere. The reason for this is the lack of centrifugal force at the poles. Imagine being an atom at the rotational equator of the earth at sea level. The speed traveled is about 1,100 miles per hour. The atom's linear velocity increases at the equator, as the placement of that atom exists further from the earth. At a certain distance, the atom reaches escape velocity removing itself from the earth's atmosphere by the means of the rotation by the earth. Actually, the atom lags behind by the means of inertia. Even though, there still is a point that the atom reaches escape velocity. At the same distance directly above the rotational axis, there is no rotational velocity to generate centrifugal force. Likewise, all points near this location have virtually no centrifugal force. Since there is nothing to push the atoms away, the gravitational force holds them to the earth.

Part 6: Post Creation Activity

For this reason, the atmosphere above the troposphere within the Polar Regions is denser.

Note: this phenomenon existed before the flood, but today, the variation intensifies, as the troposphere is thinner via lack of external water vapor within the equatorial regions. Even though, water vapor existed primarily within the confines of the "vapor ring," water vapor exerted pressure upon the entire planet. Modification of the pressure ratios occurs as the other gasses above the polar region remained somewhat intact while the equatorial vapor ring becomes deleted.

The gaseous nature of atmospheres provides another interesting phenomenon. The molecules or atoms of a gas are not confined. Their range of influence is much greater than that of a liquid or solid. Furthermore, they interlace with neighboring molecules. The linear velocity of the atmosphere spreads perpendicular affecting adjacent molecules to its path. Just as one molecule induces another to accelerate, the inverse is also true. The molecule being induced to accelerate influences the previous to decrease in velocity. This influence is locally minute to moving solid or liquid objects through the gases. On the same note, the lack of density provides a pliable substance. Multiplied by the volume of air existing within the atmosphere, it becomes a noticeable surface air mass banding phenomenon.

Actually, air banding occurs before the flood. The most distinguished tropospheric banding was between the northern and southern hemispheres. Upper, pressure of the vapor ring air mass pushes down at the equator "pinching off" the unity between the two lower tropospheric air masses. This trough is still seen today.

If we look at the upper limit of the troposphere in a cross section of the earth, we find generally an oval shape surrounding a circle representing earth. The oval, naturally, has its major axis aligned with the equator. By this, we can easily see that the upper atmosphere pushes down at the poles and that the vapor ring being an extension of the equatorial bulge. The forces forming this shape along with other factors will also provide input into dividing the resulting unprotected and thinner tropospheric atmosphere into warm and cold air masses. Armed with this information, we are ready to examine the formation of the jet stream.

Jet Stream Phenomenon

The Jet stream existed in the pre-flood atmosphere much like that observed on other planets: a thin unmoving band between major atmospheric bands. The atmospheric divisions or banding found in the gaseous atmospheres of Jupiter and Saturn has several bands via their greatness in size. Uranus has a calm undisturbed atmosphere because of its lack of differential pressure and movement between bands. Neptune's atmosphere differs from Uranus only in that a large moon sized meteor impact punctured the shell-like surface.

Chapter 21: Resulting Atmosphere

At this point we need to introduce two more factors in forming the jet stream. This is the gravitational force of the earth and the centrifugal force of the earth's spin. As noted before, the centrifugal force varies directly with the distance an object is into space and with its location in relationship to the equator. Gravity, on the other hand, varies slightly upon the surface being slightly stronger at the poles. There is a location in which the two forces clash in the upper atmosphere creating a trough. As our tropospheric atmosphere thins, this trough drops down into the troposphere. The result is that two air masses form isolated from each other per hemisphere. Let us use one set for examination. The polar air mass in the upper troposphere is isolated from the equatorial air mass, as far down as the trough reaches. The result is that the cooling air of the polar region becomes trapped from the warm air of the equatorial air mass. The polar air mass becomes colder, and the equatorial air mass becomes warmer as it is no longer fed cold air. The air that remains unified in the lower troposphere also becomes affected as the air above the two regions is now different.

Division deepens between the two masses because of the difference in density between the two. The cold air, naturally, is the denser air. This generates two independent air masses with even greater variation by attributes. These differences cause another behavioral difference to occur.

The trough that gives the jet stream its existence formulates from the difference between the polar air mass's and equatorial air mass's definition disregarding temperature and density. Let us examine the atmosphere's relationship only to gravity and centrifugal force. Centrifugal force being a primary factor separating the equatorial atmosphere from the polar atmosphere. The determining factor of the edge between the equatorial atmosphere and the polar atmosphere is the magnitude of the planet's rotation, the gravitational pull of the planet and the planet's size.

On earth, the rotational factor is not as great as seen in the larger planets. The greatest affect of centrifugal force exists between $60°$ N and $60°$ S of the equator. The reason is that the cosine of $60°$ is 0.5 or one half. This means that the fullness of the effect asserted by centrifugal force is felt at the equator ($0°$) and decreases to a half a little under the arctic circles. Recall another factor is that gravity, it is a little stronger within the Polar Regions via the earth is slightly flattened as explained in complex gravity.

For simplicity, we will use $45°$ angle. Take a plain piece of paper and fold it half horizontally and again in half vertically. Then fold it in half diagonally bringing two perpendicular edges connected to the center of the paper together. When the paper is unfolded, there should be four creases intersecting at the center of the paper. Draw a large circle inside the framework of a piece of paper using the intersection of creases as the center of the circle. This will represent a cross-section of the earth. The vertical line is the rotational axis; the horizontal line is the equator; the diagonal lines represent the division between air masses. In the top two and bottom two pie pieces, draw one-inch lines from the edge of the circle outward using the circle's center in uniform arc

Part 6: Post Creation Activity

increments. The result should be the lines seem to radiate from the surface. Upon the remaining four pie pieces draw one-inch lines parallel to the "equator" evenly spaced. This gives an exaggerated effect of centrifugal force. To be accurate, we would be somewhere between the illustrated effect and the radiant effect as gravity is still a factor. However, we now can readily observe the sharp difference between the two masses at the $45°$ angles. This is more pronounced at $60°$.

Unlike Uranus and Neptune, there is a sharp difference between the polar region and the equatorial region. These differences unify the three basic different regions (two polar and one equatorial commonly divided into two bands via the equator). This gives the trough "solidarity." These troughs are not empty, but filled with air acquiring atmospheric pressure determined by the altitude. This band moves generally faster than the air masses on either side. The air within this band fills from the outer atmosphere. Even though the outer atmosphere is moving slower in angular velocity, it is moving faster in linear velocity. The primary reason is that there is an attempted unification of each individual atmospheric band horizontally and vertically in respect to the direction of motion. This increases the linear speed as we move upward and away from the equator. Therefore, the upper air that moves into the jet stream at $60°$ latitude both northern and southern hemispheres, moves faster.

Note: the trough is not empty of gravity otherwise; it would not draw the air in from the upper atmosphere. The lines extending above the circle represent columns of air angled in response to the forces imposed upon them. The centrifugal force removes air that would otherwise occupy the region of the trough. Next, gravity provides the force in which the upper air falls into the vacant place.

Earth's tropospheric bands behave a little different from that seen on Saturn and Jupiter. The earth's jet streams are not like rigid walls but behave somewhat like a fluid tube in that it shifts, as will be seen later. In some cases on larger planets, the friction between atmospheric bands (upper and lower) and jet stream walls generates a multitude of eddies within the band. Likewise, the earthly bands themselves are not totally unified in motion. These velocities gradually change from the center of the band to the edge by the influence of angular rotation and friction from the lower troposphere but not in unison with the lower troposphere.

The next factor to examine is the nature of the winds of these air masses. As described before the equatorial air mass of the upper troposphere moves angularly slower than the rotation of the planet generating a retrograde wind. The edge of the polar cap air mass moves linearly faster than the rotation of the earth creating a wind in the direction of the earth's rotation. The reason for this is the attempted unification of the air mass within the band. The required observation is that even though the edge of the band is moving faster than the lower troposphere in the higher latitudes, it still moves slower than the angular speed of the lower troposphere at the equator. This exists because the lower troposphere moves more with the angular velocity of the planetary rotation giving

Chapter 21: Resulting Atmosphere

slower speeds as we move toward the poles. Since the buffer zone of the jet stream is only partial and there are relatively extreme differences in temperature (creating different densities), these bands interact.

Let us return to our illustration to understand the velocity conflict that occurs in a band. Imagine that the linear speed of an atom moving in the upper troposphere at 45° N is the same speed as the one moving at the equator. Let us assign this velocity to be at 90% of the linear rotational velocity of the planet at the equator. The rotational speed of the planet at 45° N is about 71% (cosine of 45°) of that at the equator. An atom moving at 90% at 45° N appears as moving faster than the planet. Imagine now that the friction between the slower polar atmospheres accelerates somewhat while the equatorial atmosphere down to 80% of the equatorial rotation speed. The polar air would still be moving slower than the equatorial air but faster than the rotation of the planet by approximately 113%. These figures are a little extreme but provide an illustration to the actual nature.

The jet stream is like a thin band of air disjoined between the major bands of air. The image is similar to a liquid escaping pressure from between two compressing solid objects. The liquid will acquire speeds beyond the motion of the two compressing objects via compression. This is true for the jet stream. Unlike the solid objects, the compressing objects partially join in the acceleration at their edges touching the stream. This creates an envelope of slower accelerated air. The outer edge of the envelope unifies with the slower external air of the jet stream. As we move inward the velocity of the air increases to the speed of the inner air. This occurs, as there is undulation (progressive wobbling) of the polar air mass. As will be seen, the jet steam is not affected uniformly creating the pulse-like effect between faster and slower moving segments within the stream itself. The laws governing this phenomenon are well defined by scientists.

Undulation of the polar air mass occurs in response to the moon's affect upon the earth. The earth wobbles. Scientists have defined this wobble as a result of the earth's inclination to rotate mutually around the moon. This phenomenon also accentuates by the closer orbital pattern of the moon. However, the mass of the moon is considerably less than that of the earth; the moon seems to be the only mass orbiting. Another factor is the gravitational pull of the moon. People on the earth feel neither of these forces, as it is so slight. As expouned earlier water responds to these forces in the form of tides. Air is much more sensitive to these forces than water because it is thinner. Air within the regions at the poles is heavier than the equatorial regions laterally. This allows the polar air to move toward the equator. Note that the change in shape of the polar air mass into an oval-like form creating a larger perimeter that requires the jet stream to stretch and move around the form.

Adding another factor involved in this undulation by the polar air mass, there are both strong and weak regions fronting at the edge of the tropical air mass. This variation

Part 6: Post Creation Activity

formulates from the differences in temperature and atmospheric pressure. This generates the Rossby waves, ripples in the otherwise uniform movement toward the equator. In essence, this breaks the dropping polar air into different segments. These waves increase as the polar air mass continues to migrate toward the equator.

Teardrop formations of the polar air mass occur for two reasons. The centrifugal force of the planetary rotation adds to the acceleration towards the equator of the lower polar air mass. This accelerates the jet stream even more in locations around the tear-like formation. The second factor is the forces that formed the jet stream are at work above trying to reform the jet stream. This decreases the amount of polar air that pours into the system by the polar air mass. Finally, it becomes a separate cyclone entity.

The Coriolis Effect (the spiraling effect) is clearly seen in the nature of the cyclones. Now, we are ready to examine cyclones, anticyclones and ultimately tornadoes. The cyclones in the Northern Hemisphere rotate counterclockwise. Conversely, the rotation in the Southern Hemisphere moves clockwise. The winds of both cyclones move toward the center of the low-pressure cell. Eventually the low fills with the warmer external air and destroys the cyclone barring external influences.

The inverse of this process is the anticyclone. Instead of the central region being low, there is a high-pressure cell. The surrounding air rotates in the opposite direction as the air mass as it pours into the lower surroundings. In essence, it is moving away from the cell's center. The image is that of air falling from a central location off to the sides again promoting an eventual leveling condition. The Coriolis Effect appears to be moving in the opposite direction of the cyclone. However, when we realize that one process in converging toward the center and the other diverging away from the center, these opposite directions are consistent with the nature of the Coriolis Effect.

Tornadoes

Tornadoes also form from the Coriolis Effect and the added intensified differentiation between cold and warm air masses.. In their case, a cold "low" cell has been broken into smaller fragments and travel into a region of high-pressure warm air. As scientists note: the tornado's cell is as a heavy cold atmospheric mass meeting an excessively warm, hence, lighter air mass. The cold air mass breaks into smaller "fronts" as it falls into the warm air mass wall. When more than one of these smaller fronts exists near each other, "fingers" formulate. The Coriolis Effect generates the rotational force. The variation of pressure forms a steeper valley to the low's ceiling. The interaction becomes more severe than a regular scenario between the forces of warm and cold air.

The result is the circling velocity is greatly increased. It becomes a tornado when the velocity reaches the point of throwing the molecules away from the center. In essence, the low increases. The air of the warm air mass feeds into the system under the edge of the storm. The image is that the edges are sucked in near the base as the top throws

Chapter 21: Resulting Atmosphere

outward the air mass material. This causes the pressure to decrease within the eye of the tornado. There is equilibrium between the vacuum of the eye and the centrifugal pulling. They mutually build in magnitude. The vacuum tries to fill itself from the air underneath. The tornado is pulled to the ground, as the external warm front can no longer meet the demand for air to fill the system.

Hurricanes function a little differently than that of a tornado. A tornado operated by a clash between the cold front from the polar region and the warm air of tropics, the hurricane forms purely from the tropics. Hurricanes form another phenomenon because of the thinned atmosphere. Like the Red Spot of Jupiter and the circular storm found on Neptune, the storm is heat related within the framework of the surrounding atmosphere. Unlike these storms, hurricanes are free to move over the surface of the planet; the heat source comes from the sun and not from a location upon the surface.

The hurricanes form about 10° to 20° above or below the equator, instead of starting at the equator and start out in the ocean. The reason that they start at these latitudes is because of the tilt of the earth's axis. This involves the scenario concerning the path that the sun makes over the surface of the earth. In this, the sun spends more time 10° to 20° above or below the equator than at or near the equator itself. Again, this is easily illustrated by looking at the basic trigonometric function of a sine wave. At 30° the sine of the angle is 0.5, and the value does not drop below 0.5 until we reach the sine of 150°. The same is repeated at 210° and 330°. When we subtract 30° from 150° or 210° from 330°, we arrive with 120°. Now, let us subtract 150° from 210°, we arrive with 60°. This interval is half of that expressed at the crest and trough of the sine wave. Translating this information to the earth's surface, the sine wave has an altitude of 23.5° or about 1,627 miles with a length of about 24,903 miles for 360°. The sun moves through this sine wave approximately 0.986° per day along the x-axis or equator. By this the ocean water above and below the equator by a few degrees interact with direct sunlight while the equatorial region remains primarily free. When the sun moves to the opposite side of the equator, it creates the greatest radiation difference. This gives the temperature variation needed to generate the storm.

Hurricanes are drawn away from the equator by cooler air, generating more temperature variation. However, the cooler air eventually destroys the storm as it cools. The image is as a sine wave starting at zero, is the formation of the storm. The crest represents the peak of the storm, and finally the storm ends as we approach the x-axis for zero. In essence, the hurricane needs two basic factors to exist, water and temperature variation. Remove either, and the storm diminishes. But, to a tornado, being over water is not necessary for its manifestation.

Tornados and hurricanes are manifestations of a weaker atmosphere. In the time of Adam to Noah, the thicker atmosphere kept the troposphere's temperatures from varying greatly between air masses. Since the Earth was closer to the sun, the thickness kept the temperatures more moderate. When we say moderate, we mean not scorching.

Part 6: Post Creation Activity

Atmospheric Layers

Finally, let us examine the layers of the atmosphere. The heat radiating from the earth via solar absorption and internal furnace gives the troposphere its warmth. As the tropospheric gases absorb this heat for their kinetic movement (maintaining a gaseous state), there is less heat available to the next atoms higher in altitude. For this reason, the temperature is constantly dropping as we rise in altitude within the troposphere. The tropopause is the location in altitude where the atmosphere starts to warm up. This is because of the photonic activity within the ozone layer (O_3) of the stratosphere. The ozone electrons are loose enough to absorb and interact with ultraviolet (short wave) light because of its molecular formation. Because of this looseness, electrons are able to release more heat. Hence, the stratosphere gains heat as we journey upward as the ozone has more sunlight that is available. The stratopause ends this gain, and the atmosphere begins to cool off again as we continue upward. The ozone process thins and other gases become predominate. Within the mesosphere, the cooling process starts again as distance from the heat source (the ozone layer) becomes greater. Finally, this cooling process ends at the mesopause. The reason for its end is that the atmosphere is thinning absorbing less of the energy from the sun. By this, the thermosphere continuously gains heat as we leave the atmosphere of the planet. In addition, clouds form under the ozone layer. The reason for this is the air is thinner in the ozone and beyond.

This brings us to the clouds forming at the top of the troposphere or the cirrus cloud formation. This is the water vapor that has continued its journey upward. Most water vapor rains back to the earth, but some does not. These clouds are temporarily trapped under the ozone layer. The reason that the ozone has the trapping effect is that the ozone molecular structure has more mass than the cloud clumping water molecules. It takes longer for the water molecules to move the ozone molecules than it does for the upward push of buoyancy of the water molecules via the troposphere. As the water molecules pass through the ozone, they warm up and breakup as a clump never to reform via thinness. By this, the ozone becomes a one-way door and troposphere loses water moister. However, the moister loss is insignificant to the comparison to the surplus still existing as the 300 feet of water left over in the ocean from the melted ice from the ice age. Even so, over thousands of years it would become significant. However, Elohim will have created a New Earth and a New Heaven long before this occurs.

Chapter 21 Quiz

1. Why are there Jet Streams in the Upper Troposphere?
   A. Thinness of the Atmosphere
   B. Turning of the Earth
   C. Air mass variation
   D. All of the Above

2. Hurricanes form where?
   A. At the Equator
   B. About 20° from the Equator
   C. About 30° from the Equator
   D. About 40° from the Equator

3. Why were there no Tornadoes during the time of Adam?
   A. Thicker Atmosphere
   B. Higher Oxygen Content
   C. There was only Hot Air
   D. The Landmass was in one Location

4. Why is there Greater Pressure in the Upper Atmosphere at the Poles?
   A. The Rossby Wave Effect
   B. Colder Temperatures
   C. Rotation of the Earth
   D. Flags need Wind

5. Temperatures Drop in the Stratosphere because:
   A. Lack of Oxygen
   B. Lack of Nitrogen
   C. Ozone
   D. Argon

6. (T/F) Today's atmosphere is thinner than that before the ice ages.

7. (T/F) Hurricanes start at the equator.

8. (T/F) Polar upper air mass is heavier than the upper air mass at the equator.

9. (T/F) Jet Streams exist between the different bands of air in the troposphere.

10. (T/F) The ozone exists only in the mesosphere and above.

Part 7

Logic of Creation

Chapter 22

The Big Ending

When we examine all the different process involved in creating the physical universe, we have to notice that it was time consuming. Yet, HE is able to restore wholeness to a body within seconds. The question that emerges is that HE could have created the entire universe within an instant, but why did HE choose not to do so? There are at least three real reasons. The primary reason is to create a universe that operates by a set of laws that gives HIM the ability to automate HIS Plan for us so that we can operate in self-awareness. Along the same reasoning, it makes us need to choose to seek HIS Face to find HIM rather than HIM already being in our presence. The second reason is that, HE knew we would fall and needed to give us a pattern of laws to illustrate HIS Nature and to establish the Sabbath progression. Three, it is to give us the ability to have dominion via the laws of nature. Finally, HE needed a place to deal with the rebellion that started in Heaven. In the Book of Revelations Chapter 21, it states that Elohim will recreate Heaven and earth without any mention of time to create. By this, we see a purpose in the "Creation Week;" it is to establish a pattern in which HE will work.

One of the interesting observations understood from the formulation of Creation is the Ten Commandments' role in forming the universe; after all, we have the Sabbath week. While the Sabbath week and Feast Days are a pattern for HIM and HIS Work, the Ten Commandments are for us. However, we need to observe both for ourselves and a witness to the world of prophetic events. The Ten Commandments are not there just to limit us from sinning. They are keys to unlock the nature of the universe- Not only in the manner of creation, but in the manner in which living can be rewarding. They are the guidelines needed for human existence to become a dream of reality. However, we all fall miserably short of these guidelines. Even so, HE makes a way.

One. Aleph is the first letter of the Hebrew alphabet. This letter has a silent sound. The picture language image is the ox. The meaning in general is the first, source and strength. Does not this describe G-d? The first commandment is to acknowledge the existence of Elohim (HaShem/YHWH). Our first phaseverse is the nulverse, which is nothing ("silent") and has no expression of itself except that it exists by the Presence of Elohim.

Two. Beth or Bet is the second Hebrew letter. It has two sounds. The picture language is the house. The general meaning is house and family. The commandment is that we should not make and pray to graven images. This may seem unrelated until

Part 7: Logic of Creation

we realize that the ultimate house form is the temple, and that our bodies are the temples of HIS Spirit. The second phaseverse is the inertverse. Inertia makes the entire universe as a single solid reflecting HIS Nature or an image of HIM but in itself has no life, which HE forms to "house" this universe.

Three. Gimmel is the third Hebrew letter. The picture language is that of a camel generally meaning lift up, pride and exalt. The Third Commandment is not to take HIS Name in vain. This means that we are to lift HIS Name above all other names in honor and respect for it is Holy or set apart. The third phaseverse is the gravverse. Within this phaseverse HE chooses a set of bethtips, and give them higher value than the rest, thereby, exalting these locations.

Four. Daleth or Dalet is the fourth Hebrew letter pictured by a door. The door represents the door, path, or way. The commandment is to remember and keep the Sabbath (Saturday) Holy. Keeping the Sabbath day for worship is the way or path of HIS followers. Yeshua said, "I AM the path and the way..." Our fourth phaseverse is the kineverse. Elohim introduces motion into the universe. The bethtips follow a particular "path" forming xergopaths throughout the universe.

Five. He or Hey is the fifth letter pictured in the Hebrew language as a window. The meaning of this letter is that of revealing, seeing beyond, and beholding. The Fifth Commandment is to honor our parents. Yeshua said that if we cannot love one and other whom we do see, how can we love HIM whom we do not see? Thereby, by honoring our parents, we see, or reveal to us, how to honor HIM. The corresponding phaseverse is the thermaverse. Juttoria, "Light," is introduced into the universe. We now have a polarity of types of acceleration, like a father and mother are two opposite sexes. The phaseverse is no longer in darkness, hence seeing is implied.

Six. Vav or Waw is the sixth Hebrew letter. The picture is that of a nail meaning in general to add, attach or pin. The Sixth Commandment is that we should not commit murder. Yeshua was nailed to kill HIM without just cause. The sixth phaseverse is the xyzenverse. Here, we see the first finite objects being created. All energy is "nailed" to individual crystals, such that, it does not destroy energy in the adjacent crystal.

Seven. Zayin is the Seventh Hebrew letter. The picture language image is that of a hammer/ weapon. The general meaning is that of weapon of war, tools and functional (use). Before the fall of man, war and weapons were not an issue, but the functions of the designs were for a specific use. The Seventh Commandment is, "do not commit adultery." The highest physical function of the human body is to procreate, just as the highest call for the soul is to know Elohim and worship HIM. The idea is that we should follow HIS Design for a man and woman. Satan has attempted to make it into a weapon to destroy those who are following Elohim. The Seventh phaseverse is the magneverse. Within this phaseverse, HE designs the purpose for each face of the xyzenthium crystal and designed the special function of magnetic interaction.

Chapter 22: The Big Ending

Eight. Heth, Het or Chet is the eighth letter in the Hebrew alphabet. The picture language for this letter is a fence. The general meaning of the picture is to set apart, protect and sanctuary. The commandment is not to steal. That which is separated, fenced off, to belong to a certain person should not be taken by another. The phaseverse corresponding to this commandment is the neutronverse. Within this phaseverse, each neutron is fenced off from the adjacent neutron; Elohim assigns each neutron a certain number of crystals.

Nine. Tet is the ninth Hebrew letter representing a snake. The primary meaning of the picture is that of surrounding. The ninth commandment is not to bear false witnesses. The idea is that we should not surround the truth with lies, thereby obscuring it. The ninth phaseverse is the plasmaverse. This is the phaseverse in which surrounding neutrons are gathered into clumps of neutrons transform into specific atomic nuclei. These nuclei are isotopes, which depict a specific chemical element.

Ten. Yod, or Yud is the tenth letter of Hebrew. It represents a hand meaning hand, work and physical action. The Tenth Commandment is that we should not covet another's belongings. Our hands should acquire our needs honestly. We should be satisfied with what is handed to us by Elohim. Within the final phaseverse, the bondverse, atoms to physically interact with other atoms to form different compounds by various bonding processes. All these processes involve atoms sharing electrons following certain rules. Once these conditions are met, they do not seek more electrons from other atoms besides that which is set in these rules. This makes any chemical compound a precise combination and a type of final product.

Why Our World?

The earth is unique, above all other planets. Not because of life, life exists elsewhere, Heaven for example. As of yet, we have no proof of life on other planets in other galaxies. But to think that because there is life on other planets in other parts of the universe that this negates Elohim's purpose for our planet would be a GRAVE mistake. There are several speculative theories on how our existence would relate to the existence of other civilizations. Rather than offering, yet, another theory, we can know from Scriptures of the Tenach, including the B'rit Hadeshah (New Testament) that if our scenario negates HIS Purposes, we are in error.

Elohim chose the earth as the final battle ground for the war between good and evil to fully express itself. We are not unwittingly victimized by this choice, but rather we are part of this choice. The battle wages even as we continue in our analysis. Despite all the misinformation and propaganda preached in the information systems of the world, there is truth and love reigning from Elohim.

To illustrate and to personalize this understanding, we will use the concept of a person dreaming a dream. In our example, we used a man; we could have just as easily used a

Part 7: Logic of Creation

woman and achieved the same results. However, the role of a woman as designed by Elohim is better portrayed for our example. Yes, there are different roles. The abusive examples that we observe in history to present was never intended by Elohim.

The Dreamer and the Dream

At night, when we dream, we are the only one aware of the dream. In fact, in our dreams, we are the center of that dream, without us the dream would not exist. Some of us are able to realize this in our dreams on occasions. When we awaken, the dream disappears. But imagine someone who is dreaming a continuous dream and is able to choose, for the dream to remain, and live in it. This is similar to the way Elohim has made physical reality. If HE wants the trees to be green, they are green. If HE wants the trees to have needles, they have needles. The whole concept behind witchcraft is a perversion of this ability. Witchcraft, weather white, gray or black, is about controlling the dream to our desires. But, that is not the purpose of the Dreamer. It is not because HE wants control. It is because control belongs to HIM. Even so, it is for our ultimate dream. What then, constitutes the ultimate dream?

As stated earlier, we will assume that the ultimate dream is that of love marital relationship. For a man, this is a dream of a woman. If it is not, it should be. We will assume that the dreamer is dreaming of love as designed by Elohim HIMSELF. He designed her in love. In the dream, He falls in love with her.

She is also in love with him. In this dream, they had so many pleasant places to experience. He gave her all kinds of gifts tailored to her being. He adorns her with so many beautiful jewelry trinkets and gave her flowers as a gift. He made sure that all of her needs were met with love. He made her co-ruler with him. Together they ruled the entire dream. The companionship this mate is to him is as "living in heaven."

Then he realizes in the dream that she is not aware of the joy, love, nor aware of the dream. All the gifts and love for her are in vain unless she is able to experience them. Her rulership was empty. He wanted her to experience all the good things that he had for her.

The desire then becomes to give her the awareness of the dream so that this mate may be aware of the joy, love, and happiness given to and expressed by her. Unlike us, Elohim is able to do this, and does. However, this required a sacrifice on HIS part. He had to give up control over her, so she could have the awareness she needed to experience love.

After imparting the mate with this awareness, their relationship continues. He gave her memory of all the time she spent in the dream unaware so that she could be aware. She reacted just as she did in the dream. They enjoyed the love and authority over the realm of the dream.

Chapter 22: The Big Ending

Unfortunately, this did not last. Whether it was one day or a trillion years, it is short in comparison to the infinite past. The mate decides that the dreamer is not worthy of her. Unlike our fallen state, it was without cause by HIM. Her own beauty blinds her. She seeks and attains many other men desiring her beauty. Then, she becomes jealous of the dreamer's position, being the source of the dream, and desires to overtake the dream. These are the actions of Lucifer. This name is Latin for Hebrew's word Heilel, root word is praise, pronounced hey-lel.

However, the Dreamer is not without recourse. It is his dream! He decides that he will divorce this mate. He forms another woman in the dream that is not his mate. She has no memory of the previous dream. This woman he courts and plans to marry her. Her image is just as the previous form. This infuriates the ex-mate. She sets out to deceive this newly formed woman. The divorce is taking away the original position and authority of Lucifer. The forming of the new woman is the human race.

He could have destroyed this ex-mate before creating the next woman, but he didn't. He chose to keep her in the dream to test the newly formed woman, whom he intends to share his dream as a mate (as before). He allows this ex-mate to speak lies to this newly formed woman. He wants a mate that desires to live the dream as it once was and in awareness. The ex-mate provides a means to test and sets this choice by the newly formed woman. This provision by the ex-mate is done by spite, and not consent.

Homosexuality is portrayed as the lure because of the prideful rebellion against the natural design. Those involved in this sin, as in all sin, prefer their deviance to the normal behavior. This is true in any sexual misconduct, they put down the norm as boring and undesirable for some easy sexual gratification. The idea that G-d is dead sounds good to some of them. The ultimate sexual experience can only be found in love within a marriage setting as designed by Elohim.

This brings in the tree of life and the tree of good and evil. Fruit from the tree of good and evil was not just an apple or peach, nor was it symbolic only of sexual errors. The fruit is "self-focused glory." Lucifer went after Eve, not because she was she spiritually weak, for this is not so. There are three reasons for this occurrence. Her formation was out of Adam and not vice versa. She was beautiful (as Lucifer was also the most beautiful). She was not commanded by Elohim personally not to eat of the tree (second hand information). Eating the fruit by her was focusing herself upon her beauty and the abilities of her body. She compared herself to Adam and exalted herself because of her body's beauty and functioning. The giving of the fruit to Adam is showing him her findings. He harkens to her words and eats the fruit. His eating the fruit is that he chose to focus on his glory that entailed his strength and tasks wrought by his hands. By this the curses placed upon them by Elohim equals the offense. She has pain in child bearing and desires his (in the case of Eve, Adam's) position. The ground cursed for his sake; he must till the ground in sweat. Now, we can continue the scenario.

Part 7: Logic of Creation

The future bride succumbs to the lies of the ex-mate. The ex-mate gloats before the dreamer and says, "I win!! I am dreaming the ultimate dream! You can't even make yourself one bride, and look at all the men and women I have! Each woman you make, I'll teach her my ways. You can't destroy us, else you have no dream." The ex-mate laughed to scorn. "When she ate my poison (lies) she became my possession. She is mine to do whatsoever I will! I have defeated that puny married for life routine."

The dreamer cut off her speech with a simple "Ahem..." She looks up and sees the new woman is still in her love for him. She then wondered, "Who is this that is with her?" and "How can she be in two places at once?" The reply is that he died for me. This speaks of Elohim coming to earth in the human form known as Yeshua and dying for our sins and rising on the third day. There will be those that reject HIS Ways and those that accept HIS Ways giving the two different outcomes. These different outcomes are not like the difference between choosing red or blue, but as making the right decision or the wrong decision.

The dreamer said, "For you there is no redemption, for you knew me intimately and rejected me. But of her, she knew me not. Moreover, no one lied to you (except you to yourself), but she was lied to by you. For her, redemption is possible. She was allowed to eat your poison. But, I have died for her to redeem her." Knowing that the authority has not been taken from her ("the initial dream"), gloats again, mocking and accusing the redeemed.

There is a day that will come, even those still living on the earth will be changed within a twinkling of an eye (I Corinthians 15:51-52). This is represented by the taking of the crown and giving it to the bride. The required observation is that in Jewish weddings the bride is wearing a crown. There is much to be learned from the Jewish wedding and the last days.

"This is MY dream; surely I will destroy you and those with you. However, this lady will become my bride, I will take the authority of the realm from you!" with those words great fear came over the ex-mate and those with her. They now realize that all their plans were nothing but self-deceiving snares for themselves.

The dreamer destroys her and those with her. He states to them, "The nightmare and torment that you created shall rise-up against you and hurt you as a burning flame." While they suffer together, they suffer separately as physical pain is felt individually.

After the Millennial Reign upon the earth and the Great White Throne Judgment (wicked beings cast into the lake of fire), a new earth and a new Heaven are created and the End is as before as in the Beginning, with the exception of individual awareness. Those who were evil will be forgotten, as there will be no place found for them. Those, justified by their faith in Yeshua HaMashiach (Jesus Christ), live in the new universe. Their lives will express the saying, "they lived happily ever after." Truly, this is the desirable outcome.

Chapter 22: The Big Ending

The Rebellion in Heaven

The Scriptures speak of three archangels, Michael, Gabriel, and the one we call Lucifer, which the Torah states as a covering cherub. In the rebellion in Heaven 1/3 of the angels fell with the covering cherub. Another interesting observation is that the Ark of the Covenant has only two cherubim covering the Mercy Seat, not three. However, the Torah does not make a direct correlation. Scriptures state that there is nothing new and the past is required by Elohim, gathered from Ecclesiastes 3:15. By this, we know that if the fall occurred in Heaven destroyed the original state of existence, and then there has to be a restoration to the original state. For our analogy of the events that took place in Heaven during the fall and the future restoration, we will use the color wheel for additive coloring. The color assignments are not arbitrarily chosen, as we will see.

Imagine each color representing one of the archangels. The red color represents the fallen archangel. Blue and green represents the two loyal archangels. It states in scriptures that his own beauty lifted up the archangel before he fell. Red is the color most consider beautiful. Without red there can be no yellow or magenta. Therefore, in this scenario, red lifts itself above the other two. Failing to realize that yellow requires green and magenta requires blue as well. Most importantly, it takes all three to create white light expressing Elohim. However, as the story unfolds, red ignores these facts for the purpose of self-exaltation. Red furthers its depravity by insulting blue and green for forming cyan a color similar in hue.

Finally, red could not tolerate being just one of three. It decides to withdraw itself and all that is associated with the color, one third of the angels, out of the equation. After all, it was they who created yellow, it was they who created magenta, and it was they that gave the other two colors the ability to create white light. White light will cease to exist, destroying the ability of Elohim to express HIMSELF fully as a natural expression within this particular realm. They thought in their hearts that Elohim will beg for their return and that they could call the terms. The terms in their minds were the exaltation of red "as" and "above" white meaning above the Most High.

The idea was that even white light could not be manifested without their input of red. In this, Elohim would lack total expression within the universe. Blinded, willingly, by the beauty of red, they could not see the obvious. The attributes of Elohim are not dictated by the dream, whether it be an earthly dream or a heavenly dream. Notice that the dreamer of a dream is in the dream and that the dreamer's existence is not dependent upon the dream. Similarly, Elohim is not dependent upon HIS Expression in Heaven for HIS Existence. The red's input to formulate white light was not a necessity to HIM. But, it was a privilege to red to help formulate the image of HIS Presence in Heaven with blue and green.

Elohim could have stopped the separation. However, love then becomes permanently imprisoned to HIS Will and not a free will offering. Red makes the first move and

Part 7: Logic of Creation

separates itself from the rest of the colors. The result looks disastrous for the remaining colors, as there are now only shades from green to blue. As red observes the results, it gloats.

After the separation occurs, Elohim is now free to offer an ultimatum for their repentance. However, from their standpoint, the whole threat seems weak- all bark and no bite. After all: the colors of yellow, magenta and white no longer able to exist and the remaining colors cannot create them without red. As noted earlier, in the Torah, we notice only two Cherubim covering the Mercy Seat, not three. It is interesting that red is the color of the blood sacrificed; the Hebrew word for earth (soil/ land), A-da-mah, comes from the word for red.

After some time, Elohim makes good of the threat. The red colors begin to see a phenomenon that they had not considered. This is the formation of another set of red. The doom and fury of the red colors were great as they saw the formulation. This new formulation is the human race. The faintness of the color represents that the work of Elohim is not yet complete.

One day, not in the too distant future, the work will be completed. The beginning will be reestablished completely. The old red attributes vanquished from their positions. They are forgotten as their vacant mansions are reconstructed for the new occupants. The only difference is that these occupants are there by their choice and willing to follow HIS Ways with their hearts in it. All their old memories will be replaced with the knowledge of the Present.

In conclusion, the purpose of attaining supernatural power (if there is such an attainment) is not for controlling others as some might suppose. The purpose is to serve others. For instance, each primary color has a role to perform in painting the universe. If red were to control the entire picture, the existence of other colors would become nonexistent. Within the human concerns, the number of primary colors equals the number of beings. By this, each person's service to the universe would be to complete the totality. Therefore, the intended actions for us supernaturally are the exact opposite of witchcraft. The greatest act is that of serving all. Ultimately, this means to serve Elohim.

Yeshua

Yeshua (Jesus, means Salvation) is HaShem (YHWH) in the Flesh. HE was born to a virgin (Isaiah 7:14) named "Mary" (in English) on the first day of Sukkot/Feast of Tabernacles, and circumcised on the Eighth Day Assembly of Sukkot. HE lived a sinless life and died for our sins on Pesach/Passover. HE rose on the third day (Mat. 12:40) and ascended to Heaven on the Bikkurim/Feast of First Fruits, and is interceding for those who accepted HIS Salvation Plan, as it is this day. HE sent HIS Ruach HaKodesh on Shavuot /Feast of Weeks or Pentecost (Implied by name, Act. 2:1-4).

Chapter 22: The Big Ending

HE is coming back for the Believers of HIS Second coming on Yom Teruah/Feast of Trumpets (Two-day long Feast, "no man knows the hour or day" Mrk. 13:32). HE is coming back to Redeem Israel by Yom Kippur, and to be crowned King on Sukkot (the time of Dedication of the Temple) rule the world for 1,000 years (Rev. 20:4). HE will perform Great White Throne Judgment on Yom Kippur (Rev. 20:11).

Repentance

Each of us needs to be saved in order to enter the Kingdom of G-d, whether Non-Jew or Jew. Salvation does not come from works, nor from intellectual achievement, but through repentance of sin and establishing a relationship by asking HIM into each of our lives. Upon receiving Salvation, you become a new creature. The Ruach HaKodesh (Holy Spirit) now exists within you. This quality marks the fact that of being "born again." The Ruach HaKodesh is an earnest down payment of the transformation that is to occur in the future. When we repent, it is a sincere promise to turn away from our contrary ways. Moreover, we desire to make HIS Ways our ways.

Repentance is a gift. Consider Esau (brother to Jacob in the First Book of Moses). He sought repentance and could not find it, even when he sought it with tears. This excludes the idea that repentance is something that can be accomplished at will.

Abominations Generally Described

Unfortunately, we have become so dull in discernment that we need the moral code written down. The following gives the generally expressed evils of the world as it is expressed in Brit Hadeshah (New Testament) Scriptures:

Revelations 9:20-21.

And the rest of the men which were not killed by these plagues yet repented not of the works of their hands, that they should not worship devils, and idols of gold, and silver, and brass, and stone, and of wood: which neither can see, nor hear, nor walk: Neither repented they of their murders, nor of their sorceries (root word implies drugs), nor of their fornication, nor of their thefts.

Galatians 5:19-21.
Now the works of the flesh are manifest, which are these; Adultery, fornication, uncleanness, lasciviousness (of no moral regard, reckless sexual behavior and being luxuriously wasteful), idolatry, witchcraft, hatred, variance, emulations (desiring to be equal or surpass, envious dislike), wrath, strife, seditions, heresies, envyings, murders, drunkenness, revellings, and such like: of the which I tell you before, as I have also told you in time past, that they which do such things shall not inherit the kingdom of G-d.

Part 7: Logic of Creation

Notice the words "and such like;" it means that there is still more. Rather than listing each of them individually, a partial list gives the general nature of these violations. Therefore, let us not think of these omissions as significant implications of acceptable behavior; rather that the focus on such things as dwelling on that which has an evil report, which has its own evil spiritual attribute. All of these sins have roots in violation of the Ten Commandments. These can be broken down to two Commandments, Love G-d with all your heart, and to love your neighbor as yourself.

Acknowledgements of Who HE is and Our Sins

While it is impossible to remember every sin committed, it is the sincerity if the heart that is important. HE wants us to desire to be right with HIM. Secondly, after repenting, do not expect that sin will stop manifesting itself, for it will. It is our willingness and desire to overcome sin with HIS Help, which is important to HIM. The following presentation provides a guideline for repentance. While it is not the absolute method to repent, it will give us an idea just how far removed that we are from living the life that HE made for us.

Acknowledge Shortcomings

Acknowledge that there is only one Elohim whose name is HaShem ("YHWH")
Acknowledge that Elohim came into the world through a Virgin (Namely Mary)
Acknowledge who HE is in the flesh. HIS Name is Yeshua (Salvation, Jesus)
Acknowledge HIS work:

> HE lived a sinless life.
> HE died for our sins (one to all the below reflecting the Ten Commandments)
> HE rose from the grave on the third day
>
> 1. Not acknowledging the existence of Elohim (HaShem/YHWH)
> 2. Having other priorities (other gods) and praying to graven images
> 3. Taking HIS Name in vain
> 4. Not keeping the Sabbath (Saturday) Holy
> 5. Not honoring our parents (for whatever reason)
> 6. For hatred/murder
> 7. Sexual immorality (physical or fantasizing)
> 8. Stealing, Cheating
> 9. Bearing false witnesses
> 10. Coveting other's belongings

Acknowledge your shortcomings (sin) to yourself, and to HIM
Acknowledge the need to repent (turn 180 degrees), and repent
Acknowledge that HE listens to prayer and pray:

Chapter 22: The Big Ending

Prayer

Father, I repent of all my sins.
Please apply Your Atoning Sacrifice to my life
By Yeshua (Jesus), the Lamb of G-d unto me
I ask You for forgiveness of my sins
Including those that I am unaware of.
I ask You to come into my life as my G-d.
In Yeshua's (Jesus') Name I pray.
Amen.

Standing Fast

Acknowledge that HE forgives those that ask.

After doing these things earnestly in faith, you assure that your name is written in the Book of Life. If we truly have repented, we will seek to walk in HIS Ways and not our own. For the initial hint of this, we need to look at our repentance and develop the relationship with HIM. There is also a need to get into fellowship with other Believers to help on this endeavor. Local Believing Congregations provide a vehicle in which Believers can develop this relationship with G-d and HIS Plan for earth (Feast Cycle).

The war is not over, in terms of our life on earth. We must continue to fight for that which is right. It is better to die fighting on the right side than to die fighting against that which is right without knowing it, leave alone knowing it. Do not worry, we will be back! HE will destroy those who seek our destruction under HIS Leadership.

Mikvah (Baptism)

As commonly believed, this symbolizes the death and resurrection of Yeshua, showing Salvation. Ultimately, it is a good illustration of this principle. It also serves as a public confession of faith (Romans 10:9).

Yeshua states that those who believe and are baptized shall be saved and those who believe not in the Gospel (of Salvation) are condemned already (Mark 16:20, John 3:18). This requirement of Baptism is the demonstration of repentance and the cleansing of sins (Matthew 3:1-6).

When Yeshua came to be Baptized, Yochanan (John) forbade Him saying that he needed to be Baptized of Him. The reason is that Yeshua did not sin. Therefore, there

Part 7: Logic of Creation

is no need of repentance. However, the retort was that all righteousness had to be fulfilled (Mark 3:13-15). This gives the dimension of obedience to the ordinances, which is true repentance and is its purpose.

There are exceptions to this rule. The primary example is the thief who died next to Yeshua. Others in similar circumstances are not required– i.e. where it is impossible or impracticable to do so.

There is also a Baptism of the Ruach HaKodesh (Holy Spirit). This is also received for the asking and the laying on hands to stir the gifts (II Timothy 1:6). In Acts 19:1-7, Paul found some disciples that did not know of receiving the Ruach HaKodesh. When they had hands laid upon them and were baptized, these spoke in tongues and prophesied. While there is no need to perform, it is good to receive the infilling of the Ruach.

Welcome

When salvation comes in our lives, we are new creatures, in that, now, we have the Ruach HaKodesh living inside us empowering us to walk in HIS Ways. We are no longer like those in the world walking in darkness. They are led by that, which seeks our and their demise. Moreover, in the world to come, our mortal bodies regenerate into incorruptible bodies without sin forever. It is also said that angels rejoice when one person becomes a new creature by a regenerated spiritual nature.

WELCOME TO THE BIG ENDING!

Epilog- Tav Judaism

Before entering into this topic, we should first recognize that Tav Judaism is not a new religion. Rather, it is the intended manifestation of Judaism. Elohim has always wanted all human beings to be in a perfect relationship with HIM, not just an exclusive group. However, the manifestation of an exclusive group within today's world became a necessity because human beings do not know HIM and live in sin.

Fragmentation of Judaism has manifested even within the knowledge of HIM. Again, sin enters the picture. However, the objective is not that of accusing any particular group of people, for we all have fallen short. We have dealt with this matter previously; therefore, we will make an attempt to not dwell on our faults. Our focus here is the examination of the nature of the relationship with HIM.

Differentiation between Jew and Gentile has been the greatest obstacle in Judaism. Yeshua came to bring both together as one people. This has not yet occurred, but in the end (Tav), it will. How this will come to be is a matter of great contention. The Jews think that the Gentiles must become Jews. Conversely, the Gentiles believe that Jews have to become Gentiles. Both are in error.

Chapter 22: The Big Ending

Immediately, we recognize two opposite patterns of Judaism- Orthodox Judaism and Christianity. Each has its incorporated some pagan concepts. The Orthodox Jews have incorporated the Kabbalah and the Christian have incorporated the pagan holidays. While these issues must be resolved, there is a deeper issue that must be addressed. First it should also be noted, there is no set of words that can bring an end to the dissention, only Elohim HIMSELF. Then why write? The purpose is to present a picture of the outcome in hope that it will help to inspire the same.

For the Jews, the initial contention was with Yeshua being the Creator in the flesh. Afterward, it was the Liberty give to the Gentiles in that they did not required to observe the Kosher laws nor Circumcision. Moreover, they did not need to observe any particular day as all days belong to HIM. The latter became an occasion for the Gentiles to incorporate pagan holidays as times of their observances (which is a mistake). The "Jewish" Feast Days are given, so that we can see the pattern of Prophecy or, in other words, the pattern in which Elohim performs HIS plan. Moreover, the Jews need to observe Kosher and Circumcision to remain Jews, as it is a Covenantal issue. However, it is not a Salvation issue. There is only one Salvation issue.

Note: the original Disciples were all Jews, as such, kept the Sabbath on the seventh day and the Feast Days. However, many who practice Messianic Judaism today (keeping the Sabbath, Kosher and the Feasts) have also fragmented the Scriptures. This is speaking of the abandonment of the Gifts of the Spirit, as most of them started out of the Charismatic Movement. This is especially true concerning the Speaking in Tongues.

As it has been painfully pointed out by many, Speaking in Tongues is the least of the Gifts. This is true, and it is also the easiest to achieve. If we neglect the least of the Gifts, how are we going to be trusted with greater Spiritual Gifts? There are many purposes to this particular Gift. As most know, it is to help us pray as we do not always know what to pray (Rom 8:26) because we do not know the total picture. It gives us the ability to speak mysteries (I Co 14:2). Moreover, it is a practice of faith that the Spirit of G-d is guiding our tongue to form words without our understanding. By this, we exercise our faith in the Living G-d. Faith without works is dead (Jam 2:17). After saying this, keeping the Sabbaths and the Spiritual Gifts profit nothing without first having a personal relationship with HIM. In fact, they are only tools to perfect our love relationship with HIM and to be a light upon the world.

There truly is not any newly formed information; we are only restoring data that which we have lost due to ignorance and sin. For an example in science, the movement of Venus past the earth facilitating the flood occurred without us knowing it for thousands of years. The neutrons, protons and electrons are only (re)discovered in recent science, but they have always existed as they are, intelligently designed from the very beginning. A tree falling in the forest without any human witness still falls and makes sound waves in the process. Truly, all these things are known to the Creator and they are a matter of HIS-Story.

Part 7: Logic of Creation

Science has been a stumbling block for believers and unbelievers alike. While those who do not believe consider the Torah as some kind of fiction, they think that everyone knows that the universe is billions of years old; therefore, the Torah is outdated. All the while, they are ignorant of the nature of time and photons. Meanwhile, some believers try to incorporate the unbeliever's theories into their own understanding. This, of course, does not work out well as there are some basic misconceptions within the unbeliever's theories. The purpose here is to show the light that is shed upon the darkness, which the unbelieving scientists have rigorously held as truth. In doing so, show the world that the Torah is not some concocted story written by Moses. The promises made by Elohim are real, and so are HIS judgments and HIS wrath are just as real. Some believe, that believing a lie will save them, thinking they are innocent, because they "did not know." Contrarily, the judgment is that they willingly believed the lie in a response of rejecting the truth; thereby, they are condemned. This book was written so that all might escape this judgment.

Note: the messianic symbol illustrated in thew back of this book initiated in the first century A.D. It combines the representation of the menorah and that of a fish; in which, the base of the menorah and the tail of the fish forms the Star of David in the center. The Star of David is perhaps the most misunderstood symbol. Many occultists have assigned the interlocking triangles as the union between good and evil, fire and water and the like. However, the initial formation came from the pre-Babylonian Hebrew for the word, David. In Hebrew, the latters are Dalet, Vav, and Dalet. Pre-Babylon Hebrew representation of Dalet is a triangle; therefore, two triangles are used in his name. The interlocking triangles cut into each other forming six Vavs. The picture language for Dalet is a door, and Vav represents a nail. The two doors are interlocked signifying their unity as one. The menorah represents the Jewish door, and the fish represents the gentile door. Both doors have three Vavs or nails. This represents the three nails used in the sacrifice of Yeshua (Jesus). In this, both doors require this sacrifice to achieve the fulfillment of salvation.

Chapter 22 Quiz

1. The Ten Phaseverses are a reflection of:
   A. The Ten Commandments
   B. The Ten Dimensions of Space
   C. The Ten Days formulating Inorganic Matter
   D. The Ten Cycles of Formation

2. Why not an instant Creation?
   A. It takes Time to Create
   B. It was a Struggle to Create
   C. A Pattern was needed for Mankind
   D. A Pattern was needed for Matter in General

3. The Ultimate Dream is:
   A. A Perfect Mate
   B. Happiness
   C. Dream's Self-awareness
   D. All of the Above

4. What Color represents the Rebellious Angles?
   A. Black
   B. Red
   C. Blue
   D. Orange

5. The Purpose of Creating this Universe is:
   A. To Restore the Beginning
   B. End Satan's Reign
   C. For Subjugating Humans
   D. To Disperse Sin

6. *T/F) The phaseverses relate to the 10 Commandments.

7. (T/F) This physical universe exists because of the rebellion in Heaven.

8. (T/F) The Dreamer wanted absolute control over the Dream.

9. (T/F) There are many ways to Heaven.

10. (T/F) Yeshua is the door to Eternal Life.

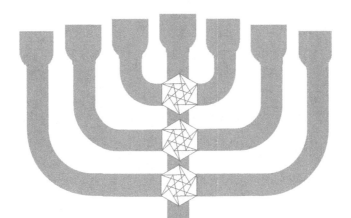

Appendix

Antiderivative

Understanding the sidestepping process of calculating an infinite number of rectangles, whose areas equal zero each, requires the understanding of manipulating sequential summations. Previously, we were adding sequential numbers. Now, we are analyzing sums calculated from the difference in x as determined by dividing a width into n segments and adding the segments together. As suspected, the total incorporates the entire original width. However, an interesting relationship exists between the segments. We find that as we subtract the lower value of x from the higher value of x for each segment that the next segment uses the higher value as the lower value for subtraction. The result of this process causes the upper value of each segment to cancel, as it becomes the lower value subtracted in the next segment. Notice also that the initial value and the finial values of x are the only values that do not cancel out. This gives the concept of subtracting the initial value from the final value to find the total of such a sum. However, subtracting zero from a final value seems meaningless; we could add zero and get the same answer. When we examine other sums starting at a number other than zero, we find that the subtractive process is correct.

When we measure from one geographical location to another, our starting point is zero and not one. So, when we measure a distance in a graph, the beginning point is zero. While it is true that an object can sit a distance from the origin (zero, or point of reference), we still have a starting point in which it cannot account for any distance that the object occupies giving a "zero" start. Secondly, the length of the object is also relative to a devised system of measurement. We can divide the length of an object into several equal sized lengths of any number. The number of units that we divide the distance into does not change the total length. Therefore, we can divide a distance into n units and the total units of the measurement will be our starting point minus the ending point of the last unit.

Our initial example of this process may seem over simplistic, but it is at the very root of the dimensional understanding of the process. If we do not start here, many will not understand the significance of the process and think it is something beyond their comprehension. The point then becomes ignored. Our initial example: if we had a rectangle three units in length at the base setting on a line five units from the origin. Our first measurement would be at five, and then we measure out one unit to six. We then subtract $6 - 5 = 1$. Our next step will be to start at six and measure out one unit more to seven. We subtract $7 - 6 = 1$. Naturally, we then measure out the last one and

Appendix

do the subtraction 8 - 7 = 1. In essence, we added the following: (6-5) + (7-6) + (8-7) or $6 - 5 + 7 - 6 + 8 - 7$. Notice that we are adding and subtracting both six and seven, leaving the subtraction of $8 - 5 = 3$.

Our next observation is that of counting from five to eight starting with five, we have four numbers or, on the graph, four locations. However, when we count the segments, we get only three. To express this relationship, we have to say that the summation of n segments or n increments, hence represented by i, i+1 (upper value), −i (the lower value) from i=1 to n equals n + 1 locations minus 1, the start, which is n. Even though it seems like we are making a simple thing complex, establishing this relationship is important in examining more complex problems as we shall soon observe.

If we add 1 five times, we get 5. If we add 3 five times, we get 15. Fifteen is three times five. Our examples of 1 and 3 are considered constants as they do not change within a given equation. The number of times we add, we will label as n. By this, we can say that the summation of any constant k added n times equals k times n or k times the summation of 1 added n times. This is illustrated in mathematics by moving the constant to the left of the sigma that represents the summation process indicating that the entire summation is multiplied by the constant k.

Our next objective is to assign exponents to the equation. This is not as bad as it may seem. For an example 17 raised to the first power (or having the exponent of 1) equals 17. This is true for any number. Since the constant k is external to the process, we do not need to raise the number to the associated exponent. This is because k is a multiple of the process. This becomes obvious as we raise the exponent beyond 1. By using 1 as our exponent, we do not alter the equation. This is the desired result.

Putting the equation together, we get the following results: $(i +1)^1 - i^1 = (n+1)^1 - 1^1$. The observation is that the resulting answer has an exponent equal to the summating equation. Despite the apparent complexity of the exponential equality of this equation, after the math is done, the equation reads the summation of ones for increment one to n equals n. This is as it should be, but one can be represented by i^0, as the result is also one. If we were seeking an answer for the summation of i^0, we would need to use such an equation. This provides us a base for a geometrical progression for more equations.

Adding any set of sequential numbers raised to the zero power is the same as adding one by the count of numerical increments within the sequence, which is n or n^1. The observation then becomes that the exponent of the increment will always produce an answer having an exponent one greater. This is because the terms of the polynomial with the highest exponent of the increment cancel out, and the polynomial answer does on cancel out. This gives us the first part of the "sidestepping" technique.

We can test this theory by increasing the exponent by one. Instead of looking the expression of (i+1)-i and (n+1)-1, we are now going to square each individual increment value. We are now solving for the difference of two squares one unit apart giving the

Antiderivative

summation $(i+1)^2-i^2$. The middle values still cancel out leaving us with the resulting value of $(n+1)^2-1^2$ which becomes $(n+1)^2-1$ as one squared is still one. We are now going to multiply the squares out and cancel. Doing the side of the summation, we get $i^2 + 2i + 1 - i^2$ and on the result side of the equation we get $n^2 + 2n + 1 - 1$. After canceling the equal valued positive and negative terms in both equations, we get the summation of $2i + 1$ from values 1 to n equals $n^2 + 2n$.

$$\sum_{i=1}^{n} i = \frac{n(n+1)}{2}$$

$$\sum_{i=1}^{n} (g_i - g_{i-1}) = g_n - g_0$$

$$\sum_{i=1}^{n} ((i+1)^2 - i^2) = (n+1)^2 - 1$$

and

$(i+1)^2 - i^2 = i^2 + 2i + 1 - i^2 = 2i + 1$

$(n+1)^2 - 1 = n^2 + 2n + 1 - 1 = n^2 + 2n$

gives:

$$\sum_{i=1}^{n} 2i + \sum_{i=1}^{n} 1 = n^2 + 2n$$

Since:

$$\sum_{i=1}^{n} 2i = 2\sum_{i=1}^{n} i \quad \text{and} \quad \sum_{i=1}^{n} 1 = n \quad \text{and} \quad 2n = n+n$$

$$2\sum_{i=1}^{n} i + n = n^2 + n + n \quad \text{Becomes} \quad 2\sum_{i=1}^{n} i = n^2 + n$$

Giving:

$$\sum_{i=1}^{n} i = \frac{n(n+1)}{2}$$

Appendix

Our next step is to divide the summation terms and substitute. Recall that the summation of 1 from values 1 to n equals n. We are going to subtract n from both sides of the equation to eliminate this variation. We are now left with the summation of 2i from values 1 to n equals $n^2 + n$. Finally, we will need to simplify the multiple of 2. Recall that the sum of segments having a value other than one is the same as multiplying the summation of one-unit lengths by this variation in units. We then say that 2 times the summation of i incremented from values 1 to n equals $n^2 + n$. We then divide both sides by 2 giving the summation of i from values 1 to n equals $(n^2 + n)/2$ or $(n(n+1))/2$. We attained this same formula from our previous example, when we added the first and last number and took that total and multiplied it by half the number of elements within the sequence. The theory works.

$$\sum_{i=1}^{n}((i+1)^3 - i^3) = (n+1)^3 - 1$$

$(i+1)^3 - i^3 = i^3 + 3i^2 + 3i + 1 - i^3 = 3i^2 + 3i + 1$

$(n+1)^3 - 1 = n^3 + 3n^2 + 3n + 1 - 1 = n^3 + 3n^2 + 3n$

Gives: $\sum_{i=1}^{n} 3i^2 + \sum_{i=1}^{n} 3i + \sum_{i=1}^{n} 1 = n^3 + 3n^2 + 3n$

$$= 3\sum_{i=1}^{n} i^2 + 3\sum_{i=1}^{n} i + \sum_{i=1}^{n} 1$$

$$3\sum_{i=1}^{n} i^2 + 3\left(\frac{n(n+1)}{2}\right) + n$$

$$3\left(\frac{n(n+1)}{2}\right) + n = \frac{3n^2 + 3n}{2} = \frac{3n^2}{2} + \frac{3n}{2} + n$$

$$3\sum_{i=1}^{n} i^2 + \frac{3n^2}{2} + \frac{3n}{2} + n = n^3 + 3n^2 + 3n$$

Antiderivative

After establishing and verifying an algebric pattern for determining summations, we are now ready to solve for the summation of squared increments. Recall that finding the summation for i, we had to examine the summation of the difference between squares. Now, we are looking for the summation of i^2 incremented from 1 to n. Just as we had to square our differences earlier, we are going to cube them. The equation now reads as the following: the summation of $(i+1)^3 - i^3$ incremented from 1 to n equals $(n+1)^3 - 1^3$. This multiplies into the polynomial expression: the summation of $i^3 + 3i^2 + 3i + 1 - i^3$ incremented from 1 to n equals $n^3 + 3n^2 + 3n + 1 - 1$. We then cancel out the two additive inverses that exist on both sides of the equation. The equation then reads: the summation of $3i^2 + 3i + 1$ incremented from 1 to n equals $n^3 + 3n^2 + 3n$. Now, we can restate the summation into the addition of three separate summations. Two of these summations we already know their outcome; they are the summation of 3i incremented from 1 to n and the summation of 1, or i^0, incremented from 1 to n. The summation of 3i incremented from 1 to n equals three times the value of the summation of i incremented from 1 to n giving $3(n(n+1))/2$ or $3n^2/2 + 3n/2$. The second summation is the equivalent of adding 1, n times, giving n.

Our next step is to perform the required algebraic math to solve for the summation of $(i+1)^3 - i^3$ incremented from 1 to n. Unlike the previous steps, this one does not work out to an easy answer. We are now working with dissimilar fractional denominators. First, we need to convert n to 2n/2 so that we can add it to $3n^2/2 + 3n/2$ giving $3n^2/2 + 5n/2$. We then need to subtract the resulting quantity from both sides of the equation. On the summation side of this equation, the terms cancel out nicely. However, on the result side of the equation, we need yet another step. We need to convert $n^3 + 3n^2 + 3n$ into $2n^3/2 + 6n^2/2 + 6n/2$, which in essence is both multiplying by 2 and dividing by 2 so that we have the same denominator of 2. Now, we are ready for the subtraction of like terms with equal denominators.

We now have	$2n^3/2 + 6n^2/2 + 6n/2 - (3n^2/2 + 5n/2)$,
which is	$2n^3/2 + 6n^2/2 + 6n/2 - 3n^2/2 - 5n/2$,
giving	$2n^3/2 + 3n^2/2 + n/2$.

We now have the summation of $3i^2$ or 3 times the summation of i^2 incremented from 1 to n equals $2n^3/2 + 3n^2/2 + n/2$.

Finally, we divide both sides by 3 to get a single occurrence of the summation of i^2. Now, we get the summation of i^2 incremented from 1 to n equals $2n^3/6 + 3n^2/6 + n/6$, which becomes $n^3/3 + n^2/2 + n/6$. Another useful solution is to factor the polynomial which is $n(2n+1)(n+1)/6$. To view the resulting mathematical solution, see next page. However, we will not be using the factored form within our analysis. Now, we are ready to return to the main presentation for the final solution of our problem. The actual math is on the next page.

Appendix

$$3\sum_{i=1}^{n} i^2 + \frac{3n^2}{2} + \frac{3n}{2} + n = n^3 + 3n^2 + 3n$$

$$3\sum_{i=1}^{n} i^2 + \frac{3n^2 + 3n + 2n}{2} = \frac{2n^3 + 6n^2 + 6n}{2}$$

$$3\sum_{i=1}^{n} i^2 + \frac{3n^2 + 5n}{2} = \frac{2n^3 + 6n^2 + 6n}{2}$$

$$3\sum_{i=1}^{n} i^2 = \frac{2n^3 + 6n^2 + 6n}{2} - \frac{3n^2 + 5n}{2}$$

$$3\sum_{i=1}^{n} i^2 = \frac{2n^3 + 6n^2 + 6n - 3n^2 - 5n}{2}$$

$$3\sum_{i=1}^{n} i^2 = \frac{2n^3 + 3n^2 + n}{2}$$

$$\sum_{i=1}^{n} i^2 = \frac{2n^3 + 3n^2 + n}{6} = \frac{n^3}{3} + \frac{n^2}{2} + \frac{n}{6}$$

or

$$\frac{n(2n+1)(n+1)}{6}$$

442

Pluto and Beyond

As noted earlier, Charon moves around Pluto at a close orbit controlling its axial spin. It spins nearly at 90°. Just as our moon always has the same side facing us, Charon always has the same side facing Pluto. Unlike the Earth, Pluto always has the same side facing Charon. An elevator could be built between the two objects without fear of destruction via rotation. Two other moons exist to Pluto that are much smaller and much further from Pluto's surface.

Nitrogen is also found on Pluto. It should be noted that the atmospheric gas has not been "contaminated" with hydrogen. In this, the nitrogen formed upon the surface by volcanic activity, and from being an internal multigalactic shell gas or internal to the surface. Even forming internal to the surface and spewing out by volcanic activity, for the lack of hydrogen to be found missing in the gas means that it formed internal to the multigalactic sphere.

Discovery of external planets, or dwarfs as scientists have chosen to classify them, has opened the possibility of many such levels of "planets." However, we do not have the data to give an exact quote on the actual count of planets beyond Pluto. There is also evidence of an external most asteroid belt that has an unknown composition except possibly for the information gained from comets. However, there are some items of information that we can extrapolate from the point of the creation of our solar system. Discovery of other asteroid belts beyond Pluto have been made. Even an oort cloud external to the final asteroid belt exists.

Methane atmospheres found in planets in Uranus and beyond are carbon based. Carbon is the sixth element, having six protons. This gives boron, beryllium, lithium and helium for hydrogen to "invade." Beyond Pluto there could be planets having frozen atmospheres of compounds formed by boron and hydrogen. Boron has one proton less within its nuclei than carbon. Others, beyond them could be planets having frozen atmospheres of hydrogenated beryllium and lithium. Perhaps in some solar systems, the furthest planets have atmospheres of only helium over oceans of hydrogen with land formations of primarily lithium and beryllium compounds.

The core material is another issue. The elements forming the planets' crust and interior are also diminished by distance. Even iron may be eliminated from the element set. This is especially true for the outer asteroid belt. However, for examining the outer

Appendix

asteroid belt, we do have two planets that have passed through it, namely Jupiter and Venus. It should be noted: sulfur is not a natural atmospheric element as it is heavier than aluminum, and it exists only in the upper atmosphere of Venus.

Each of these two planets has a peculiar phenomenon of sulfur. Jupiter has a sulfur moon that orbits close enough for friction to disperse some of its surface into a ring around Jupiter. Venus has an upper atmosphere that has sulfuric acid that never reaching the ground. Sulfur is a heavier element than that found in the rest of the atmosphere upon the Venusian's surface. Elohim used sulfur based droplets to destroy Sodom and Gomorrah. These cases can be explained by other various phenomena, overlooking the common factor. However, this factor does indicate that sulfur is one of the primary elements in the outer asteroid belt.

Life Continuum

Space: Consider the possibility that other life exists within this universe on other worlds besides this one. If this be the case:

    Is the war between good and evil restricted to our planet?
    How does it fit in with the purpose of the life on earth?

Remember the relationship of the finite universe has with the infinite universe. One, that it is as the infinitesimal universe is to the finite universe. Recall that there are many different finite locations within the finite realm. Each finite location is infinitesimally infinite within each finite location. Similarly, there are an infinite number of finite "universes" composing the infinite universe. This concept is reflected in creation. For an example, the human body is comprised of many cells, even though the actual count of these cells is finite, the size of the number is beyond our ability to count them one by one. Our life would end before we finished the count. Yet, these cells form one body.

Understanding the nature of personal awareness is deeper than the mystery of the formation of this universe. We can only understand the principle through illustrations. Consider an illustration between the ingenious puppeteer and the created puppets. In this scenario, the puppeteer desires to give the puppets the ability of being aware of the play and to feel the joy of living it. After accomplishing this feat, the play continued until the puppets desired to rewrite the play to exalt their own role. Within the exaltation of these roles, requires that other roles to diminish in glory often to the point of being extinguished. This is the knowledge of "Good and Evil." Rather than destroying the "set" and starting again HE decided to salvage the play. This is where the earth comes in and again with the birth of Yeshua (Jesus).

Now, let us bring the two concepts together. In this, we will affirm that there are several planets that have populations that have creatures in the image of the Creator. Next, we will assume that on some of these planets that the humanoids did not fail as Adam did. Contrariwise, others failed just as Adam did on the Day of Atonement. From this, we see the ultimate test occurred on the same day (1,000 year period) on all planets. The incarnation of Yeshua and death also occurred on each of the fallen planets simultaneously. By this, the death of Yeshua is a single unified experience as there is only one Elohim. But, the scenarios on other planets have no impact upon us.

Consider yet another factor to the equation. On the planets that the humanoids did not fail the test, they continue their existence within the physical universe for one week

Appendix

(7,000 years) after the creation week and are "harvested" for the "Big Ending." As this event finishes, Yeshua is incarnated 4,000 years after the fall of mankind to salvage those who desire to "return to the original play." Again, this correlates to the Feast of Tabernacles four days after Yom Kippur. HE was in the grave three days illustrating 3,000 more years to finish this physical realm before establishing the Big Ending or Life as it was in the Beginning. Ten days total equals 2*5, or double Grace (numerically).

Time: There are two more astonishing implications formulate concerning Adam and Eve. Despite the logical implications, there is a lack of absolute Scriptural backing to the following assertions. This does not mean there are no Scriptures that we can use to support the concept. It only means that there is no direct implication. Therefore, the absolute certainty of the information cannot be assured. However, in regard to being responsible to understanding, the following information is presented. Perhaps, the omission relates to our tendency to worship the creature above the Creator; consider the Catholic Church and all the obfuscation concerning Mary.

If Yeshua replaced Adam's position, then someone will replace Eve's position. The person that comes to mind is Mary Magdalene. It was her that Yeshua first appeared to when HE rose from the grave (John 20:1-18). Moreover, it was her who HE cast out "seven" devils (Luke 8:2).

The future role of Mary Magdalene is marriage. In Revelation 19:9, it speaks of the marriage supper. If Yeshua is the head of the "church" (Col 1:18) then there is no need to marry the "church" for HE is already the head of HIS Body. Then the question becomes, "Who is HE marrying and why? The answer is nation of Israel. Just as HE is the head of the "church," Mary Magdalene will be the head of Israel. Does this mean that sex is involved? Not EVEN, as written in Mark 12:25; we do not marry and have children in Heaven. Remember, that in the Garden of Eden, Adam and Eve were not to have sex. So it will be as in the Beginning. The reason for the marriage is that it will unify Israel and the "church." During this time, we will all be as those that were transformed and living in the Garden of Eden before the fall.

This marriage is to last throughout the Millennia Reign. After this, a New Heaven and a New Earth are form and the old one melts in a fervent heat (2 Pet 5:10). This new universe is to last forever. Remember that those things past will be. With that in mind: this new universe is the restoration of the original universe before the fall of Adam and before the fall of Lucifer. Mary Magdalene will then replace Lucifer as the Creators Favorite Cherub as Lucifer will be thrown into the Burning Lake of Fire (not hell). During this time, we will, by the testimony of our life on earth, judge (John 14:2) or replace the (fallen) angels (1 Cor. 6:3) in the New (or the Renewed Original) Universe.

Resources

The Torah
The Chumash
The Stone Edition
by Rabbi Nosson Scherman
Published by Mesorah Publishing, ltd.
4401 Second Avenue
Brooklyn, New York 11232

The Tenach (Holy Bible)
King James Version
A Regency Bible
Thomas Nelson, Inc.

Hebrew Word Pictures
By Dr. Frank T. Seekins
Living Word Pictures, Inc
Phoenix, Arizona

Strong's Exhaustive Concordance of the Bible
By James Strong, LL.D., S.T.D. & more than 100 associates
Abingdon
Nashville

Basic Concepts of Chemistry
Fourth Edition
Sherman/Sherman/Russikoff
Middlesex County College
Edison, New Jersey
Houghton Miffin Company
Boston, Dallas, Geneva Illinois
Palo Alto, Princeton New Jersey

Element of Physical Geography
Second Edition
By Arthur N. Strahler, Alan H. Stralher
John Wiley & Sons
New York, Chichester,
Brisbane, Toronto

Resources

Physics for Scientist and Engineers
Third Edition/Updated Version
Saunders Golden Sunburst Series
By Raymond A. Serway
Saunders College Publishing
James Madison University
Philadelphia, Fort Worth, Chicago, San Francisco,
Montreal, Toronto, London, Sydney, Tokyo

Calculus and Analytic Geometry
Fourth Edition
By Sherman K. Stein
Professor of Mathematics
University of California, Davis
McGraw-Hill Book Company
New York, St. Louis, San Francisco, Auckland, Hamburg
London, Madrid, Mexico, Montreal, New Delhi
Panama, Paris, Sao Paulo, Singapore, Sydney, Tokyo, Toronto

The World Book Encyclopedia
Field Enterprises Educational Corporation

"Isotope Information"
1. Unknown Author(s) –unable to locate original source (over updated, 2007)
    a. Holden, N. E. "Table of Isotopes," in Lide D.R., Ed., **CRC Handbook of Chemistry and Physics**, 74th Ed., CRC Press, Boca Raton FL, 1993.
    b. Wapstra, A. H. and Audi G., **Nucl. Phys**., A432, 1, 1985
    c. IUPAC Commission on Atomic Weights and Isotopic Abundances, **Pure Appl. Chem**., 63, 991, 1991

"Initial Information on..." –Hearsay, Never saw the Book
1. **Hollow Earth,** Unknown Author(s)
2. **Adam and Eve**, Rabbi Robert Benbow (unwritten) and Unknown Author(s)
3. **Bermuda Triangle and Magnetic Poles**, Unknown Author(s)

Glossary

* Words Unique to this document

Alephfield*	Energy field of physical existence, it is generated from the "true" geometric point. It supplies the three primary energies defining the physical universe.
Alephtip*	Reference angle assigned to a geometric point, it exists with or without energy assigned. Inactively simulates the geometric point in which the angle refers.
Antiderivative	Calculus term used to describe the mathematical process of reversing the derivative process. Couple of examples is an area under a curve, or the means of finding the acceleration of an object in respect to a given motion. Consider the transformation from $6x^2$ to $2x^3 + C$; C is a constant.
Antineutrino	Established within the Quark Theory, it was originally the particle was known as neutrino. Minute mass released from forming an electron, also referenced later as electron-neutrino. Within this article, the particle is named neutrino showing that it forms within the natural process of the neutron transformation.
Antiquark	The inverse charged quark in the theory of quarks. Example: an up quark has a charge of +2/3. The antiquark of anti-up has a charge of -2/3. This is not a down quark.
Asteroid Belt	Ring of rocks and small planetoids that surround the sun or given star within a general distance. Most stars, like our sun has more than one. Form when star gathers core material.
Antishell*	Galactic formulation entity, it exists outside the multigalactic shell process. Consider the neutrons represented by the space existing between four touching circles.
Asthenomoho*	Layer separation, this particular separation exists between the Asthenosphere and the molecuumsphere. Composed of lavor, vapor and liquid magma; the formation acts as a lubricant during the "dividing of the earth."
Asthenosphere	Term established in geology for a layer of semi-liquid rock just under the ultramafic layer of the planet's crust.

A-C   Glossary

Atomic Nuclei — Collection of neutrons and protons, except Hydrogen-1 having only one proton, that exists at the center of an atom. Unlike previous notions, they are crystalline.

Bermudaland* — Landmass a little over the size of Madagascar, it has about 300,000 square miles. The landmass is responsible for much of the coastal formations observed in Western USA. Some may connect it with the fabled Atlantis as magnetic energy does emit from this region.

Beryllium Rain* — Rain of beryllium formed from a cooling plasmatic state above the general atmosphere of the planet. More accurately, an elementuum breaking up as the molecuum cools into a lavor plasmatic state forming clumps of the element that are heavy enough to fall.

Bethtip* — Active alephtip (physically assimilated geometric point) charged with energy expressed by the three xevim subfields which manifest energy within this universe.

Bondverse* — Tenth and last phaseverse of Creation, this phaseverse defines the universe seen today. Within this phaseverse, atoms bond and life forms find their definitions.

Borax — Primary source mineral for Boron, it has an empirical formula of $Na_2B_4O_7 \cdot 10H_2O$. When hydroxide bonds to the forming molecule, the formula will change in configuration.

Boron Rain* — Rain formed by a collection of four boron atoms, these particles initiate the formation of borax and kernite.

Bottom (Quark) — Third generation hadronic quark, it contains a -1/3 charge, largest negative charged hadronic quark.

Brit Hadeshah — Hebrew for New Covenant, this word defines the fulfillment of the law of the Torah and the Prophets. It embodies the New Testament in Christian Bibles.

Carbon Rain — Forms from the Carbon Elementuum above the Earth when the molecuum breaks up into Lavor droplets. The initial large, in count of atoms, fell deep into the surface layers forming diamonds. The latter smaller droplets formed "seeds" that are used to form "houses" for life.

Cation — Positive ion in chemistry, the ion is either a single atom or a molecule. It is generated when a nucleus of an atom lacks one or more electrons required by its proton count.

Glossary C-D

Centrifugal Force — Energy forms by a mass, as it spins around a central location. When the mass is released from the force that holds the mass to the focal point, this energy will cause the mass to move in a straight line tangent to its path.

Charm (Quark) — Hadronic Quark of the second generation, the quark holds a charge of +2/3. It also has an attribute of strangeness.

Chexosod* — Xyzenscape (layer) of subatomic material composed of xyzenthium crystals in a magnetically checkered pattern. The net charge is zero. It has a magnetically unstable nature.

Chlorophyll — Botany term for a green substance within plants that manufactures energy from light using photosynthesis. The green coloring points to an existence before the sunlight of today; in which, red light was the only available light source from the central solar region.

Complex Gravity* — System of measurements that refute the held notion that the core of the earth is solid, but supports the concept of a shell-like object filled with plasma.

Coriolis Effect — Response in nature, it is vortical phenomena established by the rotation of a planet. Example: draining water rotates to the right in the Northern Hemisphere and to the left in the Southern Hemisphere.

Cro-Magnon — Beside the Initial historical information provided by scientists on the topic, Adam and Eve were of this race and those that were incorporated by them. However, the formulations of the racial attributes are the defunct results of the fall by Adam and Eve in the garden.

Daysod* — Alignment of xyzenthium crystals, microstrong energy covers its surface. It provides the strangeness factor found in second generation quarks.

Daysod Floor* — Interior of the electron on the side facing the nucleus designed to push the photon out of the cooled electron.

Diflohexet* — Singular stem of the diflohexius sprouts from the xentrix of the xyzenthium crystal. The image is that of a fountain spewing from the center in a tight stream only to fan out at the edge of the crystal. In an inverse manner energy pours into the "geyser" structure like a funnel. There are six of them at right angles per crystal.

D-E   **Glossary**

Diflohexius*   Internal structure of energy, it forms the xyzenthium crystal. Has six stems joined in the center at 90 degree angles. Energy flows into and out of these stems in a self-contained manner. Word formed from the words **Di**- a prefix for two, **Flow**, **Hex** for the six stems and rad**ius**.

Dividing of the Earth   Separation of landmasses, the result is the formations of continents and islands from the felsic cap. Purpose is to keep the human population dispersed throughout the Earth after the incident at the Tower of Babel.

Down (Quark)   Hadronic quark of nucleons (protons and neutrons), it has a magnetic charge -1/3. Symbolically, the letter "d" is used. Neutrons have two of these quarks, while the proton has one when smashed.

Duocollision*   Collision transference of energy established by direct contact between two bethtips, it occurs without an intermediate alephtip collecting input from their xevim.

Duoquiet*   State of a phaseverse in which neither finite motion nor infinitesimal movement exists. This condition is found in the first two phaseverses, even in the third after completion.

East (Kedem)   Naturally, east is where the sun rises. Interestingly enough, the Hebrew word, Kedem, reveals that it was also the location of the sun's setting. This, in itself, locates Eden at the Artic or Antarctic Circle. In today's directional perception, the direction would be either North or South. From the division of the Earth, the direction is north.

Eden   Centrally located within the felsic cap. The Garden of Eden location is East (see East) of this location, which is the Promise Land to Israel which is larger than the land allotted to Israel in 1948. However, Eden is much larger in area.

Electrodonativity*   Opposite of electronegativity, it is the tendency to donate electrons to other atoms in bonding. Lithium has the greatest tendency.

Electrophoton*   Photons developed within the electron's photon chamber. Generally, they are larger and less dense than nucleophotons. They provide us with light and normal heat energy.

Electrosod*   Xyzenthium crystal alignment generating a negative charged xyzenscape (layer) forming the surface of an electron.

Glossary   E-F

Elementuum* — Plasmatic vapor bonding formulation, it forms where all the atoms of a particular element or isotope are the same. These connect together forming a molecuum.

Ergitude* — Half volume of a xyzenthium-crystal-like formation dominated either by a jammerial or a juttorial force resulting of the bend seen in the wave of the electromagnetic waves propagated by photons through the space/time continuum.

Eternal Season* — Alignment of the polar axial movement producing a winter or summer that has no chance of change within the alignment parameters. This occurs as during the time before the Fall, when the Southern Rotational Axis was always facing the sun.

Exozeed* — Division of the xentrix by grouping xergopaths that split from the center forming four wedges each dimensional direction making twelve. Each exozeed contains two zeeds.

Felfraplex* — System of felfraxim, it moves as a single landmass fragment during the fragmentation process.

Felfrax* — Felsic fragment, it forms a continental landmass. Term stems from the words **Fel**sic **Fra**gment and an "**x**" to represent Continental Shield.

Felfraxim Belt* — Fragmentation of a felfrax, it forms island chains. They are caused either by friction between the edges of the two felfraxim or by friction with the ocean floor. The largest originated from the coastal regions of today's Asia.

Felsic — Geological term, it is used for the lightest crust material. The material consists primarily of iron, aluminum and silicon. Average density is between 2.6 and 2.8.

Felsic Cap* — Cap, it contains all the continents of the world. If the cap was centered over the North Pole, it would reach the Tropic of Cancer. The felsic cap forms Kadummagen.

Felsic Moho* — Thin layer or crack containing separating the felsic layer from the mafic layer. This division allows the felsic layer to travel over the mafic layer despite the friction.

Fijiland* — Landmass, it breaks off from the Australian Felfrax. It lost elevation via sinking slightly by sliding over the mafic plain collecting mafic material. It lies in the region of the Fiji Islands.

Finite Limits*	View of the physical universe, it indicates that the physical universe has both upper and lower limits. The upper singular limit is infinite, and the lower limit is an infinite count expressed in geometric points.
Force Subfield*	Any of the three energy subdivisions stemming from a geometric point. Individual pressure exerted by the xevim (pressure angles) as expressed within the physical universe.
Fractal	Patterns that repeat, it does not matter if the image is magnified or shrunk. Example: after magnifying an image of a particular square, we find that the lines are made up of squares. This is blown up again only to discover the lines forming these squares are also squares. This process could occur without ceasing in either direction.
Geometric Point	Mathematical Term that relates to the existence of an object having zero dimensions. It exists in space, but occupies no space. It is the root of physical space.
Gondwana	Larger of two shields defined in geology. These shields have an absence of sedimentary material. It was formed by Venus during its close movement over the earth during the flood of Noah.
Googolplex	Finite number beyond comprehension, it is formed by a number that is comprehendible. The comprehension begins with the googol being a decimal number having a one followed by a hundred zeroes. A googolplex is a number having a one followed by a googol zeroes.
Gravity	Jammerial collective spatial force of attraction, it forms between and within objects. Within the fibers of the physical universe, it bends the space/time continuum downward.
Gravtip*	Bethtip (active simulated geometric point) acquiring jammeria (force of attraction), it is formed in the third phaseverse. The progressive step beyond being an inertip is this form.
Gravverse*	Third phaseverse, it stems from the word gravity for a reference of understanding. Actually, gravity is a finite attribute of jammeria as a force of attraction. In this phaseverse, the inertinuum becomes discontinuous at an infinitesimal level and returns to being duoquiet. Finitely speaking the phaseverse is still continuous (see Chapter 4).

Glossary G-H

Gyro Alignment* — Today's Axial Alignment, it provides us with the four seasons. This occurs due to the stationary position the rotational axis has regardless of the Earth's position within its orbit, hence it is alignment is as a gyro.

Hadron — Type of quark. These quarks form the mesons (two hadronic quarks) and Baryons (three hadronic quark objects). It is supposed that the strong force holds them together.

Havilah — Land formation outside of Eden found in the Torah. The name means to twist circular. This planetary formation not unique to the earth; it can also be found on the moon and Mars. Today we find this formation in Tibet as the "Rooftop" of the World.

Hudson Comet* — Large planetoid smashing into the earth, it is responsible for forming the Hudson Crater and Laurasia Shield. Impact occurred while the flood covered the earth. It acquires an altered rotational axis before impact via trajectory friction. Iridium found in this impact.

Hydrogen Hydroxide — Chemical nomenclature used for water. The concept behind the formulation is that water is both an acid and an alkaline molecule giving water the pH factor of 7.0.

Hydroxide — Ion formed by one hydrogen atom and one oxygen atom. Empirical formula is OH.

Hyperatom* — Dense stars; they have a multitude of nucleons forming a single atom. Such stars are considered to be black holes, neutron stars and quasars.

Hyperons — Baryons (consisting of three hadrons) formed of the second and third generations. They have masses greater than that of any nucleon (formed up or down quarks). They contain one or more of the following quarks or their antiquarks: charm, strange, top and bottom.

Hyperpush* — Collective push by two xyzenscapes, protosod and daysod, that ejects an electron and an evaporating neutrino from the initial neutron leaving a proton.

Hypertip* — Bethtip (active simulated geometric point) charged with a double juttorial xev. These bethtips generates energy that rest at the acceleration of the speed of light. They are responsible for microstrong energy.

H-J   Glossary

Hypotip*   Bethtip charged with double jammeria xev. These rest at zero and can accelerate mass to the speed of light via gravity. Hypotips are responsible for microweak energy.

Ice Ages   Result of the Earth's movement away from the sun and the near solar centric movement by the rotational axis invoked by the passing of Jupiter.

Inertinuum*   Continuum of inertia in the inertverse, it gave no room for movement. Each geometric point has an expression of an inertip throughout the universe.

Inertip*   Bethtip existing only of inertia, its foundation is the joining of three nultips per geometric point. All xevim subfields (pressures exerted upon a geometric point) contain the expected inertial traits.

Inertverse*   Second phaseverse in creation, it is the first phaseverse that the physical universe has physical substance. Inertia defines this substance forming the inertinuum (continuous volume of inertia). Within this phaseverse, there is resistance to movement indicating the presence of the physical universe.

Israel   Location of the Garden of Eden, it is Elohim's Land.

Isotope   Unique configuration of neutrons and protons, it forms the atomic nucleus. These nuclei are considered to be radioactive when two neutrons touch.

Ion   Electromagnetically imbalance state of a particle whether it be an atom or a molecule or an atom freed electron. Ions can be positive or negative charged.

Jammeria*   Primary force of attraction, it provides the finite values for gravity and microweak forces. Within the formulation of magnetic energy, it forms the negative charge as jammeria projects itself from the magnetic source.

Jet Stream   Thin stream-like layer dividing layer, it forms an interface between the air masses, especially between the temperate and polar air masses. It has a high wind speed.

Juttoria*   Force of Repelling, acceleration reaches to the speed of light. Last force entered into the equation of the universe. Responsible for the finite qualities of heat (light) and the microstrong force.

Glossary

Kadummagen* — Primordial Shield, it is formulated from the Hebrew words for ancient and shield or the Shield of the Ancients. It is the unified felsic cap (landmass). The fragmentation of the landmass gives us our present day continents and islands. However, it does not always formulate a tectonic plate in its fragmentation.

Kernite — Mineral source for Boron, its composition is $Na_2B_4O_7 \cdot 4H_2O$. Molecular structure is subject to bonding with hydroxide. In this, fewer water molecules will be collected by the mineral. Hydroxide subtracts oxygen molecules from the formulation when encountered before the "root" molecule is complete.

Kinetic Non-Alignment* — Condition of the rotational axis, axis move around the sun neither in a gyro fashion nor in a solar centric fashion, but moves in response to an external force accelerating or decelerating the axis to a kinetic motion that is beyond the normal behavioral parameters. This polar axial alignment deteriorates by the acceleration to another axial alignment with the sun. This alignment is the gyro fashion.

Kineverse* — Fourth phaseverse in formulating the physical universe, it is, however, the second phaseverse in establishing xyzenthium crystals. Motion is established in the universe causing the universe to become monoquiet. During this phaseverse, time becomes a dimension.

Kyntip* — Bethtip (physically active simulated geometric point), it is third in its transformation acquiring kinetic energy after acquiring inertia and jammerial attributes. It is the primary expression of Elohim's movement, "fluttering," upon the waters.

Lasma* — State of matter, it occurs between plasma and molecuum. The nuclei cool from a plasmatic state just enough to collect one or two electrons giving polarity to the atom. The atoms are loosely collected by the polarity to form a mist to liquid-like substance.

Laurasia — Second shield in size, third if the initial shield is counted, it has no sedimentary covering. Determined cause of its existence is the result of the crash by the Hudson comet planetoid during Noah's Flood. Its location shatters between upper North America to northern Europe. The shield was formed by the impact of the Hudson Comet.

L-M   Glossary

Lavor* — Forth state of matter it exists between moleculum and vapor. It occurs as a result of a nucleus collecting enough electrons to be "tricked" into having enough to become an independent atom or a smaller molecule and breaks away from or "melts" the moleculum. These atoms have polarity and behave similar to lasma, the second state of matter.

Lepton — Electrons and neutrinos are examples of natural leptons forming in the first generation of quarks. Other leptons form in the second and third generation of quarks.

Levanastorm* — Secondary lighter plasmatic storms formed out of the original central plasmatic solar storm. These storms form moons that later orbit the planets. Levana means full moon in Hebrew.

Lithium Rain* — Thin layer of floating elementuum breaks into lavor droplets that fall to the Earth. Though these droplets are the furthest from the Earth, they are the lightest. They form latter and help form the upper surface.

Mafic — Layer under the felsic layer and above the ultramafic layer, it has less aluminum and more magnesium than the Felsic layer. Average density is about 3.0.

Magneverse* — Seventh phaseverse, light is separated from darkness during this phaseverse via magnetic energy. The xyzenthium crystals become fully formed. These crystals are then used to form cubical neutrons.

Meson — Byproduct of smashing nucleons, it has only two hadronic quarks in a quark-antiquark pair giving a net charge of zero.

Mesosphere — Atmospheric layer that heat absorption predominates. Its primary function is to regulate heat coming from the sun and that, which radiates from the o-zone. Cooling occurs as these gasses convert absorbed heat into kinetic energy.

Microstrong* — Neutral magnetic force, it promotes and absorbs hypertips consisting of juttorial energy. It has a strong influence at short distances but weak at large distances. It is responsible for holding electrons away from the nucleus and prevents matter from collapsing in upon itself.

Microweak* — Neutral magnetic force, it promotes and absorbs hypotips of jammerial energy. It is weak at short distances but strong at large distances. Ultimately, it holds matter together.

Glossary

M-N

Miortex* — Within the central region of xyzenthium crystal, there is a xentrix with 24 zeeds pointing to the absolute central alephtip. They collapse formulating the miortex. Miortex comes from the words mirror and vortex because it is a vortex that deflects. This process also forms the center of photons.

Mobius Strip — Thin strip of paper, or of its like, twisted one half turn and connected at the ends forming an object with only one side.

Moho — Division between the mafic and ultramafic layers, it is a layer by its own right consisting of a thin layer of gravel and rocks. The gravel and rocks formed by the movement of the mafic layer during the dividing of the earth.

Molecuum* — State of matter, it occurs between lasma and lavor. Within this state the atoms are attempting to collect all the electrons demanded by the nucleus via cooling from plasma. They grab electrons by neighboring atoms within the process causing bonding between other incomplete "electron shelled" atoms forming a mammoth molecule.

Molecuumsphere* — Dense sphere under the Asthenosphere, it provides the earth with rotation stability and most of the surface gravity. It maintains the magnetic structure of the earth via its imbalance of electron content along with the plasma underneath.

Monoquiet* — Phaseverse condition, it occurs when motion exist with the infinitesimal universe but not in the finite universe. It is similar to the idea that an electron moves within an atom in solid matter impervious to our perception.

Multigalactic Shell* — Shell formed by the central release of equilibrium by Elohim via neutron displacement resulting in a spherical layer of galaxies. Our galaxy formulated by such a process.

Muon — Second generation electron-like lepton, It is heavier than an electron and unstable. It is used to simulate an anti-proton with a positron for an anti-electron for anti-matter as in this case, it forms anti-hydrogen.

Neanderthal — Humanoid Creatures, they inhabited Nod. Before the Fall of Adam, they were the source of new life, as they lived in the birth death cycle.

Nefesh — Hebrew word, it translates into the English word for soul. It is the result of the union between body and spirit.

N

Nephilim — Giant humanoid creatures that formed by the union between the Neanderthal and the Cro-Magnon beings.

Neutrino — Scientific term used to depict the extremely small mass that is loss during the electrons' escape from the neutron.

Neutron — Initial nucleon constructed by Elohim, it is the source of all the components forming an atom. It decays into a proton after evaporating its three external most layers of xyzenthium crystals and ejecting an electron and its neutrino. The number of layers of crystal that a neutron has is 10,890.

Neutron Clump* — Initial atom forming event that occurred before protons formed. Size of the clump depended upon the clumps relationship to the expanding multigalactic sphere.

Neutron Cube* — Original form of the neutron before separation into its spherical-like form, it was many times larger than its resulting form (nee Neutronverse).

Neutron Exohole* — Absence of four xyzenthium crystals at alternate corners of the neutron cube, it determines the size of the neutron and provides stimulus for neutron separation.

Neutron Prime* — Original neutron freshly formed from a neutron cube. It is slightly heavier in mass via no loss of mass by radiation. Mass calculated to be 1.008676762 u.

Neutron Snow* — Accounts for the jump in the natural occurring isotopes between astatine and radon. Formulation occurred when the multigalactic sphere gained enough mass to draw neutrons to it instead of being drawn outward by the external mass.

Neutronverse* — Eighth phaseverse, It is the "second" occurrence of the firmament finite fractal. The first occurrence actually was in the gravverse. The universe of neutrons formulates from a cubical format into a spherical form.

Nightsod* — Inverse xyzenscape (layer) of xyzenthium crystals that form daysod, it forms the exterior of the external most layer of the neutron. The microweak faces of the crystal formulate the entire xyzenscape surface. The electromagnetic charge of the xyzenscape is neutral.

Nonquiet* — Motion existing in both infinitesimal and finite space, it is space as we know it today.

Glossary N-P

Nucleophoton*   High-energy photons, it is generated from hadronic material. Commonly, they form within the nucleus of a radioactive atom. It is an ejection of a full xyzenthium crystal from the nucleon's surface, and it is responsible for the presence of gamma and strangeness radiation.

Nultip*   First allocation of physical space, it was oblivious to physical form. Its reference to the geometric point is equal only to one disjoined xev per dimension. No energy assigned to it, just allocation by Elohim.

Nulverse*   First phaseverse, it was present only to Elohim via allocation. No matter existed; it was only angles of reference to a geometric point in a disassociated, yet, orderly fashion.

Omnipresence   Attribute of Elohim often viewed in terms of vastness. It also refers to HIS Infinitesimal Nature in that there is no place too small for HIS Presence expressed as a personal experience.

Orbital   Term used to depict the path of an electron about the nucleus of an atom. Defines electron movement within an atom as being erratic unlike the predictable planetary orbits, and defines its location statistically.

Ozone   Molecule of air, it is composed of three atoms of oxygen. Primarily, these molecules lie directly above the troposphere and form the stratosphere.

Penumbra   Solar feature of a sunspot, it is the "grayish" edge interface between the darker sunspot and the brighter fiery ocean.

Phaseverse*   Divisions formed in the creation of matter as known today. Each division depicts a state of the universe dissimilar to each other. There are ten in all, they corresponding to the Ten Commandments.

Photon   Packet of energy released by the electron in response to the energy emitted from the nucleus or impact by another photon. They form light waves; wavelengths increase in period when the juttorial ratio to jammeria is greater reaching out to the red end of the spectrum of light.

Planetoid   Defined as a small planet-like asteroid having a spherical shell shape and a rotational axis. Unlike planets, the core plasma crystallizes into dust foam-like formations leaving a thin atmosphere or void and has a thicker crust ratio to a planet.

P-Q Glossary

Plasma Physics term defining the fourth state of matter. It is the hottest form possible for atomic structures to exist. The electrons are stripped from the nucleus via extreme heat. The nucleus of the atom becomes a positive ion called cations. The electrons move freely from the nucleus like plasma. The result is a magnetic substance.

Plasmaverse* Ninth phaseverse, neutrons collect to form atomic nuclei. All matter becomes to exist as plasma as protons and electrons form giving the elements their definition. The galaxies also form during this phaseverse.

Polar Compression* Movement of the Magnetic Poles of a planet toward each other, it is in response to a stronger external magnetic force which exist at a distance. The distance gives the effect of being smaller; hence, the planet's magnetic poles bend toward it.

Polar Decompression* Attempt made by the "bar" magnet of a planet to return to its original position like that of a spring. In the case of the Earth, this was accomplished in two phases. The first phase was more dramatic as the Earth's crust moved with the magnetic poles. The second phase involved only the internal magnetic structure and finished the process.

Positron Type of antiquark, it is a lepton having a mass comparable to an electron with an electromagnetic charge of +1, instead of being -1. It is a stable man-made quark.

Preatomic* Time accounted within the universe before atoms form. This incorporates the first nine phaseverses of our universe. Atoms were not fully developed until the bondverse occurred.

Protonization* Single word usage, it is to imply the breakdown process of a neutron into a proton and lepton components. Generally, formulation of atomic nuclei from neutron clumps.

Protosod* Xyzenscape (layer) in which the xyzenthium alignment gives a surface of a pure positive charge. It is responsible for the existence of protons and the resulted positrons. It formed the sixth layer in the neutron.

Quark Fragments and formulated particles formed by bombarding subatomic matter. Most of them are mistaken for building blocks of matter; contrariwise, they are artificially formed particles of submatter.

Glossary Q-R

Quarquid* Final form of the uniquid, it forms in four resulting magnetic energy patterns. The relationship between the movement of hypertips and hypotips determines the pattern.

Quebbrix* Smallest unit of finite space possible, it is an artificial division of the uniquid used to illustrate the relationship between infinitesimal changes to its impact upon finite space.

Quub* Cubical structure initiated in the kineverse, it is infinitesimal in nature. In the central location of each edge within the structure is the original location of each kyntip. The total number of kyntips is 12, as there are 12 edges to a cube.

Quubium* Lines of moving energy (Xergopaths) dividing space into a lattice of infinitesimal cubes, it exists throughout the entire universe. A single cubical form is a quub. Many quubium lattices are possible, our focus is on the set used to form the physical elements of the physical universe.

Quuf Factor* Juttorial variable of the electron hold preventing its crash into the nucleus depicting the space/time continuum bend upward.

Rayring* Ring of small daysod crystals, it exists about 20 degrees above the electron equator internally. Because of their loss in volume without losing mass, energy emitting from them is more intense than its surrounding xyzenscapes internal to the electron's shell. It is responsible to create the miortex, which is the vortex of a photon.

Redshift Within this article, it is seen beyond that of objects moving toward or away from each other, by which we conclude that the universe is expanding. Instead, it is the photon packet relaxing as it travels through time in the void.

Rejoinion Resolve* Nuclei resolve, it is the process in which the gravity dominated neutron clump transforms itself into a magnetically crystalline dominated nucleus.

Reysh Factor* Jammerial variable of the electron's magnetic hold, it depicts the outer slope downward by the time/space continuum to the nucleus.

Ruach Hebrew word, it means Spirit or wind. The nonphysical energy, it gives life to the body. Not to be confused with the soul, which the spirit sustains. It is that which is housed by the forms of creation.

S

Glossary

Septicollision* — A collision of six bethtips traveling from the six dimensional direction upon the three dimensional axis toward a single central alephtip, this makes an involvement of seven physically simulated geometric points. Their initial existence is found in the kineverse within the physical infinitesimal space. It has a movement like pixels on a video screen.

Sevievah Ocean* — Original unified ocean, it is from the Hebrew word meaning surrounding. It surrounded Kadummagen as the only ocean. After Peleg's time, this singular ocean divides into the oceans known today as the landmasses created a sense of difference.

Shekina Glory — Attributed to the glowing quality existing when Elohim dwells with a biological inhabitant or even within an inanimate object. A couple of examples other than within the human form are the "burning bush" of Moshe (Moses) and the pillar of fire on the Tabernacle. In all cases, it is the "fire" does not devour the object.

Solar Disc* — Primordial configuration of the sun before it became the spheroid seen today. Originally it engulfed the entire solar system. The last major break formed the asteroid belt between Mars and Jupiter.

Solar Flare — Phenomenon resulting of decay in large nuclei within the material forming the fiery ocean's floor, the primary formulation is that of helium. The light weight helium is pressed to the floors surface and is violently expelled displacing the "fiery ocean matter" into the solar atmosphere.

Solar-Centric Alignment* — Alignment made with the Earth's rotational axis to the Earth's orbit; such that, one rotational pole is always facing the sun. It was the original alignment before the Fall by Adam and Eve. It is also the closeness to this particular alignment that formed the ice ages and its counter cycles.

Strange (Quark) — Second generation quark, it has a $-1/3$ charge. It has an attribute known as strangeness. The quark is larger than the first generation down quark but less than the bottom quark of the third generation of same charge.

Strangeness — Manifestation of intense radioactivity, it has a half-life longer than expected for such intensity. Result of exposed daysod layers breaking up. It is caused by the inverted layers of the second-generation hadronic quarks.

Glossary

Stratosphere
Layer of Atmosphere, it exists directly above the Troposphere. This layer contains the ozone. The electrons shared in the bonding of O_3 (ozone) interact between their nuclei causing a greater portion of the solar radiation absorbed by the nuclei to be released back.

Subfelfrax*
Fragments of the felfrax, it breaks away from the original landmass. Example: Greenland is subfelfrax formed from the North American felfrax.

Sublevels
Scientific term depicting the electrons orbital patterns within a level or stack elevation. There are seven sublevels. Perhaps, the most important sublevels are s and p, because most of the surface elements are within the first 20 elements numerically.

Submatter
Particles forming matter such as protons, neutrons, electrons or quarks and even atoms as they form molecules.

Subparticles*
Masses that form submatter, they are xyzenthium crystals and bethtips or tavtips.

Sunspot
Dark roundish regions that appear upon the surface of the sun, it occurs in its cycle of flipping its magnetic axis. They are magnetic storms with inverted polarity to the solar bar magnet; the bar magnet flips the spots polarity via attraction.

Tau
Third generation electron-like lepton, it has a larger mass than that of a nucleon.

Tav Judaism*
Final form of Judaism, it occurs when both Christians and Jews worship Elohim in the same way in Yeshua. While it looks impossible, it will happen. Doctrines will change in both.

Tavtip*
Final bethtip formulations, these develop in the magneverse. Both juttoria and jammeria xevim exchange, the result is the generation of hypertips with two juttorial xevim and hypotips with two jammerial xevim.

Tetradecahedron
Any solid object having 14 sides. The image of bethtips (active simulated geometric point) gives such a specialized version of this figure. Here, it is a cube with the corners chopped away.

Tetragrammaton
Hebrew word consisting of four letters used to represent the Name of Elohim. In Hebrew letters, it is Yod Hey Vav Hey; in Roman letters, it is YHVH, YHWH, JHVH, JHWH or IHVH.

T-U   **Glossary**

Thermaverse*   Fifth phaseverse, it is when juttoria (light, heat) is introduced and divides the uniquid into cubes that will become xyzenthium crystals.

Thermodynamics   Study of heat; it is based upon the observed phenomena in today's physical universe.

Thermosphere   Atmospheric layer in which heat increases with altitude the inverse of the mesosphere. Even so, it is an extension of the mesosphere cooling process. The atmospheric layer continues to thin allowing solar heat to express itself. It is the last atmospheric layer before reaching the external void.

Thermtip*   Bethtip having all the xevim energy present via the introduction of juttoria by which heat exists. It is the last formulation in which each bethtip are unified in formulation.

Thermtip Pulse*   Pulse of thermtips leaving the xentrix, this occurs after the xentrix compresses as much as the parameters allow and the bethtips continue their kinetic journey. It is the beginning of the formulation of the diflohexius.

Top (Quark)   Third generation quark, it has a +2/3 magnetic charge. The most massive of all quarks, it is the last quark scientists were able to formulate to complete the set of third generation hadronic quarks.

Triocollision*   Method of transferring energy in infinitesimal space between two bethtips, active simulated geometric points; it involves a central alephtip, inactive simulated geometric point, to help in the transference.

Troposphere   Atmosphere existing directly above the surface of our planet, it is the densest layer. Our weather exists within this layer. It extends approximately 40,000 feet. Primary feature is the atmospheric composition of nitrogen, oxygen and argon which are heavier than water vapor.

Ultramafic   Layer of rock under the Moho, it has the density about 3.3 (consider water is 1.0). This layer exists directly above the asthenosphere.

Uniquid*   Very thin liquid-like substance, it existed before the creation of subparticles, submatter, and atoms. Later it separates into the four quarquid patterns flowing within the diflohexius of a xyzenthium crystal.

Glossary U-X

Uniquidron*   Special energy shape, it is formulated by the energy lines between bethtips. The general formation is a tetradecahedron (fourteen faced object). The image is a cube with the corners chopped off (6 cube faces + 8 corners = 14 faces in all).

Up (Quark   Hadronic quark, it forms when scientists smashes a nucleon. A proton formulation becomes (u, u, p), this reads as two up quarks and one down quark.

Volumass*   Measure of mass, it provides us with the volume and atomic mass measurements are of an equal scale. Also able to break into sub-dimensional units as the edge of a unit cube equals one unit.

Weight   Measurement of matter, it takes both mass and gravity to determine its value. Gravity is as much a variable as mass. Consider the change in weight of an object on Mars as compared to its weight on Earth.

Xentrix*   Set of xergotips occupying the original space of the xyzenthium crystal, which shrank during the thermaverse to the central location within the crystal, theoretically the resulting volume is 1/216,000 of its original size.

Xergopath*   Path of energy, it transfers via bethtips. It has zero mass in both infinitesimal space and finite space. It comes from the words zero, erg and path.

Xergoplane*   One of many perpendicular dual planes dividing the uniquid into uniform cubes providing a means to form membranes that determine the extent of the xentrix.

Xergoplate*   Division of the xergoplane into square shapes, this is accomplished by intersecting the xergoplanes, by which the uniquid becomes divided into crystals. In essence, it is a cubical face of the xentrix. These form the initial membrane of the xyzenthium crystal.

Xergotip*   Alephtip existing at a cubical corners of the quubium within the thermaverse, these are designated by Elohim to formulate the xentrix. They are only location in the xentrix in which septicollision occurs.

Xergotip Membrane*   Initial exterior surface of the xyzenthium crystal before formation, in which juttoria was applied by Elohim. The result forms the xentrix.

X-Z

Xev* — Exchange energy Vectors are assigned to bethtips and to alephtips, only the bethtips have energy attributes present. Each of these has four angles of reference to the geometric point. Plural is xevim.

Xyzenscape* — Surface of a layer formed by xyzenthium crystals promoting a particular magnetic energy pattern.

Xyzenthium* — First particle of finite matter created, it is the smallest building block forming all submatter particles. Spherical structure with energy influences that reach out to a cubical formation. It is with an internal magnetic transfer complex within its core known as a diflohexius.

Xyzenverse* — Sixth phaseverse, the universe is divided into cubes of xyzenthium crystals.

Yeshua — Jesus is the translation in English. YHWH expressed in the human manifestation. Yeshua HaMashiach (Jesus Christ) literally means "the Anointed (Sanctioned by Elohim) Salvation" in Hebrew for HE Salvaged the Original Plan of this Creation by saving us, the "Believers" in HIM. HE will come back as the Conquering King.

Zeed* — Locations within the xentrix, it is the root in which the xergopath loops originate per exozeed. These intersections form a pyramid shape within the xentrix.

Zeedpath* — Path defined by a zeed which feeds "tavtip-2"s into the miortex (a central alephtip).

Zerosia* — Region formed between Russia and Greenland. Now, it is submerged underwater.

Quiz Answers

Chapter 1:
1. B   2. D   3. C   4. B   5. A   6. T   7. F   8. F   9. F   10. T

Chapter 2:
1. D   2. C   3. B   4. C   5. A   6. T   7. F   8. T   9. F   10. F

Chapter 3:
1. D   2. A   3. C   4. A   5. D   6. F   7. T   8. F   9. T   10. T

Chapter 4:
1. A   2. D   3. B   4. B   5. D   6. F   7. F   8. F   9. T   10. T

Chapter 5:
1. A   2. D   3. A   4. C   5. C   6. F   7. F   8. T   9. T   10. F

Chapter 6:
1. B   2. D   3. C   4. C   5. D   6. T   7. F   8. F   9. T   10. T

Chapter 7:
1. A   2. D   3. A   4. D   5. D   6. F   7. T   8. T   9. T   10. F

Chapter 8:
1. D   2. A   3. B   4. C   5. D   6. T   7. T   8. F   9. T   10. T

Chapter 9:
1. B   2. A   3. C   4. D   5. D   6. T   7. T   8. T   9. T   10. T

Chapter 10:
1. C   2. C   3. A   4. D   5. B   6. T   7. T   8. T   9. F   10. T

Quiz Answers

Chapter 11:
1. D   2. C   3. A   4. D   5. A   6. F   7. F   8. F   9. F   10. T

Chapter 12:
1. A   2. C   3. B   4. D   5. C   6. T   7. F   8. T   9. F   10. F

Chapter 13:
1. A   2. D   3. C   4. C   5. A   6. F   7. F   8. T   9. T   10. F

Chapter 14:
1. B   2. D   3. C   4. B   5. D   6. T   7. T   8. F   9. F   10. F

Chapter 15:
1. B   2. A   3. B   4. A   5. C   6. F   7. F   8. F   9. F   10. F

Chapter 16:
1. C   2. C   3. B   4. A   5. D   6. F   7. F   8. F   9. F   10. F

Chapter 17:
1. A   2. D   3. C   4 A.   5. C   6. T   7. T   8. T   9. T   10. T

Chapter 18:
1. A   2. B   3. A   4. C   5. A   6. F   7. T   8. F   9. F   10. T

Chapter 19:
1. B   2. B   3. B   4. B   5. A   6. F   7. F   8. T   9. F   10. T

Chapter 20:
1. B   2. D   3. A   4. A   5. D   6. F   7. T   8. T   9. F   10. T

Chapter 21:
1. D   2. B   3. A   4. C   5. C   6. T   7. F   8. T   9. T   10. F

Chapter 22:
1. A   2. C   3. D   4. B   5. A   6. T   7. T   8. F   9. F   10. T

Index

- A -

Abraham 23, 323, 399-400
Adam 45, 335, 392, 445-6
Africa 334, 346, 374
.... 376-7, 379-80
.... 385-6, 388
Alaska 222, 333-4, 346
.... 376, 378-80, 380-3
Alien Life Forms 443-4
Alephfield **47**, 55-7, 59, 76, 80
Alephtip **46-7**, 59, 65-70
.... 76-7, 79-81
Aleutian Islands 333, 381, 388
Aluminum 278, 280, 282-5
.... 291, 314
Aluminum Isotope **189**
Ammonia **277**, 279, 302
Antarctica 334, 344, 374-7
.... 379-80, 382-3
.... 385-7
Antiderivative 41-4
Antimatter **143**, 145, 147
Antineutrino **143**, 147
Antiquark **143**, 145, 147, 154
Antishell **266**
Arabian Peninsula 346, 376, 382
.... 385, 387
Argon 279-85, 293-4, 297
.... 318, 342
Argon Isotopes **192**
Argon Plain **280**-1, 285, 293
Asia 334, 374, 376
.... 380, 386-8
Asia Minor 382
Astatine **198**, 255-6

Asteroid Belt 275-7, 279, 284
.... 290, 310, 317-8, 344
Asthenomoho **356**, 385
Asthenosphere 355-6
Atlantis 334
Atlas Mountains 385
Atmosphere Earth 336-7, 341-4, 347
.... Ch. 20, Ch. 21
Atmosphere Otherworldly .. 277-82, 292
.... 303, 310, 345
Atomic Nuclei .... Ch. 12
Australia 334, 346, 375-6
.... 379-80, 387

- B -

Baffin Island 334, 386
Baja 384
Banks Island 334
Baptism (Mikvah) **429-30**
Baryon **143**, 145, 267
Bermudaland 334, 379, 384, 388
Beryl **291**
Beryllium 277-8, 280, 282
.... 284, 290, 291-2
.... 294, 302
Beryllium Rain **291**
Beryllium Isotope **183**
Bethtip **52**, 65, 69
.... 76-81, Part 2
Black Sea 382
Bondverse .... **32**, Ch. 13 & 15
Borax **293**
Boron 262, 275, 277
.... 280, 282-4, 290
Boron Rain **291-3**
Boron Isotopes **183-4**

B-E

Bottom (Quark) **144**-5, 147
Brazil 377, 379
Brit Hadeshah **427**

- C -

Calcium Isotopes **193-4**
Calcium 193-4, 278, 182-3
.... 293
Cameroon 379
Canada 382, 385, 386
Carbon 164, 178, 186
.... 258, 261-2, 277-84
.... 290-1, 293-5, 302
.... 306-7, 319-20
Carbon Dioxide 294-5, 303
Carbon Rain **290-1**
Carbon Isotopes **184-5**
Caspian Sea 382
Cation **240**, 352
Central America 380, 382, 384
Centrifugal Force 179, 181-2, 186-7
.... 189-91, 194, 196
.... 387
Charm (Quark) **144**-7
Chexosod **136-7**, 148-54
Chicken 320
China 380-1, 388
Chlorine 282-4
Chlorine Isotopes 191
Chlorophyll **307**
Clear to Infinity 85
Cloud (Water) 342, 347, 359
.... 362, 414
Cobalt Isotope **196**
Cold Neutrons **152-5**
Color 231-4, 425-6
Complex Gravity **311-19**
Coriolis Effect **412**
Cro-Magnon 320, **321-4**
Crystal Annihilation **132-4**

Index

Cubical Earth Illustration 172-3
Cursing of the Ground 335

- D -

Daysod **135**-38, 147, 148
.... 151-2, 154, 161, 166
Denmark 381-2, 383
Derivatives (Math) 41-4
Diflohexet **102**-3, 105-6, 129
.... 138-9, 148-9, 161
Diflohexius **102**-3, 106
.... 129, 148
Dinosaur 320
Dividing of Earth 313, 357, 362-3
.... 365, 367-8, 388
.... Ch. 18 & 19
Down (Quark) **143**-6, 154
Duocollision **76**-8
Duoquiet **69**

- E -

Earth (Creation) .... Ch. 15
Earth Cube Illustration 172-3
East (Kedem) **332-3**
Eber 23, 400
Eden 296, **305**, 320, **321-6**
.... 331-2, 357
Electrodonativity **273**
Electron 143, 145, 147, 151-2
.... 162-6, 171, 173-4
Electronegativity **273**
Electrophotons **228**
Electrosod **135**-7, 150-2
Elementless **27**
Elementuum 280-1
Elohim .... **23-24**, Ch. 1-22
Emperor Sea Mountains 375, 388

Index

Ergitude **226**-8
Eternal Season 394-7, 401
Euphrates (Perat River) 305
Eurasia 376, 379
Europe 224
Eve 320-5
Exoholes (Neutron) **134**
Exozeed **99**, 101-2, 106, 117
Expanding Sphere 38-9

- 𝔉 -

Felfraplex **373**, 379, 381, 383-4
Felfrax (im) **373**, 383-7, 379-80, 383-7
Felfraxim Belt **373**, 375-6
Felsic Cap **288-9**, 305
Felsic Moho **374**
Felsic **304**, 305-6
Fijiland **335**
Finite Limits **36-8**
Finland 381, 383
Firmament (ra-key-a) ... 177, **249-50**, 261
Flood of Noah **342-7**
Florida... 384
Fluorine 262, 277, 282-3, 293
Fluorine Isotope **188**
Force Subfield .... see Xev
Fractal 27, 29, 249-50, 297
Francium 218

- 𝔊 -

Gamma Radiation 161
Garden of Eden 51, **305**, 320, 323
....331-2
Geometric Point (Shape) **52**
Geometric Point 35-9, 41, 44, 46-7
.... 51-7, 59, 68-70
.... 87-9, 94, 76-7, 81
....104, 106

E-H

Gishon River 305
Gold Isotope **196-7**
Gondwana **344**, 373
Googolplex **37**
Gravity ....see Jammeria
..............................See Complex Gravity
Gravtip **65-8**, 71
Gravverse .... **32**, Ch. 4
Great Bear Lake 386
Great Lakes 386
Greenland **334-5**, 346, 363-4
....376, 380-6
Gulf of Alaska 380-2
Gulf of Carpentaria 375
Gulf of Mexico 384
Gulf, Libyan 224
Gulf, Persian 385
Gyro Alignment **392-4**, 397-9, 401

- 𝔋 -

Hadron **143**-7
Ham 351
Havilah **296**, **304-5**, 331, 345
.... 357-8, 363, 367-8, 374
.... 376, 378, 381, 383, 388
Hawaii............... 363, 365, 374, 375
Hawaiian Ridge 357, 363, 374, 388
Heat .... see Juttoria
.... See Thermodynamics
Helium 180, 256, 258, 262
.... 273, 281, 283-4
.... 308, 311, 336
Helium Isotopes **180**
Hiddekel River 305
Hierarchy of Energy........................**121-2**
Hollow Earth Illustration 314
Hudson Bay 334, 344, 345
Hudson Comet **344**, 345-7, 357
.... 380, 381, 383
.... 385-6

473

H-K

Hudson Crater 333, **344**
Hydrogen 165, 169-74
.... Ch. 14 & 15
Hydrogen Isotopes **179-80**
Hydroxide 273, 277, **291**-3
Hyperatom **267**
Hyperon **143**, 145
Hyperpush **162-3**
Hypertip **112**-4, 116-18
.... 121-2, 135, 148-9
.... 153, 162-4, 174
Hypotip **112**-4, 116-18
.... 121-3, 135-6
.... 148-9

- I -

Ice Ages .... Ch. 20
Iceland 335
India 334, 336, 376
.... 383, 384
Indonesia 334, 376
Indonesia-Japan Felfraxim-Belt 373
.... See Map page 333
Inertinuum 59-60, 67
Inertip **52**, 55-7, 59
Inertverse .... **32**, Ch. 3
Inuvik, Canada 364
Ion **186**, 197, 352
Iron 219, 232
.... 234-6, 263
.... 277-8, 282-6
.... 291, 293
.... 314-5, 336
.... 355, 360
Isobaric Spin **148**-2
Isotope .... **177**, Ch. 12
Israel 169, 322
.... 325, 331
.... 362, 382, 385

Index

- J -

Jammeria .... Ch. 4, **65**, 78, 86-91
.... 93-4, 98, 101-5
.... 111-2, 114, 118-22
.... 134, 138, 155
.... 161-2, 179, 201
.... 209, 215-18
.... 224-8, 238, 250
.... 252
Japan 381, 388
Japanese Trench 381, 388
Japheth 351
Java 375-6
Jesus .... see Yeshua
Jet Stream **407**, 408-12
Joseph 400, 401
Jupiter 275, 303, 311, 343
.... 358-62, 363, 364, 385
.... 398, 404, 443
Juttoria .... Ch. 6, **86**-94, 98
.... 101-6, 111-2, 114
.... 118-22, 138, 144
.... 153, 155
.... 161-2, 198-202, 209
.... 215-6, 218, 223-31
.... 236-8, 240-1, 252

- K -

Kadummagen .... **331**, Ch.17 & 18
Kamchatka 375, 380-1
Kazakhstan 383
Kernite **291**, 293
Kinetic Non-Alignment 393-4
Kineverse .... **31-2**, Ch. 5
Kola Peninsula 383
Korea 380-1
Kuril Islands 381
Kyntip **75**, 76-7, 78, 80-1, 87

Index

- L -

Lasma **243**, 260-3, 284-5
.... 290, 308-9
.... 313-4
Laurasia **344**, 373
Lavor **243**, 260, 263, 280
.... 285-6, 290-1, 294
.... 307-9, 313-4
Lepton 143, **145**-7
Levanastorm **274**-5, 310, 317-18
Lifespans **400-2**
Lithium 257, 261-2, 277-80
.... 282-4, 290-1, 293
Lithium Rain **293**
Lithium Isotopes **181-3**
Lorenz Transformation **70-2**

- M -

Madagascar 373, 376, 385, 387
Mafic **285**, 288-90, 304-5
.... 355-7, 362, 367
Mafic Havilah **367**, 383-4
Mafic Plain **289**-90, 357, 367-68
Magnesium 262-3, 278
.... 282-5, 291
Magnesium Isotopes **189**
Magnetic Compression Math **365-6**
Magnetic Decompression Math **366**
Magnetic Flip Math **365**
Magneverse .... **32**, Ch. 8
Malaysian Peninsula 379
Manganese Isotope **195-6**
Map of Kadummagen **333**
Mariana Trench 367-68, 388
Mars Wave Limit **253**
Mars 275-6, 277-78
.... 281-2, 294-5
.... 303, 311, 313
.... 317-18, 345, **359-62**
Membrane (Xergotip) **90**-1

L-N

Mercury (Element) 197
Mercury (Planet) 275-6, 277, 281-2
.... 303, 311, 319
Meson 143-4, 147, 154
Mesopause 414
Mesosphere 414
Methane **277**
Mexico 384
Microstorm 274
Microstrong Limit **224**, 225
Microstrong 86, 112, **114**, 116
.... 118-20, 120-3, 134-5
.... 138-9, 147-8, 152
Microweak 65, 112, **114**, 120-1, 123
.... 134-5, 137, 147-8
Mikvah (Baptism) **429-30**
Ministorm 275
Miortex **115**-7, 122, 124, 161
.... 221-2, 224-5, 229-30
Mobius Strip **55**
Moho **304**, 355-6
Molecuum **243**-4, 260, 262-3
.... 280-1, 284-6, 280-2
.... 301, 305, 308-9
.... 313-14, 336-7
Molecuumsphere **313**-4, 336, 354
.... 355-6, 362, 367-9
Monoquiet **69**, 80-1
Moon 249-50, 274, 304
.... 306, **309**-11, 317-19
.... 332, 334, 359
Moshe (Moses) 32, 45, 86, 322
.... 325, 400
Multigalactic Shell .... **251**, Ch. 14
Muon **145**, 147

- N -

Natural Hydrogenation **276-8**
Neanderthal 320, **321**-5
Nefesh 44-**45**

N-P

Neon 218-9, 262, 277, 282-3
Neon Isotopes **188-9**
Nephilim **321**
Neptune 276, 277-8, 302-3, 311
Neutral Plasma **242**, 337-40
Neutrino 143, 145-7, 164-6, 171
Neutron **143**-6, 151-5
.... Ch. 9, 11 & 12
Neutron Clump **181**-2, 185-6, 190
.... 192-6, 202, 256-8
Neutron Cube .... **129**, Ch. 9
Neutron Prime **167-70**
Neutron Snow **197**, 255-9
Neutron Star **266**-8
Neutronverse .... **32**, Ch. 9
New Zealand 363, 375
Newfoundland 334, 381
Nickel-60 .... **198**-200
Nightsod **135**, 136-8, 147
.... 149, 151-4, 183, 166
Nightsod Evaporation **161**-2, 169
Nitrogen 218, 261-3, 277
.... 279-84, 286, 292, 294-5
.... 297, 302, 342, 388
Nitrogen Isotopes **185**
Noah 305, 323-4, 335
.... 347, 385, 400
Nod 320-**321**
Nonquiet **69**
North (Tzafon) **332**
North America 333-4, 344, 346
.... 379-82, 384-6, 387
Norway 364, 365, 368
Nova **268**-9
Nuclei Analysis **199-202**
Nucleon **143**-6
Nucleon Flake .... **178**-9, Ch. 12
Nucleophoton **227**
Nultip **47**, 55-6, 59
Nulverse .... **31**, Ch. 2

Index

- O -

Omnipresence **35-6**
Orbital .... 173-4, **206**, Ch. 13
Orbital Limits **215-6**
Osmium Planet Illustration **314-15**
Osmo, Norway 364
Oxygen 218, 220, 232
.... 239-40, 262, 277-8
.... 279-86, 290-5, 297
.... 307, 309-11, 317-18
.... 342, 344, 358, 360
Oxygen Isotopes **185-88**
Ozone **263**, 281, 284, 295-6
.... 341-3, 347, 359, 414

- P -

Panama 333, 384
Parabola (Solution) **394-5**, 396-7
Peleg (Person) 399-400
Peleg's Earthquake .... Ch. 18-9
Penumbra **337**
Phaseverse **30**-2
Philippines 334, 376, 381
Phosphorous 278, 282-3
Phosphorous Isotope **190**
Photon Anatomy **226-31**
Pison River 305
Planetoid **309**, 319, 344
.... 345-7, 359-60, 385
Plasma .... **236**, 240-4, Ch. 12, 14-16
Plasmaverse .... **32**, Ch. 12
Pluto 275-8, 282, 301
.... 309-11, 319, 443
Pluto-Electron Illustration **173-4**
Polar Compression **354**, 363-6, 367
.... 378, 381-6
Polar Decompression **366-7**, 385-7
Polar Flip (Earth) 301-2, **365**
.... Ch. 18 & 19

Index

Polar Flip (Sun) **301**, 339-41
Polar Impact (North) **378-80**
Polar Impact (South) **376-7**
Positron 147, **151**-2
Potassium 278, 282-3
Potassium Isotopes **192**
Prayer **428**
Preatomic .... **29**-30, Ch. 2-9
Proton 135-7, 143-**146**, 151-5
.... **163**, Ch. 11 & 12
Protonization .... **168**, 170, Ch. 11
Protosod **135-6**, 137, 150-1
Pusan, Korea 381

- Q -

Quark .... Ch. 10
Quarquid **117**
Quasar **266-8**
Quebbrix **69-70**
Quub **79**
Quubium **79**-80
Quuf Factor **216-7**

- R -

Rachaf **28**, 75
Radon **197**-9, 200
.... 255-8, 318
Rayring **221**, 224, 229-30
Rebellion in Heaven **425-6**
Redshift **231**
Rejoinion Resolve **186**-7, 191-6
.... 202
Repentance **426-30**
Reysh Factor **216**
Rose Illustration **27-8**
Ruach **44**, **45-46**, 121, 306
.... 321, 426-7, 430
Ryukyu Islands 381, 388

- S -

Sakhalin Island 381
Saturn 198, 275-6
.... 276-78, 279
.... 281, 302-3
.... 307, 310
Saudi Arabia 346, 382, 387
Scandium Isotopes **194**
Seeds (Creation) **306**-7, 335
Septicollision **77**, 79
Sevievah Ocean **331**
Shekina Glory 322
Shell Sublevels 206-15
Shem 351
Silicon 278, 280-6
.... 290-1, 293, 302
.... 314
Silicon Isotopes **178**, **189-90**
Sodium 262, 248, 282-4
.... 291, 292-3
Sodium Isotope **189**
Solar Disc **274-9**, 282
.... 284, 301-3
.... 306-8, 317-18
Solar Flare **337**
Solar System 250, 257, 260-1
.... 263-4, **274-6**
Solar-Centric Alignment **392-5**, 398
Sound Movement in Core 314-17
South (Negev) **332**
South America 333-4, 377-80
.... 382-6, 388
South Sandwich Islands 383
Strange Quark **144**-7
Strangeness **147**
Stratopause **414**
Stratosphere **414**
Subfelfrax **373**, 375, 384
Subfield **86**, 90
Sublevels Shells, General .... Ch. 13

Index

Sublevel Shells, Individual **206**-8
    d **214-15**
    f, g, h and i **215**
    p **212-14**
    s **208-12**
        K (1s) **208-9**
        L (2s) **210-11**
        M-Q (3-7s) **211-12**
Submatter **30**-1, 145-6, 149, 152
Subparticle **30**
Substorms (Solar) **274**
Sulfur 277-8, 283, 303
   343-4
Sulfur Isotopes **190-1**
Sulfuric Acid **303**, 344
Sumatra 376, 379
Sun 85, 174, 209, 223
    ....241-2, 244
    ....358, 361-2
Sun (Ignition) **307-9**
Sunspots 241, **337**

- T -

Tau **145**
Tav Judaism **430-2**
Tavtip 112-18, 122-4
Tavtip-2 ....**115**
Ten Commandments **419-21, 428**
Tenach **26**
Tetradecahedron **66**
Tetragrammaton **24**
Thermaverse ....**32**, Ch. 6
Thermodynamics **236-8**
Thermosphere **414**
Thermtip **105**
Thermtip Pulse **101-2**
Time Dimension **70-2**, 80-1
Time Changes **403-4**
Tiski, Russia 363-4
Titanium Isotopes **194-5**

Tonga Trench 375
Top (Quark) **144**-5, 147
Tornados **412-13**
Tower of Babel **351**
Triocollision **76**-7, 80
Tropopause **414**
Troposphere **294-7**, 392-3, 347
    ....259, Ch. 21

- U -

Ultramafic **285**, 304
    ....355-6, 362-3, 367-8
Umbra **337**
Unified Fluttering **78-9**
Unified Knowledge **25-6**
Uniquid **68**-9, 71, 76-81
Uniquidron **79**
United States 385, 386
Up (Quark) **143**-7, 154-5
Ural Mountains 386
Uranium 218, **256**-9, 307-8
Uranus 276-8, 282
    ....302-3, 311

- V -

Vanadium Isotope **195**
Venus 175, 294-5, 303, 343-5
    ....352, 358, 361-2
Vibration **75-6**
Victoria Island 334
Void (bo-hu) **26-7**
Volumass **166**-8

- W -

Water: Creation 277-9
        Flood 342-5
        Oceans 295-6
        Today (Clouds) 408, 413-4

Index

Waters **29-30**, 32, 45
..............................67, 71, 75-6, 78
....249, 273
....279, 297
Waves .... see Vibration
Weeds **335**
Weight Illustration **164**
West (Yam) **332**

- X -

Xentrix **90**-4
Xergopath **89**, 113-15, 117
....122-5, 138-9
...................................149, 154, Ch. 7
Xergoplane **90-1**
Xergoplate **90**, 98-9, 106, 117
Xergotip **89**-93
Xergotip Membrane **90-1**
Xev (im) **56**-7, 65-6, 75
....80-1, 86-90, 101

W-Z

Xyzenscape **134**-9
....161-4, 166
Xyzenscape Sequence **136-7**
Xyzenthium (Crystals) **97**, 101-6
....161-2, 165-73
....Ch. 7, Part 2
Xyzenthium Crystal Counts 170-2
Xyzenverse .... 32, Ch. 7

- Y -

Yanchang, China 381
Yeshua **420**, 424, **426-31**
YHWH 23-4

- Z -

Zeed **99**, 101-2, 106, 115-6
Zeedpath **115-6**
Zero .... see Geometric Point
Zerosia **335**, 381

תּוֹדָה יהוה
יֵשׁוּעַ הַמָּשִׁיחַ
לַזֶה סֵפֶר:

Thank you L-RD,
Yeshua HaMashiach
(The Anointed Salvation)
(Jesus Christ)
For this Book